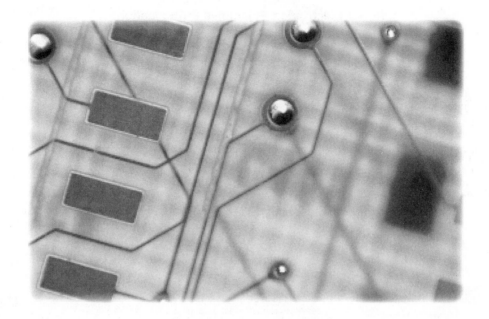

Power Electronics:
Principles & Applications

J. Michael Jacob
Purdue University
West Lafayette, Indiana

DELMAR

THOMSON LEARNING™

Australia • Canada • Mexico • Singapore • Spain • United Kingdom • United States

Power Electronics: Principles & Applications

By J. Michael Jacob

Business Unit Director:

Alar Elken

Executive Editor:

Sandy Clark

Acquisitions Editor:

Gregory L. Clayton

Development Editor:

Michelle Ruelos Cannistraci

Executive Marketing Manager:

Maura Theriault

Marketing Coordinator:

Karen Smith

Executive Production Manager:

Mary Ellen Black

Production Manager:

Larry Main

Senior Project Editor:

Christopher Chien

Art/Design Coordinator:

David Arsenault

Technology Project Manager:

David Porush

Editorial Assistant:

Jennifer A. Thompson

Library of Congress Cataloging-in-Publication Data:

Jacob, J. Michael.

Power electronics : principles & applications / by J. Michael Jacob.

p. cm.

Includes index.

ISBN 0-7668-2332-6 (alk. paper)

1. Power electronics.   I. Title.

TK7881.15 .J33 2001

621.381'044--dc21

2001032509

NOTICE TO THE READER

Contents

3
Power Parameter Calculations

4
Linear Power Amplifier Integrated Circuits

5
Discrete Linear Power Amplifier

6
Power Switches

7
Switching Power Supplies

8
Thyristors

9
Power Conversion and Motor Drive Applications

Preface

Introduction to the Herrick & Jacob series

The traditional approach to teaching fundamentals in Electronics Engineering Technology begins with a course for freshmen in DC Circuit Theory and Analysis. The underlying laws of the discipline are introduced and a host of tools are presented, and applied to very simple resistive circuits. This is usually followed by an AC Circuit Theory and Analysis course. All of the topics, rules, and tools of the DC course are revisited, but this time using trigonometry and complex (i.e., real and imaginary) math. Again, applications are limited to passive (simple) resistor, capacitor and inductor circuits. It is during this second semester, but often not until the third semester, that students are finally introduced to the world of *electronics* in a separate course or two. At this point they find what they have been looking for; amplifiers, power supplies, waveform generation, feedback to make everything behave, power amps, and radio frequency with communications examples. This approach has been in place for *decades*, and is the national model.

So what is wrong with this approach? Obviously, it has been made to work by many people. Currently, during the DC and AC courses students are told not to worry about why they are learning the material, only that they will have to remember and apply it later (provided they survive). These two courses (DC & AC) have become tools courses. A whole host of techniques are taught one after the other, with the expectation (often in vain) that eventually when (or if) they are ever needed, the student will simply *remember*. Even for the most gifted teachers and the most dedicated students these two courses have become "weed out" classes, where the message seems to be one of "if you show enough perseverance, talent, and faith we will eventually (later) show you the *good* stuff (i.e., the electronics)." Conversely, the electronics courses are taught separately from the circuit analysis classes. It is expected that the student will quickly recall the needed circuit analysis tool (learned in the DC & AC courses) taken several semesters before when it is needed to understand how an amplifier works or a regulator is designed. This leads to several results, all of which have a negative effect. First, the student sees no connection between DC & AC circuit theory and electronics. Each is treated as a separate body of knowledge, to be memorized. If the students hang on long enough, they will eventually get to the electronics courses where the circuits do something useful; second, the teachers are frustrated because students are bored and uninterested in the first courses, where they are supposed to learn all of the fundamentals they will need. But when the students need the information in the later electronics courses, they do not remember. The result is a situation under which we have struggled for decades. This is the premise upon which the concept for this new series has been developed.

The Herrick & Jacob series offers a different approach. It integrates circuit theory tools with electronics, interleaving the topics as needed. Circuit analysis tools are taught on a *just-in-time* basis to support the development of the electronics circuits. Electronics are taught as applications of fundamental circuit analysis techniques, not as unique magical things with their own rules and incantations. Topics are visited and revisited in a helical fashion throughout the series, first on a simple, first approximation level. Later, as the students develop more sophistication and stronger mathematical underpinnings, the complex AC response and then the nonsinusoidal response of these same electronics circuits are investigated. Next, at the end of their two years of study, students probe these central electronic blocks more deeply, looking into their nonideal behavior, nonlinearity, responses to temperature, high power, and performance at radio frequencies, with many of the parasitic effects now understood. Finally, in the Advanced Analog Signal Processing book Laplace transforms are applied to amplifiers, multistage filters, and other close-loop processes. Their steady-state, and transient responses as well as their stability are investigated.

The pervading attitude is "Let's do interesting and useful things right from the start. We will develop and use the circuit theory, as we need it. Electronics is *not* magic; it only requires the rigorous use of a few fundamental laws. As you (the student) learn more, we will enlarge the envelope of performance for these electronic circuits, as you become ready." Learn and learn again. Teach and teach again. Around and around we go, ever upward.

Introduction to this text

This is the first book released in what will be an integrated series of texts (Herrick & Jacob Series). It applies the circuit theory and fundamental electronics to a *wide* variety of electronics that deliver power to consumer and industrial loads. Classically, the term *power electronics* is used to mean industrial power electronics; the study of those devices and circuits used to deliver hundreds to thousands of watts to loads used in manufacturing. Focus is placed on power semiconductor characteristics, the conversion between DC and AC sources, the interface with the utility line and motor control. This is a very specific and intricate field. Those involved in the design, selection, installation, operation and maintenance of the electronics needed in a modern industrial manufacturing facility must know all of this.

Only a small percentage of technology graduates work with such industrial-strength loads. But practically every electronics circuit must manage power. Hearing aids and pagers, for example, must efficiently convert the widely varying voltage that results as their batteries discharge, to a steady level. Every cell phone, lap-top computer, and automobile must be able to clearly deliver audio power to its loud speaker. Microprocessors are everywhere; in garage door openers, microwave ovens, refrigerators, faxes, printers, copiers, and many toys. Power electronics are needed to follow their commands, to deliver current ranging from hundreds of milliamps to tens of amps to the motors, valves, heaters and lights that we use. Practically every electronics technologist's work requires an understanding of these circuits.

Yet the core electronics texts do not cover audio amplifiers with the detail required to successfully design, build, and protect a 100 watt amplifier. Audio principles (just how loud is loud enough, and what does that mean about the amplifier's gain, output level and power relationships) are never presented. Linear power supplies are covered in basic electronics texts, but the switching power supply is not. Most power supplies today are switching; a basic text, therefore, does not provide its reader with the ability to understand, design or work on most power supplies. Analog texts do not cover power switches (that's a digital thing). But logic texts and books on microprocessors are all about 0 and 1 and code; little is said about pulse width, dead time, switching frequency, H bridges, or drivers, and nothing is ever mentioned about gate capacitance, rms, or power. Those are analog topics.

The result, therefore, is a large hole in the knowledge that should be common to all electronics technology graduates. The basic existing electronics texts do not provide enough detailed information to successfully build circuits that deliver power ranging from tens of watts to thousands of watts to the wide range of loads common in everyday life. The existing *industrial* power electronics books are targeted for those few graduates destined for manufacturing. This text is designed and tested to fill that gap.

Organization of the text

Chapter 1–Advanced Operational Amplifier Principles: This chapter solidifies and reviews the fundamentals that readers should bring with them.

Chapter 2–Power Electronics Circuit Layout: This chapter explains that *how* a power circuit is built is as critical as the parts that are selected. Both protoboard and printed circuit board layout principles are explained.

Chapter 3–Power Parameter Calculations: Average, rms and power are not as simple as described in fundamental circuits courses. A few integrals are required to understand how electronic devices respond to a variety of standard waveforms. This chapter removes the magic and hand waving without becoming mired down in "the general case" that makes most engineering texts unacceptable for technology students.

Chapter 4–Linear Integrated Circuits Power Amplifiers: Every computer, phone, television, Walkman, PDA, and car now has a voice. It is expected that our electronics can talk to us. We can hear them because of the audio power ICs included in all of these products.

Chapter 5–Discrete Linear Power Amplifiers: ICs can currently only deliver about 20 W to a loudspeaker. To be louder, a discrete MOSFET amplifier is needed. The details needed to deliver a clean, low-distortion, protected signal of > 100 W are presented. Traditional texts include only a brief overview of the techniques that are explained in detail here.

Chapter 6–Power Switches: Electronics form the intelligence of the machine world, but without the ability to actually make a display light, or a disk head move, or a relay close, or a pen print, or a valve open, and so on, they are virtually useless. Power switches provide the "muscle" for digital circuits, microprocessors, and computers to actually do something with the data they have manipulated.

Chapter 7–Switching Power Supplies: All of the electronic gadgets that now form the fabric of our modern life require power. PDAs, pagers, cell phones, computers, cars, cameras, entertainment systems of all types, GPS, and a host of other systems and devices, all require voltages to be efficiently converted from one form to another. Switching power supplies are the dominant way of managing these voltages. Most traditional electronics texts do not even cover switching power supplies.

Chapter 8–Thyristors: The consumer market is becoming increasingly "unplugged", as illustrated above. But automated industrial manufacturing and even home "machines" require the control of power from the commercial power grid. Thyristors, triggers, snubbing, and proportional control are all required. Although most general texts give light coverage to these devices, they do not provide the details needed to actually design, build and safely test practical circuits to deliver power from the commercial line to a load.

Chapter 9–Power Conversion and Motor Drive Applications: These circuits form the core of most "Industrial-Strength Power Electronics" texts. But, with the foundation and techniques provided in the previous chapters, these industrial specialty power control circuits are presented as extensions of key concepts, rather than unique, involved specialized systems.

Features

Pedagogy

Each chapter begins with *performance-based learning objectives*. These are rigorously implemented throughout that chapter.

There are over 100 *examples*, each providing fully worked out and explained numerical illustration of the immediately preceding information. Every example is followed by a practice exercise that requires the reader to extend the techniques just illustrated with a different set of numbers, perhaps in a slightly different direction. Numerical results are given to allow students to check the accuracy of their own work.

The ability to find *specifications*, parts and vendors is a skill central to electronics technicians. For that reason, full data sheets are not provided in an appendix. The reader should explore the web sites of semiconductor manufacturers and vendors, developing a set of bookmarks unique to his/her own needs. However, the interpretation of these data sheets is a key part of the instruction of this text. So excerpts, captured from the actual data sheets, are included at the point in the text where they are discussed, with the relevant points highlighted.

All of the circuits are shown with *actual parts*, all power pins connected and *properly decoupled*, and other *support elements* in place. Only rarely is it necessary to simplify the circuit for clarity. Especially for technology students, it is important that the system being discussed actually work as shown. In power electronics it is often the support components that allow large currents and high voltages to be delivered to real loads though real wires and connectors. Ideal parts and schematics simplified for instructional purposes too often ignore the details of the reality of actual implementation. So the circuit fails when built, not because the student failed to understand its operation or designed it incorrectly, but because the real support components were either omitted ("they weren't in the text!") or incorrectly placed.

The *wide margins* have been reserved to serve as visual keys for important information contained in the adjacent text. Significant effort has been made to keep these margins uncluttered. The occasional figure or table supports the text beside it. Icons, notes, equations and key specifications are placed so that the reader can quickly glance along the margins to find the appropriate details explained in the adjacent text. This is very much the way a good student annotates a book as she/he studies. And, plenty of room has been left for precisely that purpose.

Computer Simulation

Electronics Workbench's *Multisim* ® and *OrCAD's* PSpice are used to illustrate the applications *and weaknesses* of power electronics simulations. There are considerable differences between the two packages. Each has its own strength and weaknesses. To rely only on one is to mislead the reader. To try to teach power electronics without a simulation package at all seriously handicaps the reader. Failures or design errors in other electronic circuits can usually be evaluated and found on the bread board. In power electronics, failures are spectacular, and happen so quickly and leave such a molten mess that little can be learned from them. Appropriate simulation is essential.

The simulations are *integrated* throughout the text, at the points in the instruction where they are needed to illustrate a point. They are *not* just included at the end of each chapter as add-ons, as if an afterthought. The appropriate icon is placed in the margin beside each example, as a key. About one third of the examples use Multisim, one third use OrCAD, and about one third are completed manually. Files for these simulations are available on the accompanying CD.

Most of the end-of-chapter problems can be solved, or the manual calculations verified by using one of the simulation software packages. The solutions manual provides the same coverage as the examples; about one third are solved manually, one third are solved with Multisim and one third are solved with OrCAD. For faculty adopting this text, these files are also available.

The education version of Multisim 6.1 and the evaluation version of OrCAD's PSpice 9.1 are used. Although later versions of each were released during the writing of this text, these do not alter the instructional value of the examples.

Multisim is a product of Electronics Workbench.
http://www.electronicsworkbench.com

OrCAD is a product of Cadence. Information is available from
http://www.orcad.com

Problems

End-of-chapter problems are organized in sections that correspond to the sections within the chapter. Problems are presented in pairs. Answers to the odd-numbered problems are given at the end of the text. Readers must solve the even-numbered problems without a target answer.

Laboratory Exercises

These are included at the end of most chapter. They have been used by over a thousand students and a variety of teachers through several revisions. They work and clearly illustrate and reinforce the principles presented in the text. Although a ±56 V, 3 A power supply and a few unique printed circuit boards are used, no specialized instruments are needed. PCB layouts are available to faculty adopting the text. Parts and special equipment used in each lab are presented in a matrix at the end of the text.

CD

The textbook version of Multisim is provided on the accompanying CD. Also the Multisim and OrCAD files used to create the examples and figures throughout the text are included there.

Supplements

Online Companion Visit the textbook's companion web site at

http://www.electronictech.com

There you will find:

Multisim and OrCAD files for the examples and problem solutions

A link to automated homework: a unique set with random values is generated each time you access the problems. Your answers are checked as soon as you enter the result of each step.

Over thirty downloadable power point presentations

Links to streamed video of the author teaching Power Electronics to his students at Purdue University's School of Technology.

Link to the author's Purdue University course web site.

Links to key manufacturers.

Sample lab results.

Art work to support any special lab printed circuit boards.

Text up dates.

Instructor's Guide

The guide contains solutions to all of the end-of-chapter problems; about one third are completed manually, one third are completed with

Multisim, and one third are completed with OrCAD. Sample lab results and printed circuit board art-work (also presented on the web site) are also provided. ISBN number: 07668-2333-4

About the author

Mike Jacob is a test and control engineer with experience in a variety of industries. He has designed a microprocessor-based packing system to load fiber bales, an IR-based rotational torque gage, and control system for tightening military tank transmission bolts, a pc-based instrumentation and control system for the testing of hydraulic steering gear on over-the-road trucks, and electronic drive and interface circuitry for laser testing of the space shuttle tiles and for measuring the blood flow rate in a portable artificial kidney. He has designed, programmed and installed automated manufacturing equipment for the artificial kidney, automotive controls and residential watt-hour meter calibration.

Mike Jacob is the current McNelly Distinguished Professor and an award-winning teacher. He has received the CTS Microelectronics Outstanding Undergraduate Teaching award as the best teacher in the Electrical Engineering Technology Department six times. He has won the Dwyer Undergraduate Teaching Award as the top teacher in the School of Technology three times. He also received the Purdue University's undergraduate teaching award (the Amoco award), the Paradigm Award from the Minority Technology Association and the Joint Services Commendation Medal (for excellence in instruction) from the Secretary of Defense. In 1999 he was listed in Purdue University's **Book of Great Teachers,** that holds the top 225 faculty ever to teach at Purdue University. He has taught at Purdue for 20 years and at a community college in South Carolina for seven years.

He is also a passionate bicycle tourist, logging 8,000 miles each year on his recumbent bike. He has ridden that bike from Seattle to Indiana, from Indiana across New England to Nova Scotia and back, and across the Great Plains, through the Rocky Mountains and down the west coast to Southern California. So keep your eyes open; you may just meet him some summer on a back road.

Acknowledgements

The author and Delmar Publishers wish to express their gratitude to the reviewers and production team that made this textbook possible. These include:

Sohail Anwar, Pennsylvania State University
Ray Bachnak, Texas A&M University
G. Thomas Bellarmine, Florida A&M University
Harold Broberg, Indiana University Purdue University Indianapolis
William Conrad, Indiana University Purdue University Indianapolis
David Delker, Kansas State University
Robert Fladby, University of Nebraska
Yolanda Guran, Oregon Institute of Technology
Robert Herrick, Purdue University
Rajiv Kapadia, Minnesota State University
Edward Peterson, Arizona State University
Hesham E. Shaalan, Georgia Southern University
Andrezej H Trzynadlowki, University of Nebraska
Norman Zhou, University of Wisconsin – Stout

The author would also like to thank Delmar Publishers for their belief and commitment to this project, particularly Greg Clayton, acquisition editor, Michelle Cannistraci, development editor, and the rest their team. Their willingness to venture down untraditional paths in pedagogy and text production has provided the essential elements necessary to convert this unique idea into a viable contribution in the spread of learning.

Avenue for feedback

No system ever performs exactly as intended. In fact the quality of the results is more a function of the quality of the negative feedback. So please contact either the author or the publisher with your suggestions.

J. Michael Jacob
  175 Knoy Hall
  Purdue University
  W. Lafayette, IN 47907
  jacobm@purdue.edu

Michelle Ruelos Cannistraci
  3 Columbia Circle
  Albany, NY 12212-5015
  michelle.cannistraci@delmar.com

Dedication

This book is dedicated to all those with the courage to faithfully press forward, every day, regardless of adversity. Remember, an attempt is only a failure if you refuse to take the lessons learned and try again. In the end, it is about the journey, not the destination.

J. Michael Jacob (Mike)   June, 2001

1

Advanced Operational Amplifier Principles

Introduction

You must master the use of power supplies, power switching, thyristors, filtering, MOSFET switches, and linear amplifiers in order to accurately apply electrical power to a load. To manage these large voltage and high current signals, however, you must first have good control over small signals. This is best done with op amps.

You should already be familiar with an op amp's basic characteristics, and its use in noninverting and inverting small signal amplifiers. In this chapter, you will strengthen these foundations. The principle of negative feedback is then expanded to create a universal analysis technique, allowing you to unravel the performance of virtually any op amp based linear circuit. You will work through amplifiers with several negative feedback loops, active-integrator feedback, composite amplifiers and a suspended supply op amp based amplifier capable of an output of over 100 V.

These op amp based tools are illustrated with both Multisim and PSpice simulations. The ability to evaluate the performance of a power electronics design before you actually turn it on gives you much more information than the pool of melted plastic and silicon and the cloud of smoke that a single miscalculation or oversight produce.

Objectives

Upon completion of this chapter, you will be able to do the following:

- Define input, output, and gain characteristics of an op amp integrated circuit, and explain the implication of each in traditional noninverting, inverting, and difference amplifiers.

- Apply these characteristics to determine the performance of a wide variety of nontraditional, linear amplifiers, including composite amplifiers with active-integrator feedback and power output stages, and a suspended supply, high voltage amplifier.

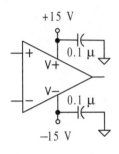

Figure 1-1 Op amp symbol

1.1 Op Amp Characteristics

The op amp's traditional schematic symbol is shown in Figure 1-1. Typically, symmetric dual power supplies are used (e.g., ±15 V). These supply voltages set the maximum amplifier voltages. The output voltage cannot be driven beyond these supplies. In addition, to prevent damage to the IC, you must assure that the voltages on each input lie between these supplies. It is also important to include decoupling capacitors (0.1 µF) connected *directly* between each power pin and ground. These assure stable, clean power and minimize the possibility that other amplifiers sharing the same power may influence the performance of this stage.

Internally, the operational amplifier consists of a high input impedance differential input followed by a very high gain voltage amplifier. The impedance looking into either input is 1 MΩ or more.

$$Z_{\text{in op amp}} \geq 1\,\text{M}\Omega$$

The gain of the op amp IC is called the open loop gain, A_{OL}. It varies radically. But under normal operation,

$$A_{\text{OL}} \gg 1000$$

Values of 100,000 to over 10,000,000 are not unusual. A simple block diagram of the op amp is shown in Figure 1-2.

$$e_{\text{out}} = A_{\text{OL}} e_{\text{d}}$$

$$e_{\text{out}} = A_{\text{OL}} (e_{\text{NI}} - e_{\text{INV}})$$

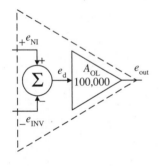

Figure 1-2 Op amp block diagram

It should seem that with A_{OL} varying from 1,000 to 10,000,000, the output would be totally unreliable. At times the output may be 1 mV. At a different time, or with a different IC of the same model, the output could be as large as 10 V with the same size input! In addition, with such large gains, the input difference must be too small to measure. Even with $A_{\text{OL}} = 1000$, an output of 10 V requires an input difference between e_{NI} and e_{INV} of 1 mV. Smaller outputs or larger gain require an even smaller input.

So, what use is the op amp anyway? The key lies in the negative input pin. A sample of the output must be *fed back* to the inverting input pin. This *negative* feedback drives the difference between the two input voltages to the required, negligibly small voltage. It takes advantage of the huge open loop gain of the IC to produce a stable, reliable output voltage, actually compensating for variations in that gain.

Look at the circuit in Figure 1-3. The output voltage is reduced by the feedback ratio, β, where $0 \leq \beta \leq 1$, and returned to the inverting input.

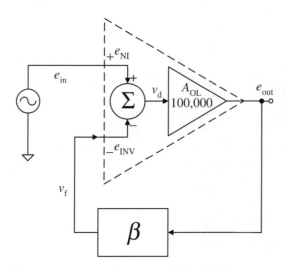

Figure 1-3 Negative feedback circuit block diagram

The output voltage depends on both A_{OL} and β. To determine this relationship, begin by finding the feedback voltage, v_f.

$$v_f = \beta e_{out}$$

$$v_d = e_{in} - v_f$$

$$v_d = e_{in} - \beta e_{out}$$

But e_{out} is set by this difference voltage.

$$e_{out} = A_{OL} v_d$$

$$= A_{OL}(e_{in} - \beta e_{out})$$

$$= A_{OL} e_{in} - \beta A_{OL} e_{out}$$

Move the e_{out} terms to the left side of the equation.

$$e_{\text{out}} + \beta A_{\text{OL}} e_{\text{out}} = A_{\text{OL}} e_{\text{in}}$$

$$e_{\text{out}} (1 + \beta A_{\text{OL}}) = A_{\text{OL}} e_{\text{in}}$$

$$e_{\text{out}} = \frac{A_{\text{OL}}}{1 + \beta A_{\text{OL}}} e_{\text{in}}$$

The closed loop gain, A_{CL}, is

$$A_{\text{CL}} = \frac{e_{\text{out}}}{e_{\text{in}}} = \frac{A_{\text{OL}}}{1 + \beta A_{\text{OL}}}$$

Practical values of feedback attenuation are

$$0.01 \le \beta \le 1$$

So, because of the very large A_{OL},

$$\beta A_{\text{OL}} \gg 1$$

This means that the 1 in the denominator can be ignored.

$$A_{\text{CL}} \approx \frac{A_{\text{OL}}}{\beta A_{\text{OL}}}$$

$$A_{\text{CL}} \approx \frac{1}{\beta} \tag{1-1}$$

The closed loop gain of an op amp based amplifier is set by the characteristics of the negative feedback, β.

$$A_{\text{CL}} \approx \frac{1}{\beta}$$

As long as the open loop gain is large enough, applying negative feedback allows A_{OL} to cancel out of the closed loop gain (A_{CL}) equation.

During this section, you have seen the following fundamental points about the performance of an op amp.

1. The voltage on the input and output pins must be kept between the limits set by the voltages on the power supply pins.

2. There is no significant current flowing into either input pin, because of the IC's very high input impedance.

3. There can be no significant difference in potential between the two input pins as long as the output is not saturated (maximum or minimum voltage as limited by the supplies). This is the result of the very high open loop gain.

4. When negative feedback is provided, the amplifier's closed loop gain is set (almost completely) by that negative feedback. As long as A_{OL} is large enough, its exact value does not matter.

Voltage Follower

The simplest amplifier you can build with an op amp is the voltage follower. Its schematic is shown in Figure 1-4. The input, e_{in}, is applied to the noninverting input. The output, e_{out}, is connected directly back to the inverting input.

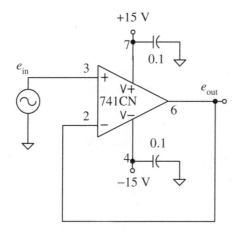

Figure 1-4 Voltage follower schematic

Assuming that the output is not saturated, the difference in potential between the noninverting and the inverting input pins is insignificant.

$$e_{INV} = e_{in}$$

The negative feedback allows the op amp to change its output to whatever voltage is needed to drive the inverting input voltage to the same voltage as the noninverting input.

Since the output voltage is the inverting input voltage,

$$e_{out} = e_{INV}$$

$$e_{\text{out}} = e_{\text{in}}$$

So, the closed loop gain of the voltage follower is

$$A_{\text{CL}} = \frac{e_{\text{out}}}{e_{\text{in}}} = 1$$

In terms of the feedback attenuation, β,

$$A_{\text{CL}} = \frac{1}{\beta}$$

But, since the feedback is not divided,

$$\beta = 1$$

Voltage follower gain

$$A_{\text{CL}} = 1$$

Either way you look at it, you get out what you put in. So, why bother? The answers are in the impedances. The op amps' input impedances are in the megohms. The voltage across that impedance is the difference between the noninverting and the inverting inputs. You have already seen that to be negligibly small. The current drawn from the input signal generator is

$$i_{\text{from signal}} = \frac{\mu V}{M\Omega} \approx 0A$$

The output of the op amp is an ideal voltage source. Up to the current limit of the op amp, the load can draw as much current from the op amp as is needed without loading down the output voltage.

Use the voltage follower as a **buffer**. Sprinkle them about liberally, anywhere you are concerned that a load may load degrade a source.

Example 1-1

multi**SIM**

   a. Calculate v_{out} with the switch open (no load), and with the switch closed (500 Ω load) for the circuit in Figure 1-5.

   b. Repeat the calculation using a voltage follower as a buffer between the voltage divider and the 500 Ω load.

Solution

   a. With the switch open, the voltage divider is unloaded.

$$v_{out\,open} = \frac{1\,k\Omega}{1\,k\Omega + 1\,k\Omega} 1\,V_{rms} = 0.5 V_{rms}$$

When the switch is closed, the 500 Ω resistor parallels the lower 1 kΩ part of the voltage divider.

$$R_{lower\,loaded} = \left(\frac{1}{1k\Omega} + \frac{1}{500\Omega} \right)^{-1} = 333\Omega$$

$$v_{out\,loaded} = \frac{333\Omega}{1k\Omega + 333\Omega} 1V_{rms} = 0.250 V_{rms}$$

So, closing the switch to connect the load to the voltage divider, cuts v_{out} in half. The load creates 50% loss.

b. Figure 1-6 shows a simulation using Multisim. The meters are all set to ac. The output of the voltage divider is 499.6 mV$_{rms}$. You calculated that it should be 500 mV$_{rms}$. There is, indeed, negligible current flowing from the voltage divider into the op amp. The load voltage matches the output of the unloaded divider.

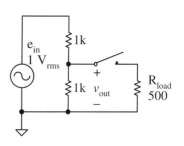

Figure 1-5 Schematic for Example 1-1

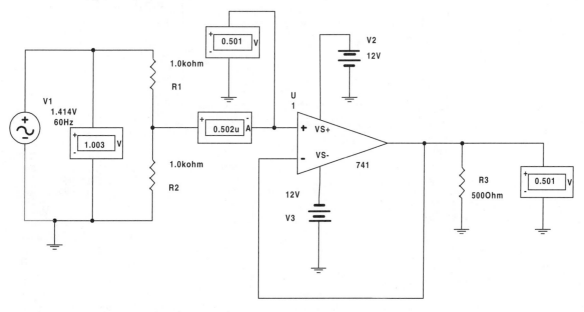

Figure 1-6 Simulation of the loaded voltage divider with a voltage follower buffer

---

Practice:

    a. What effect does the simulation Figure 1-6 predict if you lower the load to 4 Ω? Explain what happened.

    b. What happens with a 500 Ω load at 1 MHz? Explain.

Answer: **a.** $v_{out} = 98$ mV$_{rms}$ **b.** $v_{out} = 63.5$ mV$_{rms}$

Noninverting Amplifier

To increase the gain, you must reduce the amount of negative feedback. This is done by placing a voltage divider between the output and the inverting input. Look at Figure 1-7. The voltage divider, R_f and R_i, divide e_{out}, producing the feedback voltage, v_f.

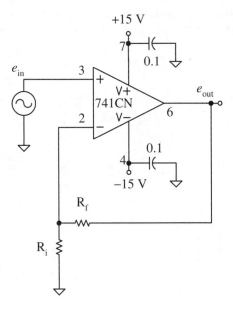

Figure 1-7 Noninverting amplifier

$$v_f = \frac{R_i}{R_i + R_f} e_{out}$$

Or,
$$v_f = \beta e_{out}$$

where β is the feedback factor.

But, according to Equation 1-1, the amplifier's closed loop gain, A_{CL}, is

$$A_{CL} = \frac{1}{\beta}$$

Substituting from the voltage divider relationship

$$\beta = \frac{R_i}{R_i + R_f}$$

$$A_{CL} = \frac{1}{\beta} = \frac{R_i + R_f}{R_i}$$

Simplifying gives

$$A_{CL} = 1 + \frac{R_f}{R_i} \qquad \textbf{(1-2)}$$

You can draw two conclusions from Equation 1-2.

The gain is positive. The output and input voltages are in phase. So this is a noninverting amplifier.

The smallest gain is 1. Reducing R_f to 0 Ω just turns this into a voltage follower.

There is a second, more powerful, way to analyze the operation of the noninverting amplifier of Figure 1-7.

1. Voltage on the noninverting input pin:

$$v_{NI} = e_{in}$$

2. Voltage on the inverting input pin:
   Since there is negative feedback, assuming that the output voltage is not in saturation,

$$v_{INV} = v_{NI} = e_{in}$$

3. Current through R_i:

$$i_{Ri} = \frac{v_{Ri}}{R_i}$$

But the voltage across R_i is the voltage on the inverting input pin, which is e_{in}.

$$i_{Ri} = \frac{e_{in}}{R_i}$$

4. Current through R_f:
    There is no current flow into or out of either input pin of the op amp because its input impedance is so large. So, the current through R_i flows through R_f.

$$i_{Rf} = i_{Ri} = \frac{e_{in}}{R_i}$$

5. Calculate the voltage dropped across the feedback resistor:
    This just requires Ohm's law.

$$e_{Rf} = i_{Rf} R_f$$

Substitute for the current through R_f.

$$e_{Rf} = \frac{e_{in}}{R_i} R_f$$

6. Calculate the output voltage:

$$e_{out} = v_{Ri} + e_{Rf}$$

$$= v_{INV} + \frac{e_{in}}{R_i} R_f$$

But you have already seen that the voltage on the inverting input pin is the same as the voltage on the noninverting input, which is e_{in}.

$$e_{out} = e_{in} + \frac{R_f}{R_i} e_{in}$$

Dividing both sides of the equation gives the closed loop gain.

$$A_{CL} = \frac{e_{out}}{e_{in}} = 1 + \frac{R_f}{R_i}$$

Noninverting amplifier gain

These six steps certainly seem more involved than using the $1/\beta$ relationship. But they work *every time* there is negative feedback, regardless of how complicated the circuit is. You will be using this technique many times.

Example 1-2

OrCAD

    a.   Design a noninverting amplifier with a gain of 100.

    b.   Verify the circuit's gain and input impedance by simulation.

Solution

    a.   Use the schematic in Figure 1-7. Pick $R_f = 100$ kΩ

$$A_{CL} = 1 + \frac{R_f}{R_i}$$

$$100 = 1 + \frac{100\,k\Omega}{R_i}$$

$$99 = \frac{100\,k\Omega}{R_i}$$

$$R_i = \frac{100\,k\Omega}{99} = 1.01 k\Omega$$

If you can tolerate a 1% error, use a 1 kΩ resistor for R_i. Otherwise you will have to use precision resistors.

    b.   The results of the Transfer Function Analysis are in the output file. They are shown in Figure 1-8.

```
V(out)/V_e_in   =   1.009E+02
INPUT RESISTANCE AT V_e_in  =   1.062E+09
OUTPUT RESISTANCE AT V(out)  =   7.694E-02
```

Figure 1-8 Probe output for Example 1-2

The simulation shows a gain of 100.9, and an input impedance of 1 GΩ. The PSpice simulation schematic is shown in Figure 1-9.

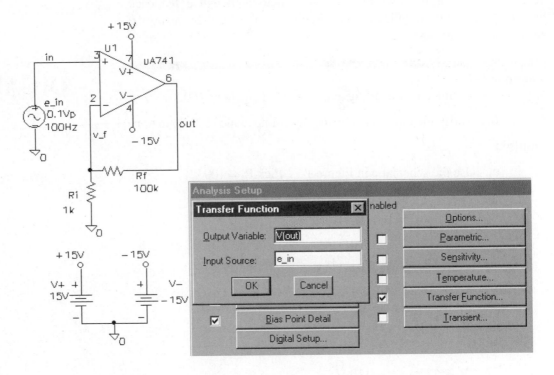

Figure 1-9 PSpice simulation schematic for Example 1-2

Practice:

a. What happens to the gain at 20 kHz? (Use Transient Analysis, not Transfer Function.)

b. Using the results of the simulation at 20 kHz, calculate A_{OL} of the 741 at 20 kHz. Remember $A_{CL} = \dfrac{A_{OL}}{1 + \beta A_{OL}}$

Answer: a. A_{CL} drops to 40. **b.** A_{OL} drops from ~100,000 to 66.

Proper layout is critical when you are building the circuit. Electromagnetic interference will induce tiny currents into all leads. These currents, flowing into the large input impedance of the op amp, produce a voltage difference that is multiplied by the open loop gain of the op amp. Even for small signals, this interference can produce very noticeable distortion at the output. Also, at high frequencies (where most electromagnetic interference exists), the inductance in the leads of a component may become dominant. The resistors and decoupling capacitors you choose become inductors to the high frequency interference causing the circuit to oscillate wildly.

Keep all leads as short as possible. Connect the resistors as close to the input pins as practical. Since the output is a very low impedance voltage source, it can easily overwhelm any electromagnetic interference generated currents. So, pull R_f back close to the inverting input pin. Look at Figure 1-10.

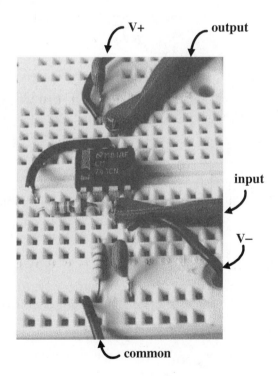

Figure 1-10 Proper layout of a noninverting amplifier (*Photo by John Zimmerman*)

Inverting Amplifier

You can reverse the phase relationship of the input and output by reversing the input connections. Just exchange the input voltage source and the circuit common, as shown in Figure 1-11(a). Notice carefully that the input connections to the *amplifier,* not to the IC, have been exchanged. A more traditional layout of the inverting amplifier is shown in Figure 1-11(b). Compare the two schematics carefully to assure yourself that they are equivalent.

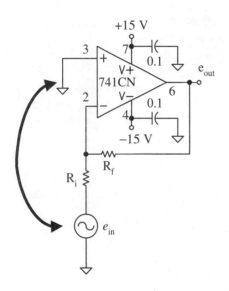

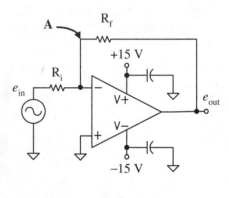

Figure 1-11 (a) Reversing inputs makes an inverting amplifier

Figure 1-11 (b) Traditional schematic

There is negative feedback from the output, through R_f, to the inverting input pin. But, establishing β is not obvious. So, to determine this amplifier's performance, apply the six steps used in the analysis of the noninverting amplifier.

1. Voltage on the noninverting input pin:

$$v_{NI} = 0\,V \quad \text{This pin is tied to common.}$$

2. Voltage on the inverting input pin:

Virtual ground

$$v_{INV} \approx v_{NI} = 0\,V$$

3. Current through R_i:

$$i_{Ri} = \frac{v_{Ri}}{R_i} = \frac{e_{in}}{R_i}$$

4. Current through R_f:

    There is no current flow into or out of either input pin of the op amp because its input impedance is so large. So, the current through R_i flows through R_f.

$$i_{Rf} = i_{Ri} = \frac{e_{in}}{R_i}$$

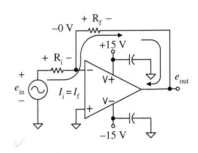

Figure 1-12   Inverting amplifier analysis

5. Calculate the voltage dropped across the feedback resistor:
    This just requires Ohm's law.

$$e_{Rf} = i_{Rf} R_f$$

Substitute for i_{Rf}, the current through R_f:

$$e_{Rf} = \frac{e_{in}}{R_i} R_f$$

6. Calculate the output voltage:

$$e_{out} = -e_{Rf}$$

Since current flows through R_f from left to right, the voltage at the right end of R_f must be more negative than its left end. But, the left end of R_f is at ground potential. So, the output voltage must go negative when the input goes positive, driving and pulling the current from left to right.

    Combining the results of steps 5 and 6,

$$e_{out} = -\frac{R_f}{R_i} e_{in}$$

Dividing both sides of the equation gives the closed loop gain.

$$A_{CL} = \frac{e_{out}}{e_{in}} = -\frac{R_f}{R_i} \qquad \textbf{(1-3)}$$

Inverting amplifier gain

    The other key issue with an inverting amplifier is its input impedance. With the noninverting amplifier of the previous section, the input voltage was connected directly to the op amp's noninverting input pin. The result of the negative feedback to the other input pin was to drive the impedance up to 2 GΩ for the 741. However, the input voltage of the inverting amplifier is connected to R_i. The other end of R_i is tied to virtual ground, that is, to ground potential. So, R_i is between the input voltage and ground potential. For the inverting amplifier, then,

Inverting amplifier input impedance

$$Z_{\text{input inverting amp}} = R_i$$

Example 1-3

Calculate the voltage at the output of the circuit shown in Figure 1-13.

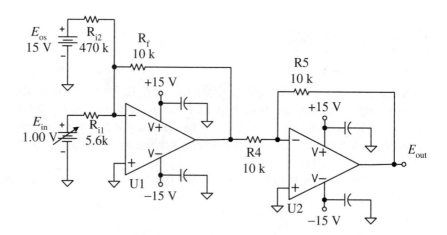

Figure 1-13 Schematic for Example 1-3

Solution

The two sources, E_{in} and E_{os}, R_{i1}, R_{i2}, R_f and the op amp U1 form an inverting **summer**. Resistors R4 and R5, and the op amp U2 are an inverting amplifier with a gain of -1. So, E_{out} is the negative of the voltage out of U1.

1. Voltage on the noninverting input pin of U1:

$$V_{NI} = 0\,V \quad \text{This pin is tied to common.}$$

2. Voltage on the inverting input pin:
   Since there is negative feedback, assuming that the output voltage is not in saturation,

$$V_{INV} = V_{NI} = 0\,V$$

3a. Current through R_{i1}:

$$I_{Ri1} = \frac{V_{Ri1}}{R_{i1}} = \frac{1\,V}{5.6\,k\Omega} = 179\,\mu A \rightarrow$$

   This current flows from left to right, entering the node at U1's inverting input.

3b. Current through R_{i2}:

$$I_{Ri2} = \frac{V_{Ri2}}{R_{i2}} = \frac{15\,V}{470\,k\Omega} = 32\,\mu A \rightarrow$$

   This current also flows from left to right, entering the node at U1's inverting input.

4. Current through R_f:
   There is no current flow into or out of either input pin of the op amp because its input impedance is so large. So, the current entering the node from R_{i1} and R_{i2} must leave the node through R_f.

$$I_{Rf} = I_{Ri1} + I_{Ri2}$$

$$I_{Rf} = 179\,\mu A + 32\,\mu A = 211\,\mu A \rightarrow$$

   This is why the circuit is called a **summer**.

5. Calculate the voltage dropped across the feedback resistor:

$$V_{Rf} = I_{Rf} R_f$$

$$V_{Rf} = 211\mu A \times 10k\Omega = 2.11V$$

6. Calculate the output voltage:

$$E_{outU1} = -V_{Rf} = -2.11V$$

The unity inverter, U2, flips this voltage over

$$E_{out} = -E_{out1} = -1 \times -2.11V = 2.11V$$

This circuit is also called a **zero and span converter**. It is used to convert the voltages out of a sensor (E_{in}) to voltages that are more meaningful. This circuit converts the 10 mV/°C from E_{in} to 10 mV/°F. The boiling point of water, 100°C, causes E_{in} = 1.00 V and an output of 2.11 V (~212°F).

These currents and voltages are verified by the Multisim simulation in Figure 1-14.

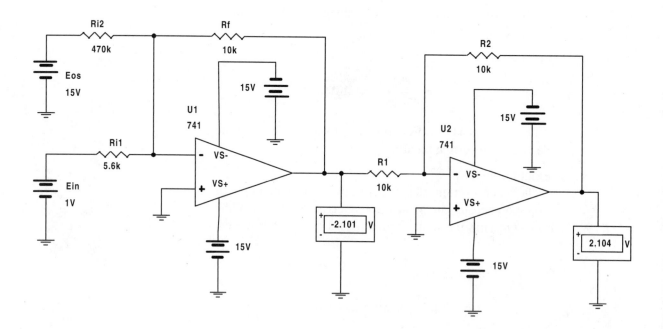

Figure 1-14 Multisim simulation of Example 1-3

Practice:

a. Calculate the output voltage when the temperature sensor is at 0°C, $E_{in} = 0$ V.

b. Calculate new values for R_{i1} and R_{i2} so that $E_{in} = -0.1$ V $(-10°C)$, $E_{out} = 0$ V; and $E_{in} = 0.4$ V $(40°C)$, $E_{out} = 5$ V.

Answer: a. $E_{out} = 0.32$ V b. $R_{i1} = 1$ kΩ, $R_{i2} = 150$ kΩ

Even though the input impedance of the inverting amplifier is set by R_i, the op amp IC still amplifies any difference in potential between its two input pins by its open loop gain. So, any interference induced onto the connections from the op amp's inverting pin to R_i and to R_f appears greatly amplified at the output. The same is true for any signal injected onto the noninverting pin (even though it is *supposed* to be ground). Just as with the noninverting op amp based amplifier, layout is important. Follow the example shown in Figure 1-15 closely.

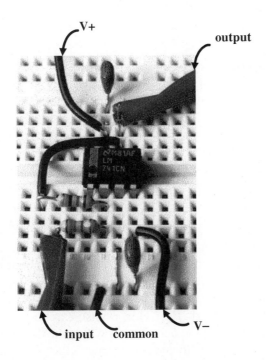

Figure 1-15 Proper layout of an inverting op amp based amplifier
(*Photo by John Zimmerman*)

1.2 A Universal Analysis Technique

During the discussion of the noninverting and the inverting amplifiers, you have seen the same six steps applied. In fact, *anytime* a linear circuit uses negative feedback, with none of the op amps in saturation, these same six steps can be used. They form a standard procedure which allows you to analyze a complex circuit that may not have a simple gain equation, or a circuit configuration that you have never seen before.

1. **Voltage on the noninverting input pin:**
   This may be as simple as recognizing that the noninverting pin is tied to ground. Or, you may have to use the voltage divider law, or some other circuit analysis technique. But since there is no significant signal current flowing into the noninverting input pin, this step is usually not too complicated.

2. **Voltage on the inverting input pin:**
   Since there is negative feedback, assuming that the output voltage is not in saturation,

$$v_{INV} = v_{NI}$$

3. **Current through each of the input resistors:**
   Usually you only have to apply Ohm's law. But, if the voltage at the inverting input is not ground, you must use the *difference* in voltage across each input resistor. There may also be current sources connected to the op amp's input pin, such as current out digital-to-analog converters, photo diodes, or sensors. So, you must know how each of these elements work.
   It is critical that you keep track of the *direction* of each of these currents. Consider placing an arrow beside each current magnitude to identify its direction. Plus and minus signs are easily confused.

4. **Current through R_f:**
   There is no current flow into or out of either input pin of the op amp because its input impedance is so large. So, the currents that flow into and out of the node at the op amp's inverting pin combine and then flow through R_f. This is just Kirchhoff's current law. What goes in must come out. This is where the direction arrows are handy, telling you how to add and/or subtract the currents.

5. **Calculate the voltage dropped across the feedback resistor:**
    This just requires Ohm's law.

$$e_{Rf} = i_{Rf} R_f$$

If the feedback element is not a resistor, then apply the current-to-voltage relationship for that element. It may be the differential equation of a capacitor or inductor, or it may be a logarithm for a diode or bipolar transistor.

Be sure to mark the polarity of this voltage drop across the feedback element.

6. **Calculate the output voltage:**
    This is just Kirchhoff's voltage law. The voltage at the output is the voltage at the inverting input pin (step 2) plus or minus the voltage across the negative feedback.

$$e_{out} = v_{INV} \pm v_{feedback}$$

Watch the polarity of the voltage across the feedback element so you know whether to add or to subtract.

That's all there is to it! Even if you have never seen the circuit before, if there is no text that gives you the gain equation, if it is an involved, multi-feedback loop monster dreamed up by some technical sadist, it is as easy as 1-2-3-4-5-6. Ohm's law, Kirchhoff's current and voltage laws, and the characteristics of the devices in the circuits are all you need.

Difference Amplifier

The difference amplifier has two inputs. Its output is the difference between these two input voltages multiplied by a fixed amplifier gain.

$$e_{out} = \frac{R_f}{R_i}\left(e_{in+} - e_{in-}\right)$$

The schematic is shown in Figure 1-16. Let's apply the six universal analysis steps to derive the equation for the difference amplifier's output voltage.

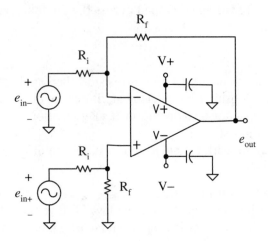

Figure 1-16 Difference amplifier

1. Voltage on the noninverting input pin:
   Use the voltage divider law.

$$v_{NI} = \frac{R_f}{R_i + R_f} e_{in+}$$

2. Voltage on the inverting input pin:
   Since there is negative feedback, assuming that the output voltage is not in saturation,

$$v_{INV} = v_{NI} = \frac{R_f}{R_f + R_i} e_{in+}$$

3 Current through R_{i1}:
   Assuming that $e_{in+} > e_{in-}$

$$i_{Ri} = \frac{v_{INV} - e_{in-}}{R_i} \leftarrow$$

Substitute for v_{INV} from step 2.

$$i_{Ri} = \frac{\dfrac{R_f}{R_f + R_i} e_{in+} - e_{in-}}{R_i} \leftarrow$$

4. Current through R_f:

   There is no current flow into or out of either input pin of the op amp because its input impedance is so large. So, the current entering the node from R_{i1} must leave the node through R_f.

$$i_{Rf} = i_{Ri} \leftarrow$$

5. Calculate the voltage dropped across the feedback resistor:

$$v_{Rf} = i_{Rf} R_f \quad - \text{ to } +$$

   Substitute for i_{Rf} from step 3.

$$v_{Rf} = \frac{R_f}{R_i} \left(\frac{R_f}{R_f + R_i} e_{in+} - e_{in-} \right)$$

6. Calculate the output voltage:

$$e_{out} = v_{inv} + v_{Rf}$$

The analysis is done. But there are several more algebraic steps still needed to produce the simplified formula. First, substitute for v_{Rf}.

$$e_{out} = v_{inv} + \frac{R_f}{R_i} \frac{R_f}{R_f + R_i} e_{in+} - \frac{R_f}{R_i} e_{in-}$$

$$e_{out} = \frac{R_f}{R_f + R_i} e_{in+} + \frac{R_f}{R_i} \frac{R_f}{R_f + R_i} e_{in+} - \frac{R_f}{R_i} e_{in-}$$

The first two terms are factors of $\dfrac{R_f}{R_f + R_i} e_{in+}$. So, factor that out.

$$e_{out} = \left(\frac{R_f}{R_f + R_i} e_{in+} \right) \left(1 + \frac{R_f}{R_i} \right) - \frac{R_f}{R_i} e_{in-}$$

Now, rearrange $\left(1 + \dfrac{R_f}{R_i} \right)$.

$$e_{out} = \left(\frac{R_f}{R_f + R_i} e_{in+} \right) \left(\frac{R_f + R_i}{R_i} \right) - \frac{R_f}{R_i} e_{in-}$$

The denominator of the first fraction cancels with the numerator of the second fraction.

$$e_{out} = \frac{R_f}{R_i} e_{in+} - \frac{R_f}{R_i} e_{in-}$$

Factor out the common fraction.

Differential amplifier output

$$e_{out} = \frac{R_f}{R_i} \left(e_{in+} - e_{in-} \right)$$

One of the major uses of the difference amplifier is in rejecting signals that are the same on each input (called common mode noise). In many industrial settings, a small voltage must be sent a long way. Along the way, noise is coupled into the wire. Often this noise is larger than the signal itself. The solution is to send two signals, equal in amplitude, but opposite in phase. Along the transmission path the noise is coupled into each wire exactly the same. At the receiver, the two wires are applied to a differential amplifier. The noise is the same on each side. So it subtracts out. The signals on each wire are opposite. The differential amplifier gives them a gain of R_f/R_i. The noise disappears and the signal is amplified. This is illustrated in the Multisim simulation shown in Figure 1-17.

The 1 V/60 Hz noise source is common to the two dc input signals. It is 100 times larger than either dc signal. These inputs are equal in amplitude, but opposite in polarity. At the output of the differential amplifier the noise has been dropped to only 3 mV, while the dc signal is 2 V. This is a improvement of over 67,000 in the signal-to-noise ratio. The tiny signals have been transmitted through a noisy environment.

But this assumes that the two input resistors perfectly match each other at R_i, and the two R_f are exactly matched to each other. What happens if you build this circuit with 5% resistors? Well, there is no formula with four different resistor values. Use the six step universal analysis approach.

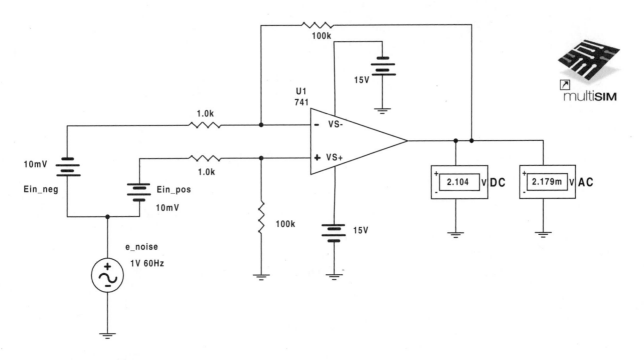

Figure 1-17 Multisim simulation of a differential amplifier rejecting noise

---

Example 1-4

Calculate the output voltage for the circuit in Figure 1-18.

Solution

With perfectly matched resistors, E_{out} should be 0 V. Even considering the nonideal characteristics of the op amp, the output should be in the 10 mV range.

To determine the effect of 5% variation in the resistors, apply the universal analysis technique.

1. Voltage on the noninverting input pin of U1:

$$V_{NI} = \frac{105\,k\Omega}{105\,k\Omega + 950\,\Omega} 5\,V = 4.955\,V$$

2. Voltage on the inverting input pin:

$$V_{INV} = V_{NI} = 4.955\,V$$

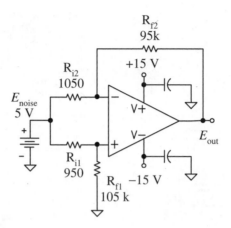

Figure 1-18 Schematic for Example 1-4

3. Current through R_{i2}:

$$I_{Ri2} = \frac{5V - 4.955V}{1050\Omega} = 42.86\mu A \rightarrow$$

4. Current through R_{f2}:

$$I_{Rf2} = I_{Ri2} = 42.86\mu A \rightarrow$$

5. Calculate the voltage dropped across the feedback resistor:

$$V_{Rf2} = I_{Rf2}R_{f2}$$

$$V_{Rf} = 42.86\mu A \times 95k\Omega = 4.071V \quad + \text{to} -$$

6. Calculate the output voltage:

$$E_{out} = V_{INV} - V_{Rf2}$$

$$E_{out} = 4.955V - 4.071V = 884mV$$

Since the output produced by the two *differential* 10 mV signals is only 2 V, E_{out} = 884 mV (from the 5 V common mode inputs) produces an error of 40%. This is caused by a 5% mismatch in resistor values. The Multisim simulation is shown in Figure 1-19.

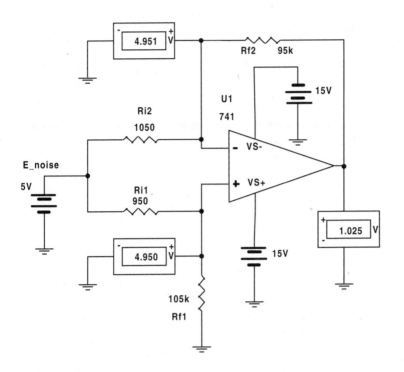

Figure 1-19 Multisim simulation of Example 1-4

Practice:

Repeat the calculations and simulation with $R_{i1} = 1050\ \Omega$, $R_{i2} = 950\ \Omega$, $R_{f1} = 95\ k\Omega$, $R_{f2} = 105\ k\Omega$

Answer: $E_{out} = -1.096$ V

Instrumentation Amplifier

The difference amplifier you saw in the previous section has two signifi-cant shortcomings. First, the input impedance seen by e_{in-} depends on the voltage at e_{in+}. So it is not fixed, and does not match the impedance at the other input. Second, and much worse, is that to adjust the gain you have to adjust both R_fs while keeping them exactly equal. Even a 1% difference can seriously degrade the amplifier's performance.

The two op amp instrumentation amplifier shown in Figure 1-20 solves both of these problems. Each input is connected directly into an op amp's noninverting input. These impedances are typically in the GΩ range. The single resistor, R_g, sets the gain. So there are no matching problems.

Example 1-5

Analyze the circuit in Figure 1-20 for all currents and voltages.

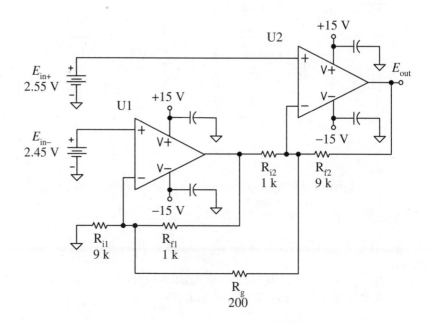

Figure 1-20 Two op amp instrumentation amplifier

Solution

   1. Noninverting input voltages:

$$V_{NIU1} = 2.45\,V \qquad V_{NIU2} = 2.55\,V$$

   2. Inverting voltages:

$$V_{INVU1} = 2.45\,V \qquad V_{INVU2} = 2.55\,V$$

3. Input resistor currents, U1:

$$I_{Ri1} = \frac{2.45\,V}{9\,k\Omega} = 272.2\,\mu A \leftarrow$$

$$I_{Rg} = \frac{2.55\,V - 2.45\,V}{200\,\Omega} = 500\,\mu A \leftarrow$$

4. Feedback resistor current, U1:

$$I_{Rf1} = 500\,\mu A - 272\,\mu A = 228\,\mu A \rightarrow$$

5. Feedback resistor voltage, U1:

$$V_{Rf1} = 228\,\mu A \times 1\,k\Omega = 228\,mV \quad +\,to-$$

6. Output voltage, U1:

$$E_{outU1} = 2.45\,V - 0.228\,V = 2.222\,V$$

Now, go back through the steps 3 through 6 for U2.

3. Input resistor currents, U2:

$$I_{Ri2} = \frac{2.55\,V - 2.222\,V}{1\,k\Omega} = 328\,\mu A \leftarrow$$

4. Feedback resistor current, U2:

$$I_{Rf2} = 328\,\mu A + 500\,\mu A = 828\,\mu A \leftarrow$$

5. Feedback resistor voltage, U2:

$$V_{Rf2} = 828\,\mu A \times 9\,k\Omega = 7.45\,V \quad -\,to+$$

6. Output voltage, U2:

$$E_{out} = 7.45\,V + 2.55\,V = 10.00\,V$$

Practice:
   a. Repeat the manual analysis with $R_g = 400\ \Omega$.
   b. Does R_g directly or inversely affect the output?
   c. Verify your calculations with a simulation.

Answer: $E_{out} = 5.50$ V

1.3 Composite Amplifier

The composite amplifier uses several amplifiers, each with its own particular strengths and weaknesses, combining them in such a way as to overcome the weaknesses of each, while exploiting their strengths. Usually this involves enclosing one amplifier within the negative feedback loop of the other.

For power amplifiers, this usually means combining an op amp and a driver. The op amp is accurate, stable, linear, and inexpensive, but can output only a few milliamperes. Power drivers are made of bipolar junction transistors, MOSFETs, or linear regulator ICs. You have to change their input voltage 0.6 V to 5 V before the part provides any output at all. The input-to-output relationship is often exponential. In fact, the two types of transistors are current output devices. Their output voltage depends not only on the input, but also on the load. Alone, they make very poor linear amplifiers. However, each is capable of providing tens of amperes of current to the load. Current limiting may be built-in, or easily added. This assures that should a failure occur, the load current does not exceed a preset limit, protecting the load and the amplifier. In fact, thermal sensing and shutdown are also available. If the power driver gets too hot, it turns itself off.

Used alone the op amp is accurate, but weak. The power driver can deliver a lot of power safely to the load, but seriously distorts the signal. Properly combined into a composite amplifier, the op amp prevents the distortion introduced by the power driver, and the power driver provides the current and protection that are outside an op amp's normal operation. And, this does not require an expensive, exotic op amp or a complex power stage. It only requires proper negative feedback.

Power Output Stage

The circuit in Figure 1-21 is a simple composite amplifier. Its role is to deliver a voltage from 0 V to 10.24 V to a load as the digital code into the digital-to-analog converter varies from 0_{10} to 255_{10}. The AD7524 outputs a current into virtual ground that is

$$I_{out} = \frac{V_{ref}}{10\,k\Omega} \times \frac{code}{256_{10}}$$

With a negative reference, the DAC sinks I_{out}.

The load's resistance is 10 Ω. So, at the highest voltage level, the load draws 1 A of current. It is also important to provide current limiting and thermal protection. The LM317 IC linear regulator does this. Its two key operating limitations are that its input voltage must always be at least 2 V greater than its output. Also, the output voltage is set 1.2 V more positive than the voltage placed on its adjustment pin.

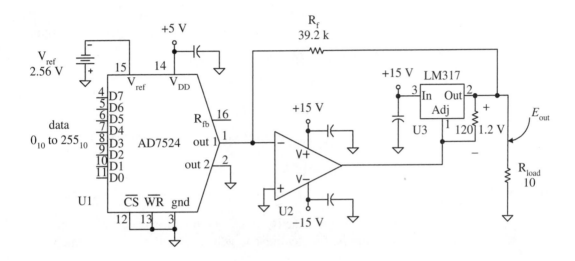

Figure 1-21 Composite amplifier with a DAC, op amp, and IC regulator

---

Example 1-6

Calculate all currents and voltages in the composite amplifier of Figure 1-21 when the data is 128_{10}. Also determine the power delivered to the load and the power dissipated by the LM317.

Solution

There is no simple gain equation for this circuit. So apply the six steps of the universal analysis technique.

1. Noninverting input voltages:

$$V_{NIU2} = 0V \quad \text{circuit common}$$

2. Inverting voltages:

$$V_{\text{INVU2}} = 0\,\text{V} \quad \text{virtual ground}$$

3. Input resistor currents, U1:

$$I_{\text{outU1}} = \frac{2.56\,\text{V}}{10\,\text{k}\Omega} \times \frac{128}{256} = 128\,\mu\text{A} \leftarrow$$

The current direction is from virtual ground at U2's inverting input, into U1, and then out U1's V_{ref} pin to the −2.56 V source.

4. Feedback resistor current, U1:

$$I_{\text{Rf}} = I_{\text{U1}} = 128\,\mu\text{A} \leftarrow$$

5. Feedback resistor voltage, U1:

$$V_{\text{Rf}} = 128\,\mu\text{A} \times 39.2\,\text{k}\Omega = 5.02\,\text{V} \quad -\text{to}+$$

6. Output voltage, U1:

$$E_{\text{out}} = V_{\text{U2INV}} + V_{\text{Rf}} = 0\,\text{V} + 5.02\,\text{V} = 5.02\,\text{V}$$

This is exactly what you should expect. When the code is set to half scale, the load voltage is half of the full scale voltage.

$$E_{\text{outU2}} = E_{\text{out}} - 1.2\,\text{V} = 5.02\,\text{V} - 1.2\,\text{V} = 3.78\,\text{V}$$

When negative feedback is applied, and the op amp does not saturate, its output goes to whatever potential is necessary to assure that there is no significant difference in potential between its inputs. In this circuit, the op amp lowered its output 1.2 V to accommodate the 1.2 V drop required by the LM317.

Notice that we did not try to go from the input directly to the output through the op amp and the LM317. Instead, using Ohm's law, Kirchhoff's laws, and the characteristics of the parts, we followed the negative feedback loop around the outer edge.

$$I_{\text{load}} = \frac{5\,\text{V}_{\text{dc}}}{10\,\Omega} = 0.5\,\text{A}_{\text{dc}}$$

$$P_{\text{load}} = \frac{V_{\text{load}}^2}{R_{\text{load}}} = \frac{(5\,\text{V}_{\text{dc}})^2}{10\,\Omega} = 2.5\,\text{W}$$

$$P_{LM317} = V_{across\ LM317} I_{through\ LM317}$$

$$P_{LM317} = (15\,V_{dc} - 5\,V_{dc}) \times 0.5\,A_{dc} = 5\,W$$

This is certainly much more current than a traditional op amp can provide, and much more power than it could dissipate. But by enclosing the power driver, LM317, inside the op amp's negative feedback loop, this composite amplifier is able to accurately deliver several watts to the load.

Practice:

Calculate the load voltage and the op amp's output voltage with data = 15_{10}.

Answer: $E_{load} = 0.59$ V, $E_{op\ amp} = -0.61$ V

Active Feedback

Composite amplifiers may also include an amplifier as part of the negative feedback loop. Look at Figure 1-22. The current through the load is sampled by R_{sense}, amplified by the difference amplifier, and fed back to U1. That op amp drives the load through the power transistor Q1.

Example 1-6

Calculate all currents and voltages in the circuit shown in Figure 1-22 when $E_{in} = 4\ V_{dc}$.

Solution

1. Noninverting input voltage, U1:

    $$V_{NIU1} = E_{in} = 4V$$

2. Inverting voltage, U1:

    $$V_{INVU1} = V_{NIU1} = 4V \qquad E_{out\,U2} = V_{INV\,U1} = 4\,V$$

3. Input (and feedback) resistors current, U1:

    $$I_{Rf\,\&\,Ri} = \frac{4\,V}{10\,k\Omega + 1.1\,k\Omega} = 360.4\,\mu A$$

At this point you have to modify the analysis technique a little. But you still only need Ohm's law and Kirchhoff's laws.

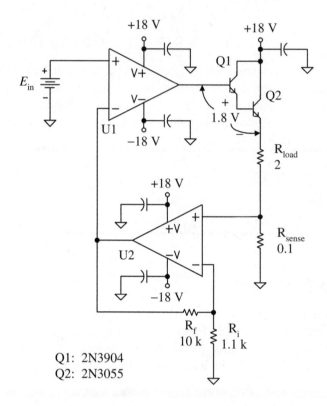

Figure 1-22 Active feedback within a composite amplifier

$$V_{\text{INVU2}} = I_{\text{Ri}}R_i = 360.4\,\mu\text{A} \times 1.1\,\text{k}\Omega$$

$$V_{\text{INVU2}} = 396.4\,\text{mV}$$

Since U2 has negative feedback, its two input voltages should be equal.

$$V_{\text{NIU2}} = V_{\text{INVU2}} = 396.4\,\text{mV}$$

This is the voltage across R_{sense}.

$$I_{\text{Rsense}} = \frac{396.4\,\text{mV}}{0.1\Omega} = 3.964\,\text{A}$$

This is the current through the load. An input voltage of 4 V produces a load current of 4 A. The composite amplifier uses its negative feedback to sense the current through the load, convert that to voltage, and apply the voltage as negative feedback to U1. Transistors Q1 and Q2 are power drivers, providing the 4 A required, but requiring only a much smaller current from the op amp.

The voltage at the emitter of Q2 is

$$V_{Q2emitter} = I(R_{load} + R_{sense})$$

$$V_{Q2emitter} = 3.96\,A(2\,\Omega + 0.1\,\Omega) = 8.32\,V$$

A power transistor may drop 1 V or more between its base and emitter leads. So for the Darlington pair of Q1 and Q2 a drop of 1.8 V is reasonable.

$$E_{out\ U1} = 8.32\,V + 1.8\,V = 10.1\,V$$

The results of a Multisim simulation of this circuit is shown in Figure 1-23. There is good correlation to the manual calculations above.

Practice:

Replace the load with an 8 Ω resistor. Determine how large of a current this circuit can provide to the load before the op amp's output runs into the 18 V power supply.

Answer:

Between 1.5 A and 2 A, depending on the actual saturation voltage of the op amp and the base-emitter drop of Q1 and Q2.

1.4 Suspended Supply Amplifier

The composite amplifiers in the previous section allow you to provide amperes of current to a low resistance load while achieving the precision of the op amp. Both the op amp and the current driver ICs or transistors are low priced. But by tucking the power driver inside the negative feedback loop of the op amp, you get the best of both parts. A feast fit for a king from popcorn priced parts!

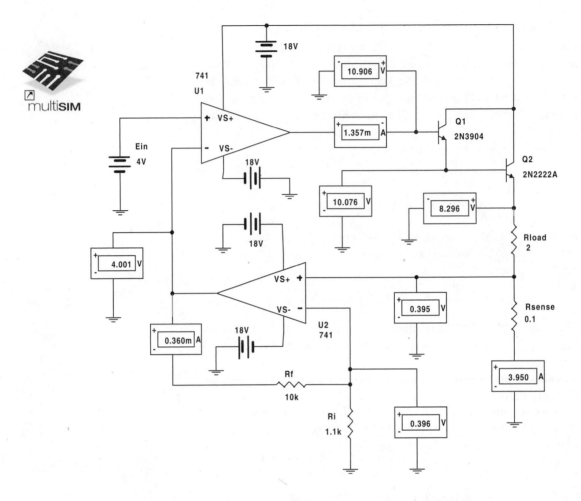

Figure 1-23 Multisim simulation of the composite amplifier in Example 1-6

But just how much power can you deliver this way? That depends on the load resistance, and the limitations of the op amp. Look at Example 1-7.

Example 1-7

For a composite amplifier, such as in Figures 1-22 and 1-23, calculate:

a.   the power you can deliver to an 8 Ω load with a sine wave.

b.   the load voltage needed to provide 100 W to the 8 Ω load.

Solution

a. Common op amps have a maximum power supply voltage of 18 V. When driven to the extreme, the output can come within about 2 V of the power supply. Also, from the simulation at the end of Example 1-6, you saw that there is a 2 V drop across the transistors' base-emitter junctions.

$$e_{\text{load p}} = 18\,\text{V} - 2\,\text{V} - 2\,\text{V} = 14\,\text{V}_{\text{p}}$$

$$e_{\text{load rms}} = \frac{14\,\text{V}_{\text{p}}}{\sqrt{2}} = 10\,\text{V}_{\text{rms}}$$

Ignoring the small drop across R_{sense}, the most voltage available to the load is 10 V_{rms}. Applied to an 8 Ω load this means

$$P_{\text{load}} = \frac{V_{\text{load rms}}^2}{R_{\text{load}}} = \frac{(10\,\text{V}_{\text{rms}})^2}{8\Omega}$$

$$P_{\text{load}} = 12.5\,\text{W}$$

This is pretty limited performance. It may be acceptable to only deliver 12 W to a table top speaker, but even the inexpensive audio systems routinely provide 25 W or more.

b. To calculate the load voltage needed to deliver 100 W of sinusoidal power to an 8 Ω load, rearrange the power equation above.

$$V_{\text{load rms}} = \sqrt{P_{\text{load}} R_{\text{load}}}$$

$$V_{\text{load rms}} = \sqrt{100\text{W} \times 8\Omega} = 28.3\,\text{V}_{\text{rms}}$$

Practice: There are 4 Ω speakers available. Repeat the calculations in Example 1-7, replacing the 8 Ω speaker with a 4 Ω speaker.

Answer: a. $P_{\text{load}} = 25$ W b. $V_{\text{load p}} = 20.0$ V_{rms}

So, a big element in delivering power to a load is to get more *voltage* out of the op amp. There are commercial op amps available with supply voltages of 100 V or more. But, their price is proportionally higher as well. A second option is to find or build a voltage driver from transis-

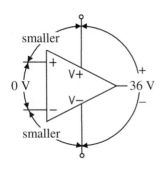

Figure 1-24 Op amp voltage limitations

tors that can boost the voltage from the op amp. This has been the approach for many years. It requires several stages of discrete transistor amplifiers. Understanding the op amp's limitations more thoroughly leads to a much simpler, cheaper, and more elegant solution. Look at Figure 1-24.

For common op amps, the maximum difference in potential between the two power pins is 36 V. With proper feedback, the inverting input pin is driven to equal the noninverting input pin. The *only* other voltage restriction is that the output and input pins be greater than the voltage on the V− pin, and less than the voltage on the V+ pin. There is no ground pin.

It is perfectly fine to apply +68 V to V+, and +32 V on V−. This means that the output pin may be at 50 V. Or, the power pins may be −32 V on V+, and −68 V on V−. This means that e_{out} could be −50 V. But how do you get the power supply values to vary? Actually, it's as simple as moving a single wire. Look at Figures 1-25(a) and (b). To suspend the power supplies, remove their connection to circuit common, and connect their common node to the op amp's output. When the output goes to +50 V, the op amp drags its supplies to +68 V and +32 V. If the output drops to −50 V, the supplies follow it down, providing −32 V and −68V to the op amp. That's it!

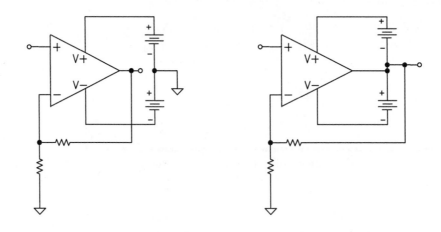

(a) Traditional grounded supplies   (b) Suspended supplies

Figure 1-25 Suspending an op amp's power supplies

Suspending the Supply

You can build the suspended supply regulator by using a zener diode and a bipolar transistor buffer. The positive side is shown in Figure 1-26. An external, ground referenced supply is required. It must be about 10 V greater than the largest output you want from the op amp. An 8 V zener is used. This sets the dc difference between the output and V+. Pick it large enough to give the op amp plenty of bias room. However, the zener also limits how close to the +56 V power supply the output can come. So, 12 V or 15 V zeners are larger than needed.

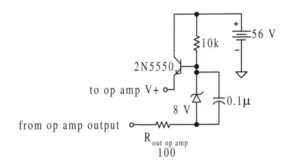

Figure 1-26 Positive suspended supply regulator

The 10 kΩ resistor limits the current through the zener. The resistor keeps the zener biased in its breakdown region (typically 100 μA), but limits the current to a level that assures the zener does not have to dissipate too much power. When the op amp's output swings up to 46 V (near its peak), the voltage at the bottom of the 10 kΩ resistor is

$$V = 46\,\text{V} + 8\,\text{V} = 54\,\text{V}$$

This leaves 2 V across the 10 kΩ resistor.

$$I_{min} = \frac{56\,\text{V} - 54\,\text{V}}{10\,\text{k}\Omega} = 200\,\mu\text{A}$$

This is plenty of current to keep the zener diode biased at its zener voltage, even as the output voltage reaches its peak.

For symmetric operation, a comparable negative suspended supply drives the op amp's V−. At the negative extreme, it is possible that the output of the op amp may go as far negative as it went positive. This puts a large voltage across the resistor.

$$V = 56\,\text{V} - (-46\,\text{V}) = 102\,\text{V}$$

The maximum current through the resistor, then is

$$I_{max} = \frac{102\,\text{V}}{10\,\text{k}\Omega} = 10\,\text{mA}$$

The resistor must be able to dissipate

$$P_{10\,\text{k}\Omega} = (10\,\text{mA})^2 \times 10\,\text{k}\Omega = 1\,\text{W}$$

The zener diode dissipates

$$P_{zener} = 10\,\text{mA} \times 8\,\text{V} = 80\,\text{mW}$$

The zener sets the dc voltage at the base of the transistor. But zener diodes are power supply parts, and do not easily pass high frequencies. And remember, the key to the suspended supply is that it drives V+ to 8 V above the output. The op amp's power pin must move as the op amp's output moves, but 8 V more positive at all times. So, the capacitor looks like a short to the higher frequencies, assuring that these signals easily pass from the output of the op amp to V+, raised by the zener's 8 V.

Many op amps cannot drive the capacitors directly. The phase shift that the capacitors introduce, when combined with the phase shift across the op amp's internal capacitor, causes the op amp to oscillate. So $R_{\text{out op amp}}$ is added to reduce this phase shift and prevent the oscillations.

The transistor buffers the zener voltage, just as you may have seen in a transistor boosted zener regulated power supply. Pick a transistor that is faster than the highest frequency signal that is to pass through the op amp. The only other consideration is the transistor's maximum collector-to-emitter specification. The transistor's emitter follows the op amp's output signal down to within 8 V of the maximum negative peak. The transistor's collector is held to the external power supply's voltage.

$$V_{CE\,max} = E_{+\,supply} - \left(V_{negative\,peak} + V_{zener} \right)$$

Using the same values above,

$$V_{CE\,max} = 56\,\text{V} - (-46\,\text{V} + 8\,\text{V}) = 94\,\text{V}$$

Pick a transistor rated above this voltage.

Example 1-8

Verify the operation of a positive and negative suspended supply by simulation. Drive the node from the op amp with a 51 V$_p$ 20 kHz sine wave. Display V+ and V−.

OrCAD

Solution

The PSpice simulation is shown in Figure 1-27.

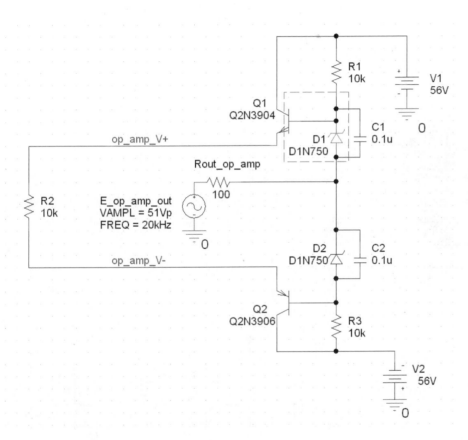

Figure 1-27 PSpice simulation schematic for Example 1-8

Several small changes are needed to simulate the circuit with PSpice. The op amp has been replaced with R2. This allows the proper continuity from V+ to V−, without having to deal with the complexity of the op amp. That comes in the next section. There are no high voltage transistors in the evaluation version of the software. But, simulating the 2N3904 and 2N3906 beyond their maximum V_{CE} does not cause the failure it would in the lab. Finally, the model of the zener diodes has a 4.7 V breakdown. This alters the results a little. So keep this in mind when you look at the Probe plot. The results are shown in Figure 1-28. The input is a 51 V_p sine wave. The positive power supply rides more positive, and V− is more negative. Even though the output swings a total of 100 V_{PP}, the *difference* between V+ and V− is a *constant*.

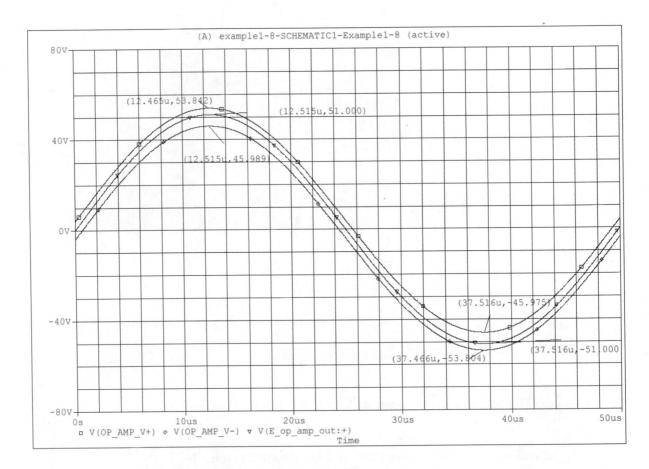

Figure 1-28 Probe of V+, the op amp's output, and V−

Practice:

    a. What is the maximum op amp output from the circuit in Figure 1-27 before the supply voltages begin to flatten?

    b. What is the effect on the maximum op amp output of changing the zener from 8 V to 15 V? Explain.

Answer: **a.** 51 V_p **b.** 43 V_p

The Amplifier

The noninverting op amp amplifier that drives the suspended supply needs one modification. And, this one change significantly alters the amplifier's performance. Look at Figure 1-29.

    The 0.1 µF decoupling capacitor is placed *between* the two power pins, instead of to common. Remember, the whole purpose of the suspended supply is to put a signal on each power pin with respect to ground. So shorting these signals to common through the decoupling capacitor is the opposite of what you want. From Figure 1-27 you saw that the *difference* between the two power pins must be the same. So the decoupling capacitor places an ac short between these pins, assuring that any noise bypasses the op amp entirely.

    The other change is the *positive* feedback created by R_a and R_b. You must assure that voltage on the noninverting input pin lies between V+ and V−. That's the purpose of these resistors. The worst case is when the output is at its peak, 45 V for the values used. Though there is some small voltage from the input signal generator, it is small. So, for simplicity, it is ignored.

$$V_{\text{NI opamp}} = \frac{R_a}{R_a + R_b} E_{\text{out p}}$$

$$V_{\text{NI opamp}} = \frac{10\,\text{k}\Omega}{10\,\text{k}\Omega + 1.8\,\text{k}\Omega} 45\,V_p = 38\,V_p$$

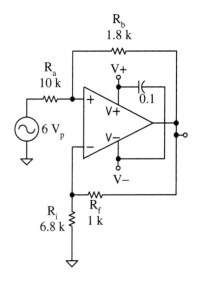

Figure 1-29 The amplifier

From Figure 1-28, you saw that the negative supply pin is about 8 V more negative than the output of the op amp, putting V− at 37.7 V when the output is 45 V.

    Be sure to keep the resistors you choose in the kΩ range. Much smaller values require too much current from the op amp at the high voltage peaks. Larger resistors combine with parasitic capacitance in the circuit to form low pass filters, limiting the high frequency performance of the amplifier.

Controlled, predictable performance requires *negative* feedback. Rf and Ri provide this. However, the positive feedback through R_a and R_b radically alters the gain equation. The derivation applies the six analysis steps and a page or two of algebra. For that last reason, the derivation is omitted here.

Suspended amplifier gain

$$A_v = \frac{R_i R_b + R_f R_b}{R_i R_b - R_f R_a}$$

Example 1-9

Apply the six steps of the universal analysis technique to calculate the output from the amplifier in Figure 1-29. Verify the results with a simulation.

multiSIM

Solution

The positive feedback requires a little algebra and patience.

1. Noninverting input voltages:

$$I_{Ra\&Rb} = \frac{E_{out} - 6\,V}{10\,k\Omega + 1.8\,k\Omega} = 84.75\,\mu S \times E_{out} = 508.5\,\mu A \;\leftarrow$$

$$V_{NI} = 6\,V + I_{Ra}R_a$$

$$= 6\,V + (84.75\,\mu S \times E_{out} - 508.5\,\mu A)10\,k\Omega$$

$$= 6\,V + 0.8475E_{out} - 5.085\,V$$

$$V_{NI} = 0.915\,V + 0.8475E_{out}$$

2. Inverting voltages:

$$V_{INV} = 0.915\,V + 0.8475E_{out}$$

3. Input resistor currents:

$$I_{Ri} = \frac{V_{INV}}{R_i} = \frac{0.915\,V + 0.8475E_{out}}{6.8\,k\Omega} \;\downarrow$$

$$I_{Ri} = 134.6\,\mu A + 124.6\,\mu S \times E_{out} \;\downarrow$$

Output voltage, U1:

$$E_{out} = I_{Ri}(1\,k\Omega + 6.8\,k\Omega)$$

$$= 7.8\,k\Omega(134.6\,\mu A + 124.6\,\mu S \times E_{out})$$

$$= 1.050\,V + 0.9719E_{out}$$

$$E_{out} - 0.9719E_{out} = 1.050\,V$$

$$0.02812E_{out} = 1.050\,V$$

$$E_{out} = \frac{1.050\,V}{0.02812} = 37.34\,V$$

The Multisim display is shown in Figure 1-30. Notice that the supplies are suspended. The simulated output voltage correlates well with the manual calculations. Notice that the voltage on the two inputs are equal and between

$$V+ = 37.56\,V + 8\,V = 45.56\,V$$

and $$V- = 37.56\,V - 8\,V = 29.56\,V$$

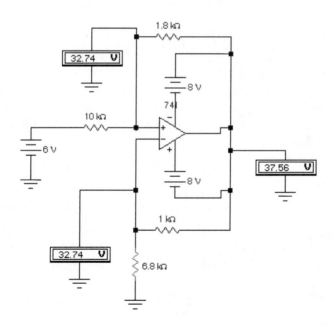

Figure 1-30 Simulation for Example 1-9

Practice:
    a.   Verify the output by using the gain equation.
    b.   Calculate a new value for R_f to give a gain of 10.

Answer: **a.** $A_v = 6.27$ **b.** $R_f = 1.082$ kΩ

Tolerances, Tolerances, Tolerances!

Look carefully at the second part of the Exploration above. Changing R_f from 1 kΩ to 1.08 kΩ changes the gain from 6.3 to 10. That is, a 8% change in one resistor changes the gain 59%! Why is the amplifier so sensitive? The gain equation has a negative sign in the denominator.

$$A_v = \frac{R_i R_b + R_f R_b}{R_i R_b - R_f R_a}$$

As $R_i R_f$ falls close to $R_f R_a$, the denominator approaches zero. As the denominator get smaller and smaller, the gain increases radically. For a stable amplifier,

$$R_i R_b > R_f R_a$$

The larger the inequality, the more stable the gain.

    Since you cannot buy (at least for production) resistors that are exactly what you specified, it is important to take into account this variation. To build a stable amplifier with 5% resistors, even if R_i and R_b are small and R_f and R_a are large,

$$0.95 R_i \times 0.95 R_b > 1.05 R_f \times 1.05 R_a$$

5% resistors   Or, $R_i > \dfrac{1.22 R_f R_a}{R_b}$ for 5% resistors.

Using 1% resistors

$$0.99 R_i \times 0.99 R_b > 1.01 R_f \times 1.01 R_a$$

1% resistors   Or, $R_i > \dfrac{1.04 R_f R_a}{R_b}$ for 1% resistors.

    Mathematically, you can set the gain to any value you want by adjusting R_f or R_i. But in practice, the range of R_f and R_i are constrained to assure that the denominator of the gain equation does not go to zero, causing the amplifier to oscillate.

For the circuit in Figure 1-28, with 5% resistors

$$R_i > \frac{1.22 R_f R_a}{R_b} = \frac{1.22 \times 1 k\Omega \times 10 k\Omega}{1.8 k\Omega}$$

$$R_i > 6.778 k\Omega$$

Since $R_i = 6.8\ k\Omega \pm 5\%$, this amplifier will be stable. But, with R_i this close to the limit, it is reasonable to expect wide variation in the gain, as different resistors (within the 5% range) are used.

Example 1-10

OrCAD

Use PSpice's Worst Case Analysis to determine the maximum and the minimum gains from this amplifier built with 5% resistors.

Solution

Enter the schematic just as you saw it in Figure 1-30. However, use a 1 V_{dc} input, instead of the 6 Vdc power supply. This way, the output voltage matches the gain.

The results of the analysis is at the bottom of the Output File. The nominal gain is 6.26, just what is expected from using the gain equation. The Worst Case Analyses discovered that by increasing R_a and R_f by 5% each, and decreasing R_b and R_i by 5% each, the output voltage goes to 119.7.To find the minimum gain, select the MIN button in the Function section of the Worst Case Analysis. The results from the end of the two Output files are shown in Figure 1-31.

R_Ra R_Ra R .95 (Decreased)				R_Ra R_Ra R 1.05 (Increased)			
R_Rb R_Rb R 1.05 (Increased)				R_Rb R_Rb R .95 (Decreased)			
R_Rf R_Rf R .95 (Decreased)				R_Rf R_Rf R 1.05 (Increased)			
R_Ri R_Ri R 1.05 (Increased)				R_Ri R_Ri R .95 (Decreased)			
RUN MINIMUM VALUE				RUN MAXIMUM VALUE			
ALL DEVICES 3.4209 at V_Ein = 1				ALL DEVICES 119.7 at V_Ein = 1			
(54.578% of Nominal)				(1.9097E+03% of Nominal)			
NOMINAL 6.2679 at V_Ein = 1				NOMINAL 6.2679 at V_Ein = 1			

Figure 1-31 Output file results for Worst Case Analysis of Example 1-10

To set the tolerance of each resistor, double click on that resistor and add the tolerance in the appropriate line, as shown in Figure 1-32. Be sure to add the tolerance for each resistor.

Set up a DC Analysis. This properly completed dialog box is also shown in Figure 1-32. The Worst Case Analysis dialog box in Figure 1-32 is set to determine the maximum output.

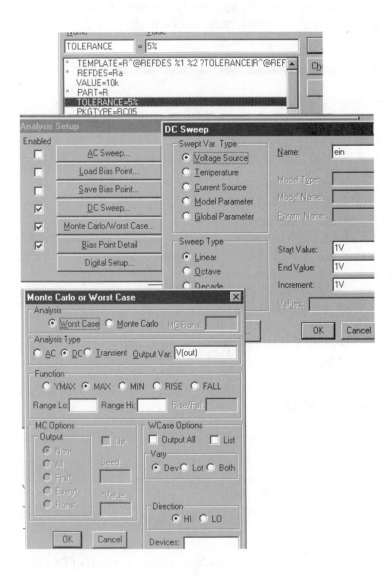

Figure 1-32 Dialog boxes needed to run the Worst Case Analysis for Example 1-10

Again, the nominal gain is 6.26. The analysis found a minimum gain of only 3.4, half of the design specifications.

So, with 5% resistors the gain varies from 3.4 to 120.

Practice:

   a. Repeat the Worst Case Analysis with $R_i = 15 \, k\Omega \pm 5\%$.

   b. Repeat the Worst Case Analysis with 1% resistors.

   c. Evaluate these efforts to lower the gain variation.

Answer:

   a. Gain varies from 1.52 to 1.96, too small to be of value.

   b. Gain varies from 5.32 to 7.68, still over 15% variation.

Composite Amplifier

The suspended supply amplifier can output as large a voltage as your application needs. It only requires an op amp, seven resistors, a capacitor, two transistors, and two zener diodes. All told, that is only a few dollars worth of parts. But, even if you use expensive, precision resistors, the results vary widely. This is just not an acceptable result for a circuit you plan to use in production. The solution is to enclose the suspended supply amplifier within the negative feedback loop of a composite amplifier, as shown in Figure 1-33.

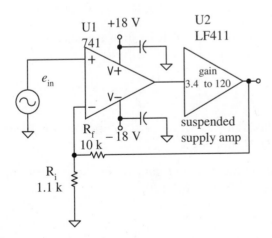

Figure 1-33 Composite amplifier with the suspended supply amp

The entire suspended supply amplifier – op amp, resistors, capacitor, transistors, diodes, and widely varying gain – is represented by the triangle. The op amp used within the suspended supply amplifier (U2) must be considerably faster than the outer op amp (U1). Otherwise, U2's output responds more slowly than its input, introducing a phase lag. At some frequency this lag causes the feedback to be in-phase, rather than being negative feedback. The amplifiers oscillate. So a LF411 is used within the inner (suspended supply) amplifier.

Example 1-11

a. Calculate all voltages and currents for the circuit in Figure 1-33, assuming $e_{in} = 4.5$ V$_p$ and a 3.4 gain from the suspended supply amplifier.

b. Repeat the calculations with a gain of 120 from the suspended supply amplifier.

Solution

a.

$$v_{NI} = e_{in} = 4.5\,\text{V}_p$$

$$v_{INV} = v_{NI} = 4.5\,\text{V}_p$$

$$i_{Ri} = \frac{4.5\,\text{V}_p}{1.1\,\text{k}\Omega} = 4.091\,\text{mA}_p \downarrow$$

$$e_{out} = 4.091\,\text{mA}_p \times (10\,\text{k}\Omega + 1.1\,\text{k}\Omega) = 45.4\,\text{V}_p$$

With a gain of 3.4 from the suspended supply amplifier, the output of U1 is

$$e_{U1out} = \frac{45.4\,\text{V}_p}{3.4} = 13.3\,\text{V}_p$$

This value is within the abilities of the 741 running from ±18 V supplies.

b. Assuming that the gain of the suspended supply amp increases from 3.4 to 120,

$$v_{NI} = e_{in} = 4.5\,\text{V}_p$$

$$v_{INV} = v_{NI} = 4.5\,\text{V}_p$$

$$i_{Ri} = \frac{4.5\,V_p}{1.1\,k\Omega} = 4.091 mA_p \downarrow$$

$$e_{out} = 4.091\,mA_p \times (10\,k\Omega + 1.1\,k\Omega) = 45.4\,V_p$$

This is *exactly* the same output you calculated in part a, when the amplifier had a much smaller gain.

With a gain of 120 from the suspended supply amplifier, the output of U1 is

$$e_{U1out} = \frac{45.4\,V_p}{120} = 0.378\,V_p$$

So, the 741 op amp changes its output to whatever voltage is needed to assure that the difference between its input pins is insignificant. If the encircled suspended supply amplifier has a small gain, the 741 outputs a large signal. If the suspended supply amplifier's gain goes up, the negative feedback allows the 741 to mirror the change by proportionally dropping its output.

The 741 (outer amp) gets the overall gain right. The suspended supply amplifier (inner amp) produces the high voltage needed at the output.

Practice:

Set up a simulation of the suspended supply amplifier combined with a 741 op amp to form a composite amplifier. Set the outer amplifier's R_i = 1.1 kΩ and R_f = 10 kΩ. Monitor the output voltage and the voltage out of the 741. Apply an input of 4.5 V. Run the simulation three times, to determine the output voltage from each op amp when the suspended supply's R_f = 900 Ω, 1 kΩ, and 1.2 kΩ.

Answer:

R_f = 900 Ω,	$E_{out\ 741}$ = 10.6 V,	$E_{out\ LF411}$ = 45.4 V
R_f = 1 kΩ,	$E_{out\ 741}$ = 7.3 V,	$E_{out\ LF411}$ = 45.4 V
R_f = 1.2 kΩ,	$E_{out\ 741}$ = 0.8 V,	$E_{out\ LF411}$ = 45.4 V

Figure 1-34 is a complete schematic of the composite amplifier, using a 741 for the output amplifier, a LF411 for the high voltage inner amplifier, and the suspended supply.

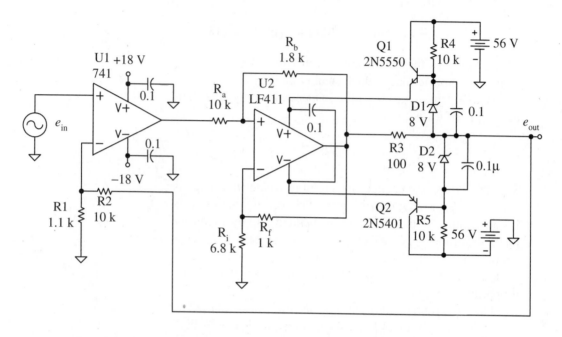

Figure 1-34 Complete high voltage, suspended supply, composite amplifier

Summary

To properly manipulate and apply power to a load, you must first be able to accurately handle small signals. This is most effectively done with an operational amplifier integrated circuit. The op amp IC has a very large input impedance. This means that there is no significant signal current flowing into its input pins. The output voltage is the open loop gain, multiplied by the difference in potential between the two input pins. Since the open loop gain is usually several thousands or more, as long as the output voltage is reasonable (not saturation), the input difference must be insignificant.

To take advantage of these two characteristics, an op amp is operated with negative feedback. This is often done by dividing the output voltage with a voltage divider (β). The output voltage is driven to whatever voltage is necessary to make the inverting input (through the negative feedback) equal the noninverting input. Since the output voltage is divided to make it equal to the input voltage, the output voltage must be larger than the input. This gain is set, almost exclusively, by the negative feedback voltage divider. So, the closed loop performance of the op amp is determined by the negative feedback voltage divider ($1/\beta$), rather than the op amp IC's characteristics (A_{OL}).

You saw three traditional op amp based amplifiers; the voltage follower, the noninverting amplifier, and the inverting amplifier and summer. The voltage follower has a voltage gain of 1 with a very large input impedance. Its main function is as a buffer, to keep a load from reducing the voltage from a source. The noninverting amplifier also has a very large input impedance. Its gain is

$$A_{\text{noninverting amp}} = 1 + \frac{R_f}{R_i}$$

The inverting amplifier grounds the noninverting pin. The inverting input pin is then driven to 0 V (virtual ground) through the negative feedback. Since the input pin is at ground potential, the input impedance is

$$Z_{\text{input inverting amp}} = R_i$$

The gain for an inverting amplifier is

$$A_{\text{inverting amp}} = -\frac{R_f}{R_i}$$

Any amplifier with negative feedback can be analyzed by calculating the following: voltage on the noninverting pin, voltage on the inverting pin, current through each element tied to the inverting pin, current through the feedback pin, voltage across the feedback element, and output voltage. This technique was applied to the difference amplifier, and a two op amp instrumentation amplifier.

The composite amplifier employs several amps, combined to take advantage of the strength of each while overcoming each IC's weaknesses. The accuracy of an op amp can be combined with the power and protection of a voltage regulator by driving the regulator with the op amp, and connecting the load to the regulator. The key is to tie the negative feedback to the load. This encloses the regulator inside the op amp's feedback. Active feedback can also be implemented to produce a voltage in–high current out amplifier. The six analysis steps apply to composite amplifiers too.

High power usually requires voltages much larger than is traditional from an op amp. The suspended supply amplifier allows the output of the op amp to drive the common of its supplies, dragging them up and down as needed to create an output of 50 V_p or more from a simple op amp IC. Positive feedback is applied to keep the input pins within the supplies. This changes the gain equation, but the six analysis steps still apply. Resistor tolerance can radically alter the amplifier's gain. So care must be take to assure that the amplifier is stable. A composite amplifier produces an inexpensive, stable, accurate, high voltage amplifier.

Layout is as important to the correct performance of an amplifier as the selection of its components. Keep leads to the inputs as short as possible, beware of parallel components, and be sure to decouple the power pins directly at the IC. This is important in the small signal amplifiers you saw in this chapter. When you combine these op amps with high voltage, high current, high temperature power stages, even a small amount of interference will devastate performance. Where you place the parts and how you interconnect them are critical. Circuit layout is not a rare art form, or black magic. There are a few techniques that, when followed rigorously, assure you a circuit that works when you build it just as it was designed. These techniques are presented in a step-by-step sequence in the next chapter, and applied to both a breadboard and a printed circuit board implementation.

Problems

Op Amp Characteristics

1-1   Calculate the input difference in potential ($e_{NI} - e_{INV}$) for an op amp amplifier with an output voltage of 10 V and an open loop gain of
   a.   2000.
   b.   30,000.

1-2   Locate specifications for the LM741 and the LF411 (National Semiconductor). For each determine the worst case
   a.   input impedance.
   b.   open loop gain.
   c.   gain bandwidth.

1-3   Assuming $\beta = 0.05$, and $A_{OL} = 2000$, calculate the closed loop gain using the *approximation*.

1-4 Repeat Problem 1-3 using the *exact* equation for closed loop gain. Compare the results of the exact equation with the approximation.

Voltage Follower

1-5 Repeat the simulation for Example 1-1 and Figure 1-6 using
   a. a 741 with the source at 500 kHz.
   b. a LF411 with the source at 500 kHz.
   c. Compare the results.

1-6 Repeat the simulation for Example 1-1 and Figure 1-6 with the source voltage set to 9 V_{max} and the power supplies set to ±5 V. Explain the results.

Noninverting Amplifier

1-7 Calculate all indicated currents and voltages for the circuit in Figure 1-35.

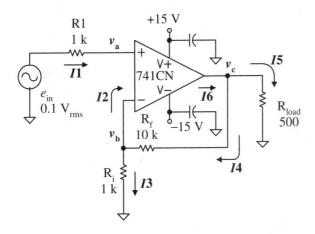

Figure 1-35 Noninverting amplifier for Problem 1-7

1-8 Design a noninverting amplifier with an input impedance of 10 kΩ and a gain that is adjustable from 5 to 20. Verify its input impedance and gain with a simulation. Set $e_{in} = 0.1$ V at 1 kHz.

Inverting Amplifier

1-9 Calculate the currents and voltages for the circuit in Figure 1-36.

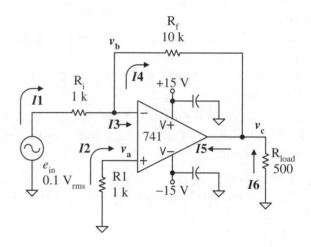

Figure 1-36 Inverting amplifier for Problem 1-9

1-10 Design an inverting amplifier with a gain of −30 and an input impedance greater than 3 kΩ.

1-11 Design a zero and span circuit (Figure 1-13) that converts an input range of 0.8 V to 4.4 V into an output range of 0 V to 5 V.

1-12 Calculate the dc and ac currents and voltages for the circuit in Figure 1-37.

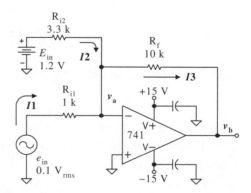

Figure 1-37 Inverting summer schematic for Problem 1-12

Universal Analysis Technique

1-13   Calculate all of the currents and voltages for the circuit in Figure 1-38.

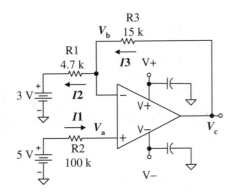

Figure 1-38 Schematic for Problem 1-13.

1-14   Calculate all of the currents and voltages for the circuit in Figure 1-39 for: **a.** $I1 = 20$ mA **b.** $I1 = 4$ mA.

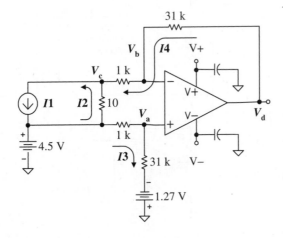

Figure 1-39 Schematic for Problem 1-14

Composite Amplifier

1-15   **a.** Design a composite power amplifier that outputs 15 V into a 10 Ω grounded load when the input is 2.56 V_{dc}.

      b. Calculate all currents and voltages in your design when the input is 1.28 V_{dc}.

      c. Can your design be used for an ac input? Explain.

1-16   Modify the design of an active feedback amplifier in Figure 1-22 if it is necessary to tie the 2 Ω R_{load} directly to ground, and place the 0.1 Ω R_{sense} between the load and the transistor.

1-17   **a.** Calculate all indicated dc voltages and currents in the circuit in Figure 1-40. The op amp in the feedback is called an integrator.

      b. Is the capacitor charging or discharging?

      c. What is the dc output voltage when the capacitor stops charging?

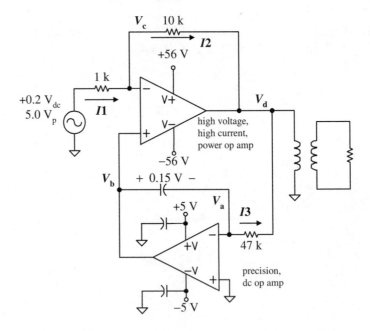

Figure 1-40 Integrator active feedback

1-18 Search data books or the World Wide Web to find op amps to be used in the circuit shown in Figure 1-40. The power op amp must be able to output at least 500 mA. The precision op amp must have an input offset voltage less than 50 µV, and input bias currents less than 1 nA.

Suspended Supply Amplifier

1-19 For the circuit in Figure 1-34, calculate the current through each component and the voltage at each node, for a 3.5 V_{dc} input. For dc inputs you may ignore all capacitors and assume that the drop across R3 is 0 V.

1-20 Repeat Problem 1-19 for an input of −4.0 V_{dc}.

1-21 Calculate new resistor values for the suspended amplifier to provide an output of up to 100 V_p, with a nominal gain of 20. Use 1% resistors.

Suspended Supply Amp Lab Exercise

A. Suspended Supply
  1. Build the circuit in Figure 1-41. Set $\pm V_{HI} = \pm\,18$ V.

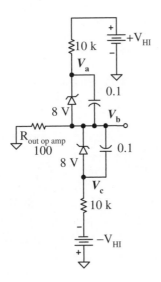

Figure 1-41 The zener part of the suspended supply

2. Apply power and measure the three dc voltages (V_a, V_b, V_c). Do *not* continue until these voltages are correct.

3. Add the transistors and the 10 kΩ emitter resistor shown in Figure 1-42. Connect the input (at the left of the 100 Ω resistor) to ground. Set $\pm V_{HI} = \pm 18$ V.

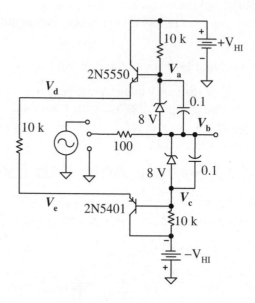

Figure 1-42 The completed suspended supply

4. Measure and record the five voltages indicated in Figure 1-42 with a digital multimeter (V_a, V_b, V_c, V_d, V_e). Assure that they are all correct before continuing.

5. Replace the ground connected to the 100 Ω resistor with the signal generator. Set the generator to 10 V_p at 1 kHz. Display waveforms at v_d and v_e on the oscilloscope. When they are correct, accurately record them.

B. The Amplifier
1. Build the amplifier as shown in Figure 1-43.

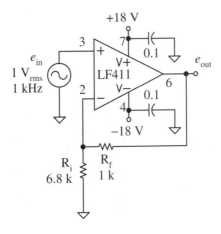

Figure 1-43 Simple noninverting amplifier

2. Measure R_f and R_i. Using these values, calculate the output magnitude you expect.

3. Display the output and input on the oscilloscope. Measure and record the output magnitude and dc with a digital multimeter. Do not continue if your measured values are more than 5% different from those calculated in step B2.

4. Add R_a and R_b as indicated in Figure 1-44. Be sure to measure and record their value.

5. Using the measured values of the four resistors, calculate the output magnitude you expect.

6. Display the output and input on the oscilloscope. Measure and record the output magnitude and dc with a digital multimeter. Do not continue if your measured values are more than 5% different from those calculated in step B2.

7. Connect the amplifier to the suspended supply as shown in Figure 1-45. Be sure to remove one of the op amp's decoupling capacitors and connect the other *between* the power supply pins.

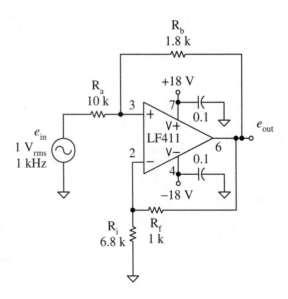

Figure 1-44 Four resistor feedback amplifier

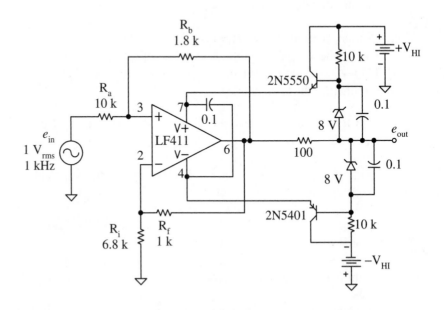

Figure 1-45 Suspended supply amplifier

8. Set $\pm V_{HI}$ to ± 18 V.

9. Apply power. Display the output and input on the oscilloscope. Measure and record the output magnitude and dc with a digital multimeter. Do not continue if your measured values are more than 5% different from those calculated in step B2.

10. Display the op amp's two power supply voltages on the oscilloscope. Carefully record these signals.

11. Measure and record the ac and dc difference in potential between V+ and V− with a digital multimeter.

12. When you have the circuit from Figure 1-45 working correctly, set $\pm V_{HI}$ to ± 56 V.

13. Repeat steps B9 through B11.

14. Increase the signal generator's amplitude while monitoring the wave shape at the output. Set the amplitude to just below the point at which the output begins to clip.

15. Repeat steps B9 through B11.

2

Power Electronics Circuit Layout

Introduction

How you build the circuit you want to test is as critical to its perform-ance as the parts you choose. Properly placed and correctly intercon-nected, your circuit components give you performance that closely fol-lows your calculations and simulation. However, careless placement or poor wiring assures that the large signal from the output contaminates the input, causing the gain to drop, or inducing oscillations everywhere.

The rules for printed circuit board (PCB) layout parallel those for good breadboarding. It is particularly frustrating to somehow manage to get the circuit to work on the breadboard, but then when the PCB is de-signed and built, oscillations reappear or the signal is just plain ugly.

Fear not, however. Proper layout is not black magic. It simply re-quires that you abandon bad habits, strictly follow a proven series of steps, and take little for granted. You define the mechanical details, identify the critical signal path, and place those components first. Con-nections are made only after all parts are in place. Wires, traces and pads must all be sized to handle the current you expect. Proper power connection and star connecting the common returns must also be assured to keep low level input signals and large power outputs apart. Step by step, the examples in this chapter take you through the proper bread-board and PCB layouts of a power amplifier.

Objectives

In this chapter, you will complete the layout of a power electronics cir-cuit on a breadboard and a printed circuit board. You will be able to:

- Define the mechanical requirements of the circuit.

- Identify the critical path.

- Place the components correctly.

- Select the correct interconnector sizes and route signal, power, and circuit common connections.

- Discuss other concerns.

2.1 Mechanical and Thermal Details

Before you can begin to place the parts on the breadboard or the printed circuit board there are several mechanical and thermal questions you must answer.

Determine board size first.

1. How big is the board?

    For a prototype testing arrangement, this is not usually of major concern. You are going to lay the breadboard on a bench top and cluster the support equipment and test instruments around it. Space and physical configuration are not constraints. But, even at this early stage, it is important to plan ahead. Normally, you prototype and test a circuit in stages. It is important to know how large the complete circuit will eventually be. Do you need to leave space on the breadboard for additional circuitry? What does that future circuit do (that affects where you put the extra space)? Are there to be other boards? If so, you need to plan now where on the test bench they go, and how the signals and power are to flow back and forth. A scale drawing or a photocopy of your bare board with blocks drawn for the locations of all the key elements prevents you from having to rebuild the circuit just to get a little more room as the project grows.

    When it comes to creating the printed circuit board, a well designed breadboard provides a good starting point. Besides knowing the size of the finished board, find out if you can adjust this as the layout evolves. It may make your layout much simpler if you could just add ¼″ along the bottom edge for a common return trace. Since your board's size affects the entire system, there may be many people involved in granting you that extra space. So determine the absolute maximum dimensions as well as the preferred size at the beginning.

    You also need to know what is around your board. Relays generate noise, light bulbs increase the temperature, and openings to the outside affect where you put the heat sinks and the controls. Do not forget that the circuit is three dimensional. Find out how much space *above* the board you are allowed. It is very discouraging to find that the board works well, but the heat sink or capacitor is too tall to fit into the case. So, get all of this information before you start to lay out the circuit board.

2. How is the board to be mounted?

    It may seem a little strange to talk about mounting a breadboard. After all, it's just going to lie on the bench while you test it. But you must attach leads for the power, common return, signal, and load. Test instruments must be connected all over the place. These appendages make the board very unwieldy. Moving one probe, or the slightest bump, shifts everything, perhaps even flipping the board over. Bare connections short out, inputs slip off, amplifiers go to the rail, high voltage is shorted to the input of a low voltage amplifier, and massive amounts of short circuit current char the board before you can turn off the power. So, secure *everything*. Tape the board to the bench. Tape the leads and probes so that their motion is greatly restricted. It takes a few extra minutes, but saves hours of work destroyed in less than a second by a slip. Do not apply power until you are sure nothing can move.

    When you design the printed circuit board, plan for a generous amount of strong mounting. Mechanical failure due to vibration or shock is common, so provide good support. Decide *now* where the mounting hardware is going. You need to know so that it does not interfere with the placement of components or traces.

    Do not plan to carry the weight of the populated board on its connectors. They are designed to pass the signals, not to be a mechanical brace.

3. What connectors are you going to use and where do they go?

    Bringing signals or power onto your protoboard in a simple wire is fine. You do not need to use special connectors. Plug the wire into a hole immediately adjacent to the component to which it is to be connected. Avoid the use of busses, especially for the high current power lines and common returns. Bus resistance of 0.1 Ω is normal. With supply or common currents of 1A, this creates 100 mV of drop. If this same common bus is used for low level amplifiers, the difference in potential along the common bus may be bigger than the input signal! It overwhelms the signal, destroying gain or creating oscillations. Busses also tend to route signals all over everywhere. This seriously increases the chances of low level signals picking up interference. So, as much as practical, connect external signals directly to the points where they are to be used.

    Keep low level inputs and high level outputs as far apart as possible. On a breadboard, this is not difficult to do. If you do not keep

Be sure that the breadboard or PCB is firmly anchored.

Avoid breadboard busses.

Keep inputs and outputs well separated.

these signals apart, interference coupled from the output back into the input may make the board oscillate.

Beware of alligator clips!

Use alligator clips very cautiously and only when no other options are available. Generally they are far too big. They often have a lot of exposed metal, threatening to short out your circuit. Moreover, they slip off. Usually a wire plugged directly into your board is a better choice. If you must use a clip, use a mini-grabber, with a small, spring-loaded tip that retracts back inside of an insulating sheath.

Do *not* solder wires directly to the PCB. Use a proper connector.

Though tempting, it is usually a bad idea to solder wires directly to a printed circuit board. Boards must be installed and removed from the system many times during system prototyping. Once in the field, easy maintenance demands that boards must slip in and out quickly.

Placement of the connectors strongly affects the performance of your electronics. It is critical to keep the large output signals as far from the small susceptible inputs as possible. The connector placement also dictates component and trace placement. Which pins within a connector carry which signals also influences interaction. Keep large and small signals well separated. Place common returns and supply voltages between signals that may interact. But be aware of pad spacing and potential differences. Connections for ± 56 V supplies must be kept at least 0.06 inches apart. So they should not be in adjacent pins in the connector. Even if the connector is rated to tolerate >100 V between pins, adjacent edges of the pads on the PCB may end up too close together. Finally, verify that the connector you are using is rated to handle the current that must pass through it. Voltage drops in the connector, as it responds to mechanical vibration, poor seating, contamination, fungus, etc., create problems that are very hard to track down once the board is shipped to the customer.

Plan controls (switches and pots) now.

4. What other connections to the "outside world" are needed?

Switches, potentiometers, LEDs, and other indicators are often part of your circuit. When laying out the breadboard be sure to make them easily accessible. Too often much time is wasted trying to adjust a potentiometer that is buried deep within the layout. You just cannot seem to get your screwdriver in the slot. "Oops, I must have shorted something out with its shaft, or knocked something loose," can be heard. Confirm switch operation before you put it into the breadboard. Hours of confusion can be avoided by knowing that the switch is normally open, not normally closed.

You may be able to attach these controls directly to the printed circuit board, and let them stick through the front panel. Good planning at this point in the printed circuit board layout means you do not have to add connectors to access these parts, and do not have to solder wires directly from the front panel to the PCB. A good breadboard layout serves well as a mock-up for the PCB.

5. What points are needed for testing?

The point of breadboarding your circuit is to test its performance. So be sure, as you plan the layout, that you can easily get at all of the circuit's nodes. Running long leads through your circuit to get at a spot buried deep within the board provides another opportunity for interference. It is better to allow space for the test probes to connect directly to the node of interest. Also, anticipate that you will want to change components. Keep this in mind when deciding where each part goes.

Designing for testability starts at the beginning of the circuit layout.

Printed circuit boards may be tested three ways: manually, in-circuit, and functionally. When testing a PCB manually, avoid connecting your probes directly to a component or a pad. The leverage provided by the probe can easily degrade its solder joint as you pull on the probe. Pushing a needle tipped probe against a solder pad also damages that pad. Instead, plan ahead. Include test points, and perhaps even pins, at the nodes you want to measure.

In-circuit testing uses an automated, spring-loaded, bed-of-nails to access nodes on your board. Again, do not probe the solder pads or components directly. Provide test points. Often, the automated tester has predefined locations for its probes. So, you must know, at the beginning, where you must put these test points.

Functional testing runs the board through connectors, powering and stimulating the board, then measuring its responses. It is often desirable to have a variety of intermediate signals available at the connector, as well as the input and output. Find out exactly what signals are needed for functional testing and where their connectors need to be. This way, you can incorporate them into the layout. Trying to add test connectors after the circuit is configured is much more difficult than doing it correctly from the start.

6. What are the thermal characteristics of the circuit?

For both the breadboard and the printed circuit board, there are two separate but closely related issues: which parts of the circuit are

Figure out where the heat is going, and protect the sensitive components and the board.

sensitive to heat, and which parts on and near the board generate heat?

At the very least, begin by keeping the heat sensitive parts away from those that create heat. Typically, heat sinks are moved to one edge of the board, while the sensitive electronics are clustered on the other side. Do not forget that the breadboard and the printed circuit board themselves can be damaged by the heat dissipated through the heat sinks. It is not unusual for the heat directed down by the heat sink to literally melt a plastic breadboard. Fiberglass printed circuit boards expand seven times more vertically than horizontally or laterally. This puts a great deal of stress on the solder joints in the heated area. Eventually those joints fail, and the board malfunctions. Unless the printed circuit board is specially designed to transfer heat, you must prevent it from heating.

Anticipate the effects of thermal expansion and contraction.

It is particularly important to consider heat flow if a fan is added to the system. Arrange the boards in the system and the parts on each board so that cool, external air first flows across those parts most affected by heat. The fan should then force the air out of the enclosure past the heat sinks. Also, do not forget that heat rises. So the sensitive parts should be on the bottom, and the heat generating elements at the top.

If the system is going to be operated across a wide range of temperatures, consider arranging the flow of air so that it avoids the temperature sensitive components entirely, passing only across the heat generators. Also, military grade parts may be appropriate.

Example 2-1

Define the mechanical and thermal elements needed to build the circuit in Figure 2-1 on a solderless connection breadboard.

1. Address each of the six items above.

2. Indicate the general area for the power transistors and the op amp and its elements, but do not begin placing parts yet.

3. Leave room for several TO92 (small plastic case) transistors and a couple of resistors and diodes to be placed near the power transistors. These are not shown in the schematic, but will be called for as the design grows.

4. Consider how the load and the support equipment can be arranged on your test bench.

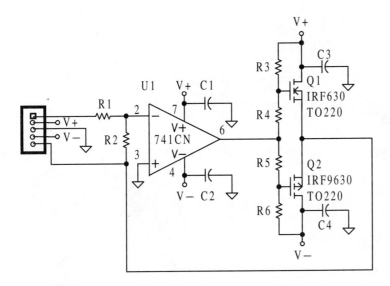

Figure 2-1 Power amplifier to be laid out

Solution

1. How big is the board?

   A standard solderless connection breadboard is to be used. Although the circuit in Figure 2-1 fits on a small board, future development indicates that there are other parts to be added. So a larger board should be chosen.

1. How is the board to be mounted?

   The test bench layout is shown in Figure 2-2. The breadboard is taped to the bench, as close as practical to the power supply and to the load. This allows you to keep the wires that carry the high current to the transistors and to the load as short as practical. The input signal comes in on the left and the load is on the right.

   A single point common return has been established at the common connection of the power supply. The function generator, voltmeter, and oscilloscope commons all connect there. The other leads for these instruments are also taped to the bench.

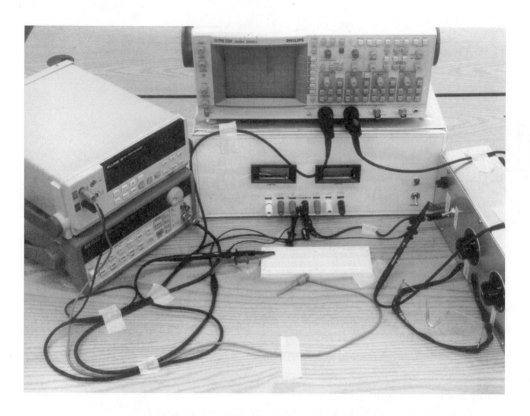

Figure 2-2 Breadboard and test bench layout
(*photo by John Zimmerman*)

3. What connectors are to be used?

For prototype testing, connections can be made directly to the breadboard. Look at Figure 2-2 again. The input comes from the function generator through a BNC-to-mini-grabber cable. Power can be provided thorough simple wires without connectors. The same is true for the signal to the load.

4. What other connections are needed?

There are no switches, potentiometers, or indicators to be considered. If there were, a good plan would be to place them on the side of the breadboard closer to you, and place the fixed parts, when practical, mostly on the side closer to the power supply.

5. What points are needed for testing?

Look at Figure 2-1 again. It is reasonable to measure at U1 pin 6, the junction of R3-R4, and the junction of R5-R6. So, when it comes time to place parts, be sure to make these places easily accessible from the front edge of the bread-board. Similarly, you may need to change any of the resistors, so keep R1 and R2 convenient as well.

6. What about heat?

The transistors must be mounted on heat sinks. This is shown in Figure 2-3. If the transistors are to be electrically insulated from the heat sink, then be sure to use a mica wafer between the transistor and the heat sink, and a nylon nut and bolt. Connecting the transistor directly to the heat sink means that the heat sink is at ±V. Provide thermal insulation at the bottom of each heat sink to protect the breadboard.

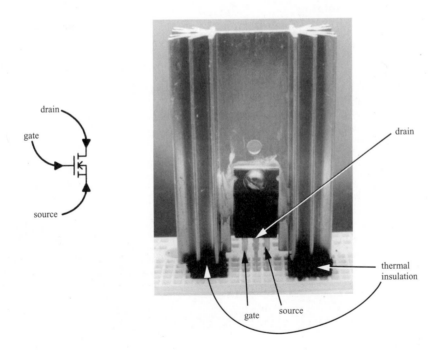

Figure 2-3 Transistors with their heat sinks
(*photo by John Zimmerman*)

Example 2-2

Begin the printed circuit board layout. Use a 2-to-1 scale. The breadboard is a reasonable mock-up.

1. The board is to be 3″ long by 2″ wide.

2. Place 1/8″ diameter mounting holes in each corner, 1/4″ from the edge.

3. Place the SIP 5 connector in the center of the 3″ edge, aligned with the mounting holes. Pin-signal assignment is defined in Figure 2-1.

4. There are no front panel controls or indicators.

5. Test points are needed for U1 pin 3, U1 pin 6, junctions of R3-R4, R4-R5, R5-R6, V+, common, and V−.

6. Locate the heat sinks as close as practical to the long edge opposite to the SIP connector.

Solution

The mechanical layout is shown in Figure 2-4.

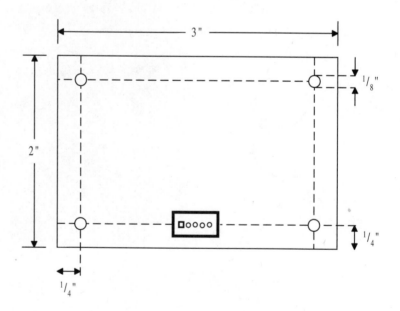

Figure 2-4 Mechanical details for Example 2-2

2.2 Part Placement

Once you have defined the mechanics of the board, it is time to place the parts. Where you place them is one of the major factors in how well the final circuit performs. Parts carrying small signals or connected to high impedance nodes (op amp inputs) must be kept close together. They must also be kept as far away from large signals as possible. Adjacent, parallel parts encourage electromagnetic coupling. On the other hand, low impedance nodes, such as those from the output of an op amp, or the emitter or source of a transistor, can easily overwhelm small noises. Therefore, it is wise to spread out parts driven from low impedances.

For all of these reasons, you must understand the circuit's operation before you can do a good job of placing the parts. A PCB layout program *cannot* auto-place parts as well as you can. Begin with the schematic. Identify the **critical path**. This is the path that the signal follows from input to output. The critical path for the example amplifier is shown in Figure 2-5. The signal out of U1 splits. Positive voltage passes through R4 and Q1. The negative part of the signal goes through R5 and Q2. These two parts of the signal recombine at the transistors' sources and are sent to the output.

Study the schematic, then identify the critical path.

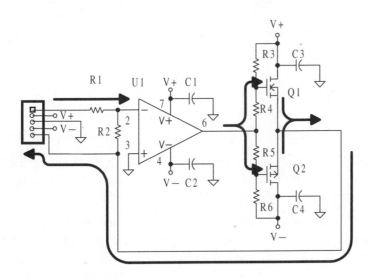

Figure 2-5 Critical path definition

Keep small inputs and large outputs well separated.

This is also the time to decide which pins in the connector carry which signals. Place the small input and the large output signals as far apart as practical. Put circuit commons and power supply voltages between the input and output signals.

The parts in the critical path are laid down first. This assures that they are in the best location. After all, they are the most important. For the circuit in Figure 2-5 these parts are: pin 1 of the connector, R1, U1-pin2, U1-pin6, R4 and R5, Q1 gate and Q2 gate, Q1source and Q2 source, and finally connector-pin5.

The signal from connector-pin1, R1, and U1-pin2 is small. And the impedance into U1-pin2 is very high. So these parts must be kept as close together as possible. On a breadboard this means that you must trim the leads of R1 short. Connect the input source into the hole immediately adjacent to R1. The other end of R1 is plugged into the hole next to U1-pin2.

On a printed circuit board, you must space component pads apart by one pad diameter. This is shown in Figure 2-6. Placing pads closer puts too much copper near the solder joint. When you try to solder to these pads, the solder cools too quickly, forming a cold solder joint.

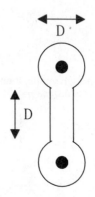

Figure 2-6 Minimum pad spacing

Example 2-3

Place the components in the critical path (from Figure 2-5) onto your breadboard. Remember to keep the power transistors well separated from the op amp and its resistors. Also allow space for future expansion of the circuit, and do not forget to plan access to the test points at U1-pin 6, R3-R4, and R5-R6.

Solution

Figure 2-7 shows the layout of the critical-path components on the breadboard. There are several points you need to notice.

4. Only the components in the critical path are placed now.

5. From the test instrument layout in Figure 2-2, the signal generator is on the left and the load is on the right. So, R1 and U1 are placed on the extreme left, and Q1 and Q2 are on the right.

6. It is fine to run the *output* of the op amp the length of the breadboard. This signal is large, and driven from a low impedance. So, it can overcome noise coupled into that node.

Figure 2-7 Breadboard layout of the critical path
(*photo by John Zimmerman*)

4. Connections to the *input* of the op amp are kept as short as possible. Here the signal size is very small and the impedance may be very large–an invitation to noise.

5. The positive voltages are on the right end of the supply. So Q1 is placed on the right and Q2 is placed to the left.

6. The two transistors are placed in the center set of the five holes in their rows. This allows the inputs to enter from the near edge of the board, and the high current outputs to exit from the holes behind the transistors. This way the two sets of signals do not have to cross.

7. U1 has been reversed so that its pin 6 is on the near side of the breadboard. This reversal took some thought. Although the IC may seem to be upside down, now the output can leave pin 6 and be routed to R4 and R5 without having to criss-cross the feedback signal. The feedback signal can be sent from the back of the transistors, to the left, through a resistor to pin 2. Even though it is premature to place these connections or even R2, its a good idea to plan ahead.

8. Efforts have been made to lay out the circuit around Q1 to look like the circuit around Q2. This symmetry helps during testing.

9. There is plenty of room between U1 and the transistors, and on the near side of the board in front of the transistors for future expansion.

Example 2-4

Place the critical-path components onto the printed circuit board.

Solution

The initial layout is shown in Figure 2-8.

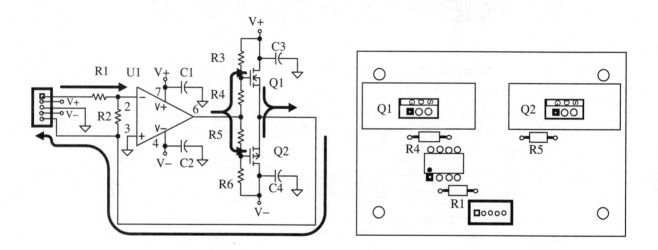

Figure 2-8 Printed circuit board critical-path layout

Outlines have been drawn around the two transistors to indicate the heat sinks. Without these outlines it is easy to place a part under the sink. It would also have been tempting to move the transistors closer to the top edge. But there the sinks would block the mounting holes. This placement is tentative. You will have to jiggle parts a little as the design progresses.

The structure of the board is now in place. Next, identify the other parts that touch the critical path. These go onto the board next. For the circuit in this example, these are R2, R3, and R6. Resistor R2 is tied to the inverting input pin of the op amp. This node has a very small signal and large impedance. So, R2 should be as close to U1-pin 2 as practical. The other resistors are connected to nodes which are driven by the output of the op amp, a fairly large voltage, low impedance source. So their placement is not as important.

Example 2-5

Place R2, R3, and R6 on the breadboard.

Solution

The placement of these components is shown in Figure 2-9

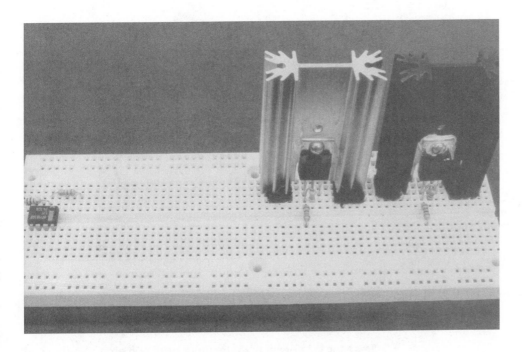

Figure 2-9 Placement of other parts on the critical path
(*photo by John Zimmerman*)

Notice that R2 is pulled as far back toward U1-pin 2 as possible and plugged into the hole immediately adjacent to R1. Now, look at R3 and R6. They connect between the gate and drain of Q1 and Q2. These two pins are only 0.1″ apart. So the resistors stand on end. The long, bare lead from the top bends down and then into the hole next to the gate pin. Since that point is one of the test points, these bare leads make a good place to which to connect a probe. This technique of standing a resistor on end works well for manually placed prototypes. But on a production PCB, this arrangement cannot be done with an automated placement machine and may cause thermal drift problems.

Example 2-6

Add R2, R3, and R6 to the printed circuit board.

Solution

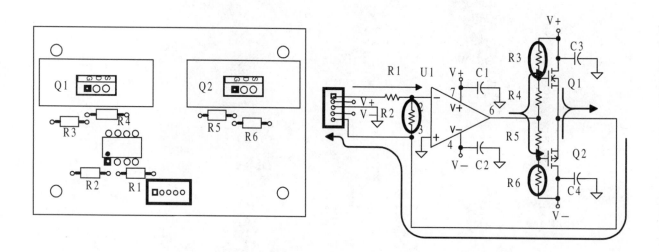

Figure 2-10 PCB placement of other components on the critical path

Resistor R2 connects to the input of the op amp. So that end of the resistor should be as close to U1-pin 2 as possible. The other end is driven from the outputs of the transistors, a large signal from a low impedance. So, a long trace to the left end of R2 does not cause a problem.

The gates of the two transistors are very high impedance. Try to keep R3 and R6 reasonably close to the gates. However, since the op amp drives these resistors, it should not be a problem if R3 or R6 must be moved a little way from the gates.

All of the remaining parts can now be moved onto the board. For the circuit in the example, only the decoupling capacitors are left. They should be close to the power pins that they are decoupling.

You may want to nudge parts a bit to get them onto a standard grid if the parts are to be automatically inserted. Check to see that parts are not side by side. This looks good and makes fabrication easy. But current flowing through one part produces magnetic fields that couple current into the part beside it. When all of the parts are on the board, even though you have not connected any of them, most of the work is done.

Example 2-7

Place the four decoupling capacitors on the breadboard.

Solution

The capacitors have been placed in Figure 2-11.

Figure 2-11 Decoupling capacitor placement on the breadboard (*photo by John Zimmerman*)

The decoupling capacitors for the op amp connect directly to pin 4 and pin 7 of the op amp. The other end of each capacitor goes into the inside bus on the top and the near side of the breadboard. The decoupling capacitors for the two transistors are placed behind the transistors. They can be seen in the backside view. Each connects directly to the drain. The other end of each capacitor goes to the inside bus. On this breadboard, the busses are broken at the center of the board. This allows you to have a low voltage common return, for the op amp, and a high voltage common return for the transistors. Its reasonable to expect that the transistors put much more noise on their common return than does the op amp. So, being able to keep these two common returns separate helps control noise and oscillation.

Example 2-8

Finish placing components onto the printed circuit board.

Solution

The PCB layout with the decoupling capacitors is shown in Figure 2-12. At this point, just get these parts close to where you think they should be. Expect that they will have to be moved a little as the traces go down.

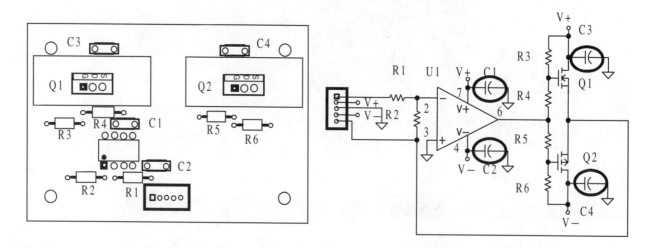

Figure 2-12 Printed circuit board with all parts in place

2.3 Interconnections

It is now time to connect the parts. But, before you grab some wire and start hooking stuff up on your breadboard, or start running traces on the printed circuit board, consider several fundamentals. Of initial importance is the size of the interconnection. For a breadboard, this means what gage wire should you use? On the PCB, you must decide how thick the copper is to be. Knowing that, you can then choose the trace width.

Wire Gage

When using a solderless connection breadboard, it is common practice to use 24 AWG or 22 AWG solid wire. Why? That is what works well with the little spring clips buried in the breadboard. Is that OK? We can hope. If it is *not*, then you should select a different breadboarding technique. You may have to manually solder properly sized wires from point to point between the parts. Forcing larger gage wire into the holes of a solderless breadboard destroys the clip. Besides, the holes and clips are sized to accommodate the current that 22 AWG can properly carry. Driving more current into the solderless connection breadboard degrades its performance. The result is that your testing results may be caused by the misuse of the breadboard not by your circuit.

There are two issues that determine what gage wire you need; voltage dropped by the current as it flows through the wire, and continuous-duty rating. Table 2-1 indicates the resistance per inch for each of the standard wire gages. Although these resistances seem tiny, when you try to deliver several amperes of current to a remote load, the losses quickly become considerable.

Table 2-1 Wire resistance per unit length

AWG	$\mu\Omega$/inch
0	8
2	13
4	21
6	33
8	52
10	83
12	132
14	210
16	335
18	532
20	846
22	1,345
24	2,139
26	3,401
28	5,408
30	8,600
32	13,675

NFPA No. SPP-6C, National Electrical Code ®, copyright © 1996, National Fire Protection Association, Quincy, MA 02269

Example 2-9

Select an appropriate wire gage to deliver 15 V to a 4 Ω load that is 8″ from the amplifier, and the power supply common. You may lose no more than 100 mV across the wiring.

Solution

The current sent to the load is

$$I_{load} = \frac{15\,V}{4\,\Omega} = 3.75\,A$$

This current may drop no more than 100 mV across the wiring resistance.

$$R_{wiring} = \frac{100\,mV}{3.75\,A} = 26.67\,m\Omega$$

It is 8″ from the amplifier out to the load, and another 8″ back from the load to the power supply common. So the resistance/inch is

$$\frac{\Omega}{inch} = \frac{26.67m\Omega}{16inches} = 1670\frac{\mu\Omega}{inch}$$

So, even with 3.75 A of current, you can use 22 AWG wire to connect the load to the amplifier and to common, as long as you are willing to lose 100 mV, and you can keep the load within 8″ of the source.

Practice

Now that you have tested the amplifier, you want to install the 4 Ω speakers across the room. Running 22 AWG along the wall means that the speakers are 30′ (one way) from the amplifier. When the amplifier outputs 15 V, how much voltage is actually delivered to the load? Suggest a more appropriate wire gage.

Answer

V_{load} = 12.1 V. There is a 19% loss of voltage across the wires. The wire must dissipate 8.7 W. It will get hot! You must use 6 AWG gage wire to lower the loss to 100 mV, and keep the wires cool.

Table 2-2 Maximum current carrying capacity

AWG	MIL-W-5088 Ampere	Underwriters Lab Ampere
30	-	0.2
29	-	0.4
26	-	0.6
24	-	1.0
22	9	1.6
20	11	2.5
18	16	4.0
16	22	6.0
14	32	10.0
12	41	16.0
10	55	-
8	73	-
6	101	-
4	135	-
2	181	-
0	245	-

The second concern is the trade-off between how hot the wire gets and how big (and heavy) the wire is. For example, in aerospace applications every ounce is critical. The equipment is designed to tolerate wide temperature variation, often has multiple redundant systems, and is closely monitored and well maintained. For these applications it is quite acceptable to let the wire get warm. Maximum current-carrying capacity for wires used in aerospace applications are indicated by MIL-W-5088, and are listed in Table 2-2.

Consumer electronics are at the other extreme. The owner sets up the equipment, turns it on , and forgets it. It must run safely even though books and coats are piled on top of it, or its wires run under a rug. Dust and grunge build up and even clog air intake vents. But under even the worst cases of neglect or outright abuse, it must continue to perform *safely*. The wiring must not get hot enough to present a fire hazard. The Underwriters Laboratory issues standards for wiring in consumer equipment. These are also shown in Table 2-2.

Using Table 2-2, the 3.75 A from Example 2-9 needs only a 22 AWG wire in an aerospace application. But you should use an 18 AWG wire if that same current is to flow in a consumer product. Neither column of Table 2-2 predicts that a 6 AWG wire is needed for a 30′ run of wire. So use these two tables carefully. Consider current, distance, acceptable loss, and eventual application.

Example 2-10

For the circuit board from Examples 2-1, 2-3, 2-5, and 2-7 place the wiring to complete the circuit. Connect the parts in the following order: 1) parts in the critical path   2) parts touching the critical path   3) other parts   4) power and common return

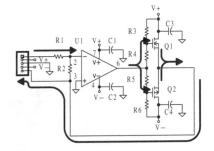

Solution

Figure 2-13 shows the connection of the parts touching the critical path. The input from the signal generator connects to a pin in the hole immediately adjacent to the left of R1. The output of the op amp runs along the near side of the board to R4, then R5.

Figure 2-13 Critical path wiring on the breadboard (*photo by John Zimmerman*)

The back side of the breadboard is shown in the photograph on the right. The source of Q1 is connected to the source of Q2 and then run along the back edge of the breadboard to R2. This keeps the input to the transistors and their output as far apart as practical. The load is connected to the transistors on the edge furthermost from the input. The only other components in this circuit are the decoupling capacitors. One end of each is already plugged into the IC or transistor power pin. The other end of each capacitor is plugged into a common return bus. Look for these in the photographs of Figure 2-13.

This leaves only the power and common return connections. Look at Figure 2-14.

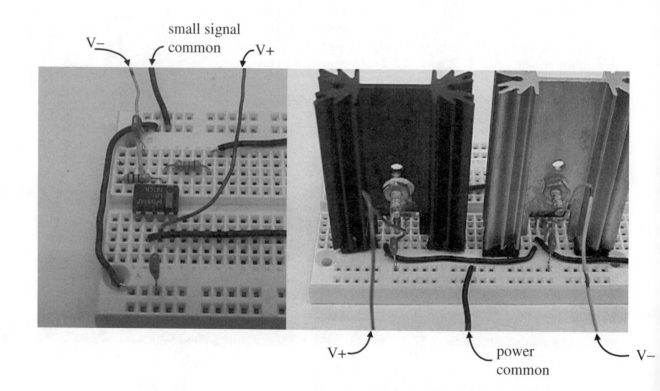

Figure 2-14 Power and common return connections
(*photo by John Zimmerman*)

Even though the small signal op amp and the power transistors use the same supply voltages, separate wires are run *directly* to each. This reduces the possibility that the major variation in current from the power transistors is coupled back into the op amp along a shared power bus. With separate power wires, the massive capacitance of the power supply sits between the transistor power pins and the op amp power pins. The same is done for the common return lines. The op amp needs the cleanest common line you can provide. So do not contaminate it with current from the power stage.

Figure 2-15 shows the interconnection of the entire test system. Look carefully at how the supply leads are kept short and separate, as are the two common returns. The signal enters from the left. The load is connected at the extreme right. Finally, current from the load returns *directly* to the supply common in its own lead. This assures that signals created by the large load return current cannot get back into the small input signal.

Figure 2-15 Complete breadboard with test instruments (*photo by John Zimmerman*)

Traces

When drawing traces on your printed circuit board, use the widest trace that fits. The wider the trace, the lower its resistance, and the smaller the resulting voltage drop. As with wire gage, size determines how hot the trace gets. The standard MIL-STD-275 provides two graphs that allow you to determine the relationship between current, trace temperature rise, copper thickness, and trace width. This is shown in Figure 2-16.

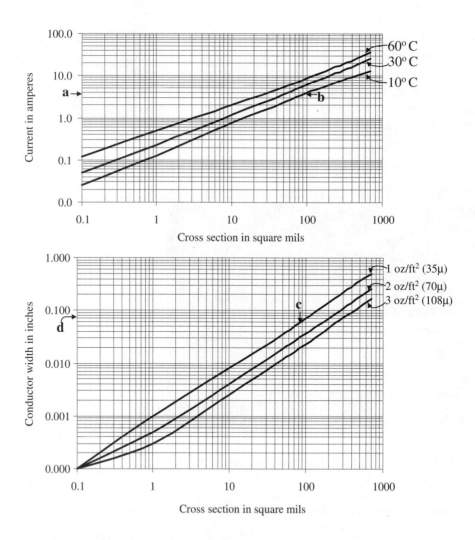

Figure 2-16 Trace size and current carrying capacity (*from IPC-D-275*)

As a general rule, for small signal traces carrying less than 750 mA, set the width to 0.025″ (25 mils).

Example 2-11

How wide should the output trace be for the PCB layout from the previous examples? Assume that there may be as much as 3.75 A of current, and that 1 oz/ft^2 copper cladding is used on the board.

Solution

The other information that you need is "How much temperature rise are you willing to tolerate in the trace?" Since you do not know the eventual use of this board, it is wise to create as small a rise as practical. So, assume that a 10°C rise is as much as you can accept.

Look at the upper graph in Figure 2-16. Find the current level, 3.75 A on the vertical axis (point a). Move from that point horizontally to the right until you intersect the 10°C line (point b). Now go down to the proper trace for the copper thickness: 1 oz/ft^2 on the lower graph (point c). Finally go back to the left to the lower graph's vertical axis (point d). A 0.070″ (70 mil) wide trace made of 1 oz/ft^2 thick copper rises only 10°C when carrying 3.75 A of continuous current.

Practice

What is the temperature rise of a 25 mil wide trace made from 1 oz/ft^2 thick copper that has to carry 2 A?

Answer About 20°C.

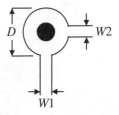

The trace width you just selected may need to be modified, based on the size of the pads to which it runs. If the traces leading into a pad are too wide, they pull the heat away when the component is soldered to the pad. This causes a cold solder joint. The guideline is

$$W1 + W2 + \ldots \leq 0.6D$$

as illustrated in Figure 2-17.

Figure 2-17 Trace and pad size

Figure 2-18 Trace junctions

When you solder the components to the board, flux is left behind. Flux is mildly corrosive. If you do not clean it all from the board, it will eventually eat through any trace with which it is in contact. Any traces meeting at less than 90° form a trap. Flux gets stuck in these acute angles, is not washed out during the cleaning process, and eventually (long after you have shipped the board) eats through the trace, causing a failure. Since it is impossible to make a junction that is exactly 90°, even when you try to make two traces meet at right angles, one angle is slightly more than 90° and the other angle is less. That spells trouble. The solution is to bevel each junction. Look at Figure 2-18.

When a trace abruptly changes direction, the resulting point transmits electromagnetic energy. So, instead of forming right angles as you lay down the trace, create arcs. If the software package you are using does not allow arcs, then at least go around a 90° bend in two 45° steps.

Example 2-12

Following these rules, insert the traces for the critical path on the printed circuit board whose parts are shown in Figure 2-12.

Solution

The critical path traces are shown in Figure 2-19.

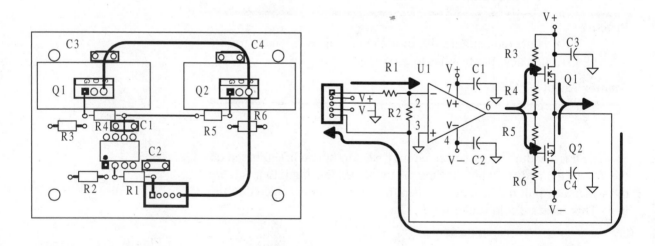

Figure 2-19 Critical path traces

> Traces are as wide as possible without violating the $W1 + W2 \leq 0.6D$ rule. Also, notice that the right angles have been converted to arcs.

Drawing the critical path first assures that it gets the best route on the board. Next, run traces to the ends of the components that touch, but are not in, the critical path. For the example circuit in Figure 2-19, these are R2, R3, and R6. These traces are shown in Figure 2-20. First, connect the end of the component that touches the critical path. Then place the traces that go to the other end of these secondary components. If the other end goes to power or the common return, rather than to a component, wait to run that trace until you are running the power and common traces.

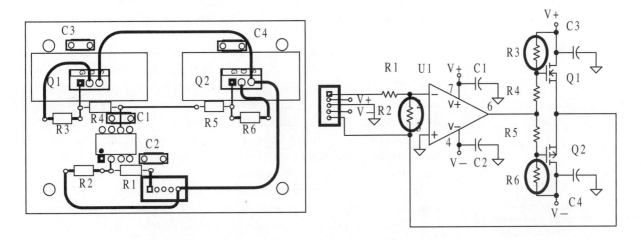

Figure 2-20 Traces run to other parts touching the critical path

Notice, to accommodate the trace from the right end of R6 to the drain of Q2, it is necessary to move the trace carrying the output signal from Q2's source. Since this is a large signal, driven from a low impedance, a little extra length does not degrade the signal.

This example has resulted, so far, in a single-sided board; components on one side and all of the traces on the other. This is simplest, and assures the best route for the signal. On a multilayer board, the trace travels from one layer to another through a **via**. But these vias are vulnerable. Mechanical stress and thermal expansion weakens them. Long term reliability is greatly improved if you can keep the entire signal route on one side of the board. If you must send the signal from bottom to top, or top to bottom, do it very sparingly.

With all of the signal traces complete, it is time to run power. If you must use a multi-layer board, put power, and common return, on the component side (or in a layer of their own). Since these runs are strongly driven to a fixed potential, they make good ac commons. In fact, snaking ±V and the common return all over the board may actually serve to shield the most sensitive signal runs.

Run separate power branches to the low level and to the power amplifiers.

When possible, run the power to the decoupling capacitor first, then to the IC or transistor. This gives the capacitor a chance to short out any noise on the power line before it gets to the amplifier. Similarly, if the amplifier has a sudden need for charge, it is available from the decoupling capacitor. That charge does not have to travel down the power bus. This keeps glitches created by a part from getting out, onto the power traces.

Keep the traces to the decoupling capacitors as short as practical. All conductors have inductance, as much as 25 nH per inch for 22 AWG wire. This means that a single inch of trace to a 0.1 μF capacitor has more inductive reactance than there is capacitive reactance in the capacitor at 4 MHz. Noise spikes can easily generate harmonics at and above this frequency. So if you cannot connect the decoupling capacitor directly to the power pin, it may be worse to have it connected with a long trace than to leave it out altogether.

Run separate common return branches to the low level and to the power amplifiers.

Finally, keep the power and the common return traces that go to the small signal amplifiers separate from those that go to the power transistors. When the main transistors ask for more or less current, small variations in the power supply voltages result. Daisy chaining the power lines sends this variation to the small signal amplifier before it gets back to the connector. This increases the possibility of oscillations. Running the power and common return to the transistors and the small signal amplifiers in separate leads (a star configuration) reduces interaction.

Example 2-13

Following all of these guidelines, route V+ and V− on the example printed circuit board.

Solution

The result is shown in Figure 2-21.

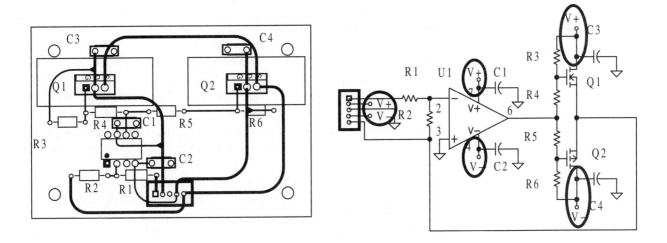

Figure 2-21 Power traces run on the example printed circuit board

As you begin to run these traces, several conflicts appear. But, since you have already placed and routed the critical components, as long as you do not make major changes to the critical path even poor decisions do not hurt performance.

The V+ trace is split, with separate branches run to the small signal op amp and to the power transistor. In this example, it is not practical to run the power trace to the decoupling capacitor before it goes to the IC or transistor. As long as the decoupling capacitor is close to its amplifier, that compromise is reasonable. In order to keep the op amp and the power transistor V+ on separate branches, R5 has been shifted to the left. This allows the V+ trace to pass under R5, at right angles. This right angle configuration ensures minimum magnetic coupling from the current flowing in the V+ trace and the signal passing through R5. Since the voltage from the op amp through R5 to Q2 is as large as the output, and is being driven from the low impedance of the op amp, sliding R5 a little is not a problem.

Because the V− trace passes directly under R6, the trace from R6 to Q2's drain has been changed to take advantage of that long run. Bevels have also been added to all of the places where traces meet in a right angle.

So, at last it is time to run the common return traces. You should consider several common branches, all starring together at the input connector. Low level signal amplifiers, like the op amp, need a very clean common. Any noise appearing there may be interpreted as part of the signal and amplified. Certainly, if the load or the power transistors have a common return, keep it separate from the small signal amplifier common, until it gets back to the connector. Otherwise, variations in the load common return current is coupled into the small signal amplifier common, causing variations in the load return current, that is coupled into the small signal amplifier, and so on and so on.

Since the common return acts as a shield, you may snake it about liberally over your board. In general, this reduces interference. If you have the time and the tools, you may even want to spread the common return across most unused areas of the board. Just be careful that you do not get it too close to other traces or components. If you are working with a double sided (or multi-layer) board, then the component side is the ideal place for a ground plane. It also may be convenient to route ±V on the component side.

Example 2-14

Add the common return traces to the example PCB.

Solution

Figure 2-22 shows the routing of the common return traces.

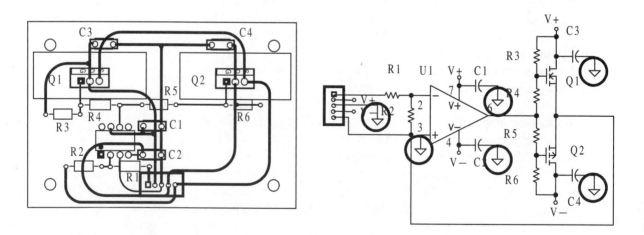

Figure 2-22 Routing common returns

The common returns play two separate roles in this circuit. The more critical is the connection to the noninverting input of the op amp. This point serves as a reference for the amplifier and must be as clean as possible. All of the other returns connect to decoupling capacitors.

While looking at the PCB layout after routing ±V, it seems that the decoupling capacitor returns can all be grouped down the center of the board. Getting to the return of C1 initially seems like a problem. But by moving C1 to the right a little, it straddles the V+ trace, and is aligned with C2 and the connector. In fact this helps, because, now the V+ trace to the op amp can be run to the decoupling capacitor before it goes to the op amp.

This only leaves the common return line to pin 3 of the op amp. Look back at Figure 2-21. With V− running from the connector to pin 4 of the op amp, there seems to be no good way to get even a narrow trace from pin 3 to the common return connector pin. But if you cannot find a place for the trace, move the trace that is in the way. In Figure 2-22, the V− trace from the connector to the op amp has been moved, run under R2, and then down under the IC to approach pin 4 from the inside. This then leaves room for an amplifier common return branch to go from the connector to pin 3.

You have seen, as the PCB layout unfolds, that it may be necessary to rethink, and jiggle parts and traces. Identifying the critical path, laying down those parts first, and running their traces before other traces assures that the adjustments needed to complete the design do not compromise the performance of the circuit. As you gain experience by laying out boards and by analyzing how experts lay out similar boards, you may be able to see some of the changes several moves ahead. If not, that is all right. It is normal to complete a board, let it sit for awhile, then see a better scheme.

Placement of the test points has been saved for the end. Their positions have not been specified, so precisely where they are placed is not important. Therefore placing them was delayed until all of the other items are in place. However, if locations for the test points were specified, then they must be placed *first*. The more flexibility you are given in the test point location, the later in the process you can wait to place them. Test points are needed at U1-pin 6, R3/R4, and R5/R6. These test points are aligned across the center of the board, shown in Figure 2-23.

Do not forget the test points. When you place them depends on how important their location is.

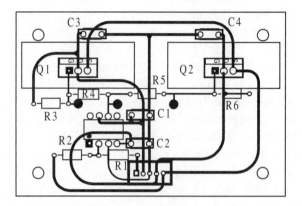

Figure 2-23 Test points added

2.4 Other Considerations

In the layout example, traces have been placed wherever it is convenient. Spacing between traces is kept as large as practical. This reduces the possibility that a signal on one trace can couple interference into an adjacent trace. This is a particular concern for long traces.

The other consideration in determining trace spacing is the difference in potential between the traces. At some point, as this voltage rises, the board no longer provides adequate insulation. Arcing occurs if the traces are too close. Table 2-3 lists the worst case recommended separation, for traces lying on the top or bottom of a board made of standard materials, with no protective coating, used at a variety of altitudes.

Table 2-3 Electrical conductor spacing (*from IPC-D-275*)

Voltage between conductors (dc or ac peak)	Minimum spacing (inches)
0 - 50	0.025
51 - 100	0.060
101 - 170	0.125
171 - 250	0.250
251 - 500	0.500
greater than 500	0.001/volt

You certainly cannot go wrong using these conservative guidelines. If your layout is especially tight, refer to the standard from the Institute of Interconnecting and Packaging Electronic Circuits (www.ipc.org) for more detailed guidance.

Keep in mind that these spacings are not from center-to-center. They are from the closest adjacent points, edge-to-edge. This is particularly important when considering pads to which parts are soldered. The transistors in the suspended supply amplifier, Figure 1-33, come in TO92 packages. The leads are placed on 0.100″ spacing, center-to-center. With ±56 V supplies, the output may be driven as negative as −46 V. This places the base of the upper transistor at −38 V. The collector is at +56 V, a difference of 94 V from the base. Using the normal TO92 footprint puts the adjacent edges less than 0.060″ apart. So you must provide a different footprint, and consider how this impacts automated assembly. When dealing with multilayer, or even two-sided boards, do not forget that high voltages can also cause breakdown through the board. Standard board materials can tolerate as much as 10 V of difference for each 0.001″ (mil) of thickness.

These clearance requirements point to a common, unwise practice in routing traces. It is tempting to try to sneak a trace *between* two IC or transistor pins. With very careful placement, it is possible to get a very narrow trace between the pins of an IC. However, there is *not* room for the trace, the pads, and 0.025″ clearance on each side of the trace. Even if you are not concerned about arcing, running a trace that close to an IC pin invites interference. Even a little coupling into an amplifier's input pin can be devastating.

Do *not* run traces between the pins of an IC or transistor.

On two-sided, or multilayer boards, vias provide connection between the layers. But as the board twists, flexes, and expands and contracts in response to temperature changes, these vias are stressed. They are mechanically the most vulnerable part of the layout. Remember, you should place the parts in the critical path first. When you begin routing the board, the traces along this critical part are the first to go down. So the critical path gets the best real estate. There should be no reason to put a via in the critical path. If you do, you are building in a future failure. Vias may be used to carry power and the common return. There are usually many opportunities to build in redundancy along these traces by adding a number of vias. So if one via does fail, there are several others also carrying the supply or return.

The proportion of metal connected to a pad strongly affects the quality of the solder joint. If the traces coming into the pad are too wide, the solder freezes on top of the metal when you try to solder a component to the pad. The solder does not penetrate the surface while it is molten. The solder just sticks the parts together. A metallurgical bond is not made. You have just created a cold solder joint. That is the reason for the rule

$$W1 + W2 + \ldots \le 0.6D$$

A common, extreme example of violating this precaution is trying to solder components directly to a trace. The connections of C1 and C2 to the common return in Figure 2-23 are flirting with this problem. A far better solution is to run the trace close to the pads, and then send a small side trace over to the pad. This is shown in Figure 2-24.

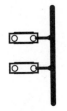

Figure 2-24 Connection to a trace

Another common mistake is to be tempted by all that great expanse of ground plane. It seems like such a good idea to arrange the heat producing parts so that their cases can be bolted directly to these large metallic regions. What a cheap and readily available heat sink! The problem is that the standard printed circuit board materials expand in thickness seven times more per degree increase in temperature than they do in the plane of the board. This exerts considerable forces on the vias and thru-hole parts solder connections. In response to cycle after cycle of heating up and cooling down, eventually these connections break. An intermittent occurs. It is down in the middle of the board, between the top and bottom where you cannot see it. Worse, it only happens when the board warms up. The fault goes away as the board cools, when you turn off power to try to fix the board.

Heat is the enemy of your board. Do everything that you can to move heat *away* from the board by properly placing and using heat sinks.

Summary

A good layout assures that even a marginal design gives you its best possible performance. A poor layout dooms even the best of electronics designs to failure. This is true when you first prototype the circuit on a breadboard, and true on its final printed circuit board configuration.

It is tempting to jump right into arranging and wiring the parts. But you must first clearly understand the mechanical details. They define the playing field. Be sure to determine how big the board is to be, how it is to be mounted, what connectors are to be used, and where they have to be located. Other mechanical details include how and where controls are to be mounted, where in the circuit and where on the layout test points are needed, and (in power circuits one of the most critical problems) what parts are going to generate heat and how you are going to remove it.

Only after you have answered all of these questions can you begin to place parts. Look at the schematic. Identify the critical path that the signal takes from the input to the output. These are the most important parts and are placed first to assure that they get the best location. Keep the small input and the large output well separated. The smaller the signal, and the higher the resistances tied to a node, the closer together the parts should be placed. On a PCB, however, pads can be no closer than one pad diameter, to assure a good solder joint. As the signal grows and as nodes are driven by amplifiers, things may be spread apart. Here is your opportunity on a breadboard to allow space for future development. Once the parts in the critical path are in place, move in those touching that path. Finally, position the other parts. While you are placing parts, do not run any wires or traces.

For breadboarding, wire gage depends on the allowable voltage drop across the wire and the acceptable temperature rise. On a printed circuit board, you can determine trace size using a set of graphs along with the current to be carried by the trace and the allowable increase in temperature. Run the traces in the same order as you placed the components; parts in the critical path, parts touching the critical path, other parts. When running power, try to provide a separate branch for the low signal parts and another for the power parts. It is particularly important to create a star configuration for the common return lines, keeping the small signal amplifiers' common away from the power circuit until they are starred together at the supply (or connector).

On printed circuit boards, be sure that the sum of the widths of the traces entering a pad is less than 60% of the diameter of the pad. Otherwise, cold solder joints may result. Avoid right angles. Where two traces meet, bevel the junction. When a trace changes direction, use an arc to avoid radiating interference.

Traces should be kept as far apart as possible to avoid interference. The minimum spacing is governed by the voltage difference between the traces. Vias make connection between layers of your board. But they are the weakest mechanical link in the signal processing chain. Do not use a via in the critical path. Keep all of the signal routing on one side. Sending power and common returns through vias is not a problem if you are able to provide several vias along a trace. This redundancy improves reliability when the board flexes and expands as it heats up. Try hard to keep heat away from the board. Never use a plane of metal on the board as a heat sink.

The amount of heat that you must get rid of and the amount of power you can deliver to a load are determined by the average value and the root mean squared value of the signal that is being processed. You are already familiar with these values of steady dc and for a sinusoid. But most practical power electronics circuits use much more complicated wave shapes: half sinusoids, pulse width modulated rectangles, phase angle modulated sinusoids, triangles, and trapezoids. During the next chapter you will see how to calculate the average value, the root mean squared value, and the average power delivered by these practical waveforms.

Problems

2-1   Lay out the circuit in Figure 2-25 on a breadboard. The input amplitude is less than 2 V. There is to be a 10 Ω load connected between the output terminal and the power supply common.

2-2   Design the printed circuit board layout for the circuit in Figure 2-25. The mechanical specifications are:

a)   The board is to be 3″ long and 2″ wide.

b)   Place 1/8″ diameter mounting holes in each corner, 1/4″ from the edge.

c)   Place the SIP 5 connector in the center of the 2″ edge, aligned with the mounting holes. You may determine the pin assignment.

d)   There are no front panel controls or indicators, but be sure to place the potentiometer R2 so that it can be adjusted from the top long edge (with the SIP to the right and the heat sink to the left as viewed on the component side view).

e)   Test points are needed at the wiper of R2, U1 pin 7, U2 pin 2, U2 pin 7, Q2 gate, +5 V, and +56 V.

f)   Locate the heat sink as close as practical to the short edge opposite to the SIP connector.

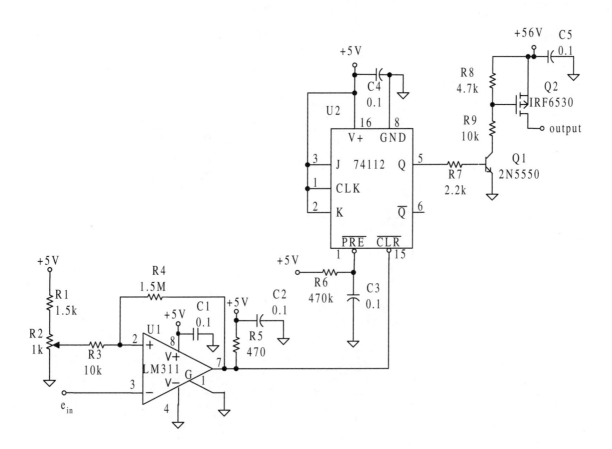

Figure 2-25 Over-temperature power control for Problems 2-1 and 2-2

3

Power Parameter Calculations

Introduction

Practical power electronics circuits use waveforms that are much more complicated than a simple dc or sine wave. An audio amplifier often processes its signal in two parts. The power that the electronic components must dissipate comes from a dc supply. So the power transistor has a constant voltage on one side, an inverted sine on the other, while the current through it is a half sine.

Power from the utility line is delivered as a sinusoid. The electronics then passes part of this to the load, turning on for some portion of each half cycle. A heater or light responds to the rms value of this partial sinusoid. DC motors and power supplies produce an output proportional to the signal's average value. The current through the electronic switch is part of a half cycle, but the voltage across the switch is $2 V_{dc}$.

Transistor switches pass power to the load in rectangles. Filtering converts the current into a ramp. How do you calculate the current from the supply, and the power dissipated by the transistor?

In this chapter, you write the equations for signals used in delivering power. Then you integrate these over a cycle to determine their average value. A more accurate comparison among signals is based on their root-mean-squared (rms) values. The chapter ends with the computation of average power from a variety of current and voltage wave shapes. All of these shapes, and average, rms, and power equations are gathered into a table you can use throughout the rest of this book.

Objectives

By the end of this chapter, you will be able to:

- Determine the equation for dc, sine, half-wave rectified sine, full-wave rectified sine, rectangle, ramp, and trapezoid signals.

- Calculate the average value of each of these signals.

- Calculate the root-mean-squared value of each of these signals.

- Define instantaneous and average power.

- Calculate the average power delivered by a variety of waveforms.

3.1 Common Waveforms

Before you can determine the effect a waveform has on the load or on the electronics circuits processing that signal, you must be able to write its equation, as a function of time (*t*). Once you have that equation, you can then integrate it to determine the average value, or square it before integrating it for the root-mean-squared value, or multiply it by another waveform before integrating it to calculate the average power.

DC

V_p

0

Figure 3-1 DC waveform

The simplest signal is pure dc. This is the wave shape of the voltage from the power supplies used to bias the electronic circuits. As shown in Figure 3-1, dc is a constant value, never varying from its peak.

$$v(t) = V_p$$

Sinusoid

The most common test signal is the sinusoid. Its shape is not affected by circuits made from linear components. You only have to determine the effect the circuit has on the signal's amplitude and phase.

Power is delivered by the utility company as a sine wave. The voltage is stepped up to thousands of volts. Power is more efficiently transmitted as a large voltage because as the voltage goes up, the current is decreased. The lower the current sent through the transmission lines, the smaller the losses as that current flows through the wires' resistance. Power electronic circuits then manipulate the magnitude and/or frequency of the power company's sine wave to allow you to dim the lights, change the motor's speed, or turn the heat down.

It is easiest to represent the waveform as a function of angle, rather than a function of time. If you must use time as the independent variable, remember that

$$\theta = 2\pi f t$$

θ = the angle, measured in radians
2π = the length of a single cycle in radians
f = the signal's frequency
t = the independent variable, time

Written in terms of the independent variable, θ, the equation for the simple sinusoid shown in Figure 3-2(a) is

$$v(\theta) = V_p \sin\theta \qquad\qquad\qquad \text{**Simple sine wave**}$$

Amplifiers and transformers may invert the sine wave, as shown in Figure 3-2(b). The equation of an inverted sine wave is

$$v(\theta) = -V_p \sin\theta \qquad\qquad\qquad \text{**Inverted sine wave**}$$

or $\qquad\qquad v(\theta) = V_p \sin(\theta + \pi)$

or $\qquad\qquad v(\theta) = V_p \sin(\theta - \pi)$

Remember, θ is the independent variable. The phase shift of 180° is accounted for by the constant $\pm\pi$ radians.

Purely capacitive circuits cause the current to lead the voltage by 90° ($\pi/2$ radians). Look at Figure 3-2(c).

$$i(\theta) = I_p \sin\left(\theta + \frac{\pi}{2}\right) \qquad\qquad \text{**Sine wave leading by 90°**}$$

Purely inductive circuits cause the current to lag the voltage by −90° ($-\pi/2$ radians). This is shown in Figure 3-2(d).

$$i(\theta) = I_p \sin\left(\theta - \frac{\pi}{2}\right) \qquad\qquad \text{**Sine wave lagging by 90 °**}$$

Figure 3-2 illustrates each of these typical sinusoids.

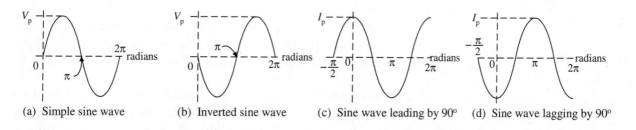

(a) Simple sine wave (b) Inverted sine wave (c) Sine wave leading by 90° (d) Sine wave lagging by 90°

Figure 3-2 Sine waves with different phase shifts

Rectification is used to convert the utility line sine wave into a steady, dc voltage. In the simplest version, a single diode is used to pass the positive half-cycle to the filter, and block the negative half-cycle. The voltage delivered to the load, and the power dissipated by the rectifier are calculated from that half-wave rectified sine wave.

Class AB audio amplifiers process each half of a sine wave separately. The positive half-cycle goes through one set of transistors, and the negative half-cycle goes through a different set. These two halves are then recombined at the load. So, for these transistors, the signal is processed as two half-wave rectified sine waves. The current from the power supply, the power delivered from that supply, and the power that the transistors have to dissipate all are based on the half-wave sine. The signal is shown in Figure 3-3.

$$v(\theta) = \begin{cases} V_p \sin\theta & 0 \le \theta < \pi \\ 0 & \pi \le \theta < 2\pi \end{cases}$$

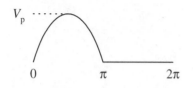

Figure 3-3 Half-wave rectified sine wave

The equation has two parts. From the beginning to π radians, the signal is a sine wave, and that simple equation applies. After the midpoint, the voltage is constantly zero. At first, this awkward equation may appear difficult to work with. But you are going to integrate it to find the average value, square it then integrate it to get the rms value, or multiply it by some other signal and then integrate the product to find the average power. In all three cases, the key operation is an integration. Integration is just a sum. So, in all of these calculations you just perform two integrations: the first using $V_p \sin\theta$ with limits from 0 to π. The second part of the calculations is from π to 2π. But that is zero and falls out.

$$\int_0^\pi V_p \sin\theta \, d\theta$$

or

$$\int_\pi^{2\pi} V_p \sin\theta \, d\theta$$

Full-wave rectifiers pass power to the load on both half-cycles. But the second half of the sinusoid is inverted, so that it too arrives at the load as a positive voltage. This is shown in Figure 3-4.

Like the half-wave sinusoid, the equation for this signal must be written in two parts. The first is just the equation used before. For the second part, look back at Figure 3-2(b). From π radians to 2π radians, that signal is a positive half-cycle. So use that for the equation for the second half of the full-wave rectified sine wave.

$$v(\theta) = \begin{cases} V_p \sin\theta & 0 \le \theta < \pi \\ -V_p \sin\theta & \pi \le \theta < 2\pi \end{cases}$$

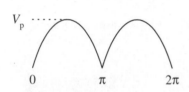

Figure 3-4 Full-wave rectified sine wave

Rectangular Waves

Switching power supplies are widely used. They allow you to step voltages up or down. They are typically 90% efficient. This means that very little of the input power, that often comes from a battery, is wasted. It also means that there is very little heat generated. So you do not need high wattage parts, big heat sinks, or lots of cooling. Switching power supplies also run at high frequencies. This means that the components can be small, reducing size and lowering cost. All of these advantages come from the fact that the transistors in the switching power supply are driven either hard *on* or hard *off*. The rectangular wave is the key signal in all switching power supplies.

This same principle is applied to dc motor drives. Passing a rectangle to the load is easily done by a transistor switch. Varying the width of that rectangle changes the average value, the dc, passed to the motor. This same technique of varying the width of a rectangle has been extended to synthesize sine waves for driving synchronous ac motors, and even speech and music.

Figure 3-5 shows a rectangular wave. The wave goes high at $t = 0$, and stays at V_p for the pulse width. It then drops to 0 V for the rest of the period. The duty cycle, D, is a unitless fraction. It indicates how much of the period the rectangle is high.

$$D = \frac{t_{on}}{T}$$

where

D = duty cycle
t_{on} = the time the signal is high
T = period

So the signal stays on for t_{on} before dropping back to zero.

$$t_{on} = D \times T$$

The equation for this wave, too, is divided into two parts, the on time (t_{on}) and the off time.

$$v(t) = \begin{cases} V_p & 0 \le t < DT \\ 0 & DT \le t < T \end{cases}$$

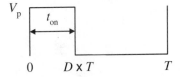

Figure 3-5 Rectangular wave

Triangular Waves

All of the switching applications described previously pass power to the load in rectangles. Often that step is passed on to an inductor within a filter, or to an inductive load, such as a motor. The inductor opposes a rapid change in current, allowing the current to gradually build up. The result is that current from a switch may be converted into a triangular shape.

The simplest triangle is shown in Figure 3-6. This is a straight line.

$$y = mx + b$$

y = the dependent variable (the height at some instant)
m = the slope
x = the independent variable (distance along the horizontal axis)
b = the y intercept (the height at the beginning)

Converting these into quantities to describe the electrical signal,
$y = i(t)$ Current is plotted on the vertical axis.
$x = t$ Time is plotted on the horizontal axis.
$b = 0$ The signal starts at 0.

The slope, m, is $m = \dfrac{\Delta y}{\Delta x} = \dfrac{\Delta i}{\Delta t}$

The signal starts at 0 A and rises linearly to I_p over the period of the wave, T.

$$m = \frac{I_p}{T}$$

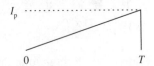

Combining these substitutions gives the equation for the ramp.

$$i(t) = \frac{I_p}{T} t$$

Figure 3-6 Ramp

Remember that I_p and T are constants, and t is the independent variable.

Example 3-1

a. Write the equation for a ramp that has a peak current of 3 A and a frequency of 150 kHz.

b. Calculate the current 5 μs after the beginning of the signal.

Solution

a.
$$i(t) = \frac{I_p}{T}t$$

$$T = \frac{1}{f} = \frac{1}{150\text{kHz}}$$

$$T = 6.67\,\mu\text{s}$$

$$i(t) = \frac{3\text{A}}{6.67\mu\text{s}}t$$

$$i(t) = 449.8\frac{\text{kA}}{\text{s}}t$$

b. At $t = 5\,\mu\text{s}$

$$i(5\mu\text{s}) = 449.8\frac{\text{kA}}{\text{s}}\times 5\mu\text{s}$$

$$i(5\mu\text{s}) = 2.25\,\text{A}$$

Practice: Write the equation for a 52 kHz ramp with a 0.5 A peak. Calculate the current at 12 μs.

Answers: $i(t) = 26\dfrac{\text{kA}}{\text{s}}t$ $i(12\mu\text{s}) = 312\text{mA}$

The trapezoid shown in Figure 3-7 is a more practical, though a bit more complicated, waveform. At $t = 0$, the current makes an abrupt jump up to I_{min}. It then ramps up, reaching I_p at the end of the duty cycle, DT. For the rest of the period the current is zero.

$$i(t) = \begin{cases} \dfrac{I_p - I_{min}}{DT}t + I_{min} & 0 \le t < DT \\ 0 & DT \le t < T \end{cases}$$

As with the rectangle, this equation is written in two parts. The first part is just the slope intercept equation for a line, $y = mx + b$. The slope is the ramp starting at I_{min}, and ending at I_p. At $t = 0$, the current jumps to I_{min}. That is the intercept, b.

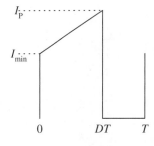

Figure 3-7 Trapezoid

Example 3-2

a. Write the equation for a trapezoid with a frequency of 500 kHz, a 40% duty cycle, a peak current of 2 A, and a ripple of 300 mA$_{pp}$

b. Find the current at 400 ns, and at 1 μs.

Solution

a.

$$i(t) = \begin{cases} \dfrac{I_p - I_{min}}{DT} t + I_{min} & 0 \le t < DT \\ 0 & DT \le t < T \end{cases}$$

$$I_p = 2\,A$$

$$I_{min} = I_p - I_{pp} = 2\,A - 0.3\,A$$

$$I_{min} = 1.7\,A$$

$$T = \frac{1}{f} = \frac{1}{500\,kHz} = 2\,\mu s$$

$$i(t) = \begin{cases} \dfrac{2\,A - 1.7\,A}{0.4 \times 2\,\mu s} t + 1.7\,A & 0 \le t < 0.4 \times 2\,\mu s \\ 0 & 0.4 \times 2\,\mu s \le t < 2\,\mu s \end{cases}$$

$$i(t) = \begin{cases} 375\dfrac{kA}{s} t + 1.7\,A & 0 \le t < 800\,ns \\ 0 & 800\,ns \le t < 2\,\mu s \end{cases}$$

b. At 400 ns,

$$i(400\,ns) = 375\frac{kA}{s} t + 1.7\,A = 375\frac{kA}{s} \times 400\,ns + 1.7\,A$$

$$i(400\,ns) = 1.85\,A$$

$$i(1\,\mu s) = 0\,A$$

Practice: Write the equation for the current and find the current at 1 μs if $f = 52$ kHz, $I_p = 4$ A, $I_{pp} = 0.8$ A, and $D = 70\%$.

Answer: $i(1\,\mu s) = 3.77\,A$

3.2 Average Value

The average value is also referred to as the dc value. When you measure a signal with your dc meter, its average value is indicated. DC motors and power supplies respond proportionally to the dc or average value of whatever signal is applied.

For a steady signal, as shown in Figure 3-8, the average value is its peak value as well.

$$V_{dc} = V_{ave} = V_p$$

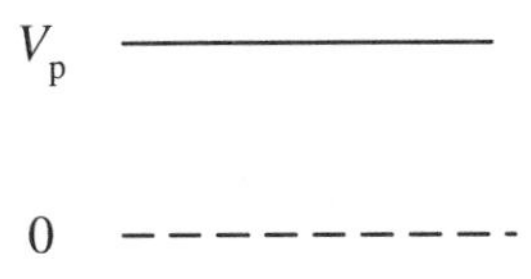

V_p _____

0 ‐ ‐ ‐ ‐ ‐ ‐ ‐ ‐ ‐ ·

Figure 3-8 DC signal

Rectangular Waves

The average value of the rectangular wave is almost as obvious. If a signal is at 10 V for half of the time and is 0 V for the rest of the time, intuitively, its average value, or average height is 5 V_{dc}. If that signal is 10 V for only 20%, and 0 V the rest of the time, then the average value drops to 2 V_{dc}.

Mathematically this is

$$V_{ave} = DV_p$$

To find the average height, calculate the area under the curve. Then divide that total area by the total length. For the rectangular wave in Figure 3-9, the area under the curve is

$$area = V_p \times DT$$

Its average value is

$$ave = \frac{area}{length} = \frac{V_p \times DT}{T}$$

$$V_{ave} = DV_p$$

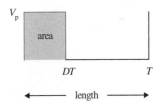

Figure 3-9 Rectangular wave

This approach works fine for a simple shape whose area is easily calculated. You must divide more complicated waves into many, many little rectangles, calculate the area of each, then add all of these areas together. That is exactly what the integral does. So, the general equation for the average value is

$$V_{ave} = \frac{1}{T}\int_0^T v(t)\,dt$$

For the rectangular wave,

$$v(t) = \begin{cases} V_p & 0 \le t < DT \\ 0 & DT \le t < T \end{cases}$$

The two parts of the equation are integrated with two separate integrals. The first has limits from 0 to DT, and the second goes from DT to T.

$$V_{ave} = \frac{1}{T}\left(\int_0^{DT} V_p\,dt + \int_{DT}^T 0\,dt \right)$$

The quantities V_p and 0 are constants. They can be brought out of the integral.

$$V_{ave} = \frac{1}{T}\left(V_p\int_0^{DT} dt + 0\int_{DT}^T dt \right)$$

The second integral is multiplied by 0. So, its value is 0.

$$V_{ave} = \frac{V_p}{T}\int_0^{DT} dt$$

Another way to interpret the integral is as the reverse of the derivative. The integral of the derivative of t is just t.

$$V_{ave} = \frac{V_p}{T}\left(t\,\Big|_0^{DT} \right)$$

The calculus is done. Now all you have left is a little algebra. First, apply the limits.

$$V_{ave} = \frac{V_p}{T}(DT - 0)$$

$$V_{ave} = \frac{V_p}{T}DT$$

$$V_{ave\ rectangle} = DV_p$$

Average value of a
rectangular wave

This certainly seems like a lot of work to get back to what you knew intuitively. But this verifies that the rigorous mathematical steps produce a result that makes sense. So, now you can apply this integration to more complicated signals and be confident that the result (when done correctly) does indeed give you the average value.

The program MATLAB® allows you to perform symbolic integration. This can be used to verify the derivation of the average value of a rectangular wave. The steps are shown below. Explanation is in the notes in the margin.

Defines these four letters
as variables. x is the independent variable.

```
» syms V D T x

» (V/T)*int(diff(x),0,D*T)

ans =
   V*D
```

$$\frac{V}{T}\int_0^{DT} dx$$

Sinusoids

The average value of the sine wave, shown in Figure 3-10, is also intuitive. The area between the curve and the horizontal axis is the same for the first half and the second half of the wave. From 0 to π radians, the area is positive (above the horizontal axis). For the second half, from π radians to 2π radians, the area is negative (below the horizontal axis). The area of one half cancels that for the second half. On the average, the area is zero. So the average height is also zero.

$$V_{ave} = 0$$

The integration should give the same result.

$$V_{ave} = \frac{1}{T}\int_0^T v(t)dt$$

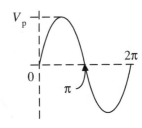

Figure 3-10 Simple sine wave

But, the sine wave is defined in terms of angle, θ, not time.

$$V_{ave} = \frac{V_p}{2\pi}\int_0^{2\pi} \sin\theta\, d\theta$$

Taking the integral gives

$$V_{ave} = \frac{V_p}{2\pi}\left(-\cos\theta\Big|_0^{2\pi}\right)$$

When evaluating the result at the two limits, be careful not to lose a sign.

$$V_{ave} = \frac{V_p}{2\pi}\{-\cos(2\pi)-[-\cos(0)]\}$$

$$V_{ave} = \frac{V_p}{2\pi}[-1-(-1)]$$

$$V_{ave} = \frac{V_p}{2\pi}(-1+1)$$

Average value of a sine

$$V_{ave\ sine} = 0$$

The calculus matches intuition again!

   Since the average value of a sine wave is zero, if you try to drive a permanent magnet dc motor with a full sine wave, the motor's shaft does not move. During the positive half-cycle, the shaft tries to turn one direction. The negative half-cycle sends the shaft the other direction. Repeated hundreds of times per second, the motor cannot overcome the starting inertia. So it just sits there, even though you may be applying a large sinusoidal voltage.

   To get dc from ac, you rectify it. A single diode gives a half-wave sinusoid, as shown in Figure 3-11. The equation developed in the last section is

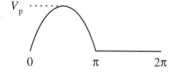

Figure 3-11 Half-wave rectified sine

$$v(\theta) = \begin{cases} V_p \sin\theta & 0 \le \theta < \pi \\ 0 & \pi \le \theta < 2\pi \end{cases}$$

Since the entire signal is above zero, this wave has a nonzero average value.

$$V_{ave} = \frac{1}{2\pi}\left(\int_0^\pi V_p \sin\theta\, d\theta + \int_\pi^{2\pi} 0\, d\theta\right)$$

The first integral contains a constant, V_p. So V_p can be brought out in front of the integral. The second integral is zero.

$$V_{ave} = \frac{V_p}{2\pi} \int_0^\pi \sin\theta \, d\theta$$

The integral of $\sin\theta$ is $-\cos\theta$.

$$V_{ave} = \frac{V_p}{2\pi}\left(-\cos\theta\Big|_0^\pi\right)$$

Evaluating the result at the two limits gives

$$V_{ave} = \frac{V_p}{2\pi}[(-\cos\pi)-(-\cos 0)]$$

$$V_{ave} = \frac{V_p}{2\pi}[-(-1)-(-1)]$$

$$V_{ave} = \frac{V_p}{2\pi}2$$

$$V_{ave\ half-wave\ sine} = \frac{V_p}{\pi}$$

Average value of a half-wave rectified sine

Example 3-3

   a. Calculate the average value of a half-wave rectified sine wave from the 120 V_{rms} line voltage.

   b. Verify your calculation with a simulation.

multiSIM

Solution

   a. First, you must determine the peak value of the line voltage.

$$V_{p\ sine} = 120\,V_{rms}\sqrt{2} = 169.7\,V_p$$

$$V_{ave\ half-wavesine} = \frac{V_p}{\pi} = \frac{169.7\,V_p}{\pi}$$

$$V_{ave} = 54.0\,V_{dc}$$

   b. The Multisim schematic is shown in Figure 3-12. The result is in good agreement with the theory. This is *not* always true with the results of simulation. It is a good idea to verify that the simulator gives correct results with a simple circuit before going to more complicated ones.

Practice: Intuitively, what should be the average value of a full-wave rectified sine wave?

Answer: Since there are twice as many humps as with the half-wave sine, it seems that the average value of a full-wave rectified signal should be twice that of a half-wave rectified sine.

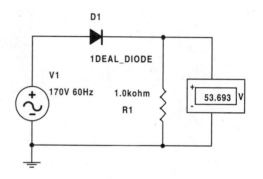

Figure 3-12 Multisim simulation

The derivation for the average value of a full-wave rectified sine wave follows closely what you just saw.

$$v(\theta) = \begin{cases} V_p \sin\theta & 0 \le \theta < \pi \\ -V_p \sin\theta & \pi \le \theta < 2\pi \end{cases}$$

$$V_{ave} = \frac{1}{2\pi}\left(\int_0^\pi V_p \sin\theta\, d\theta + \int_\pi^{2\pi} -V_p \sin\theta\, d\theta \right)$$

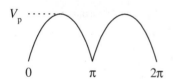

Figure 3-13 Full-wave rectified sine

There are two integrals, one from 0 to π for the first half-wave, and the other from π to 2π for the second half-wave. Each term has the constant V_p. Bring each V_p all the way out to the front. Also notice that the second integral is negative because of the negative sign associated with its V_p.

$$V_{ave} = \frac{V_p}{2\pi}\left(\int_0^\pi \sin\theta\, d\theta - \int_\pi^{2\pi} \sin\theta\, d\theta \right)$$

The integral of the $\sin\theta$ is $-\cos\theta$.

$$V_{ave} = \frac{V_p}{2\pi}\left[\left(-\cos\theta\big|_0^\pi\right) - \left(-\cos\theta\big|_\pi^{2\pi}\right)\right]$$

With all of these negative signs, be very careful. Evaluate each result at the two limits.

$$V_{ave} = \frac{V_p}{2\pi}\{[(-\cos\pi)-(-\cos0)]-[(-\cos2\pi)-(-\cos\pi)]\}$$

$$V_{ave} = \frac{V_p}{2\pi}\{[(1)-(-1)]-[(-1)-(1)]\}$$

$$V_{ave} = \frac{V_p}{2\pi}\{[2]-[-2]\} = \frac{V_p}{2\pi}4$$

$$V_{ave\ full\ rectified\ sine} = \frac{2V_p}{\pi}$$

Average value of a full-wave rectified sine

This is just twice the average for the half-wave sine. That makes sense.

Triangular Waves

The simple triangle wave is shown, again, in Figure 3-14. You can graphically imagine the average value as a line. Half of the area under the curve is above that average value line, and half of the area is below the average value. For this linear ramp, the average value should be half way up the ramp. That is

$$I_{ave\ ramp} = \frac{I_p}{2}$$

The equation for the ramp is

$$i(t) = \frac{I_p}{T}t$$

Its average value is

$$I_{ave} = \frac{1}{T}\int_0^T \frac{I_p}{T}t\,dt$$

Figure 3-14 Ramp

Both I_p and T are constants. So their fraction can be brought out of the integral.

$$I_{ave} = \frac{1}{T}\frac{I_p}{T}\int_0^T t\,dt$$

$$I_{ave} = \frac{I_p}{T^2}\int_0^T t\,dt$$

The integral of t is $\frac{1}{2}t^2$.

$$I_{ave} = \frac{I_p}{T^2}\left(\frac{1}{2}t^2\Big|_0^T\right)$$

Evaluate this equation at the two limits.

$$I_{ave} = \frac{I_p}{2T^2}\left(T^2 - 0^2\right)$$

Average value of a ramp

$$I_{ave\ ramp} = \frac{I_p}{2}$$

This matches the initial thoughts.

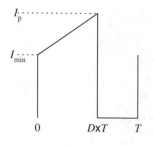

The other triangle wave is the trapezoid. It is shown in Figure 3-15. Its equation is

$$i(t) = \begin{cases} \dfrac{I_p - I_{min}}{DT}t + I_{min} & 0 \le t < DT \\ 0 & DT \le t < T \end{cases}$$

Remember, I_p, I_{min}, D, and T are all constants.

Figure 3-15 Trapezoid

$$I_{ave} = \frac{1}{T}\int_0^T i(t)\,dt$$

$$I_{ave} = \frac{1}{T}\left[\int_0^{DT}\left(\frac{I_p - I_{min}}{DT}t + I_{min}\right)dt + \int_{DT}^T 0\,dt\right]$$

By applying the steps you have seen in previous examples,

Average value of a trapezoid

$$I_{ave\ trapezoid} = D\left(\frac{I_p - I_{min}}{2} + I_{min}\right)$$

This seems right. The first term within the parenthesis is the average value of the ramp. The second part is for the pedestal upon which the ramp sits. These must be reduced by D (<1), since the trapezoid is only on for that part of each cycle.

MATLAB can be used to verify the integration. The result of the integration is easier for the software to simplify if any constants are distributed within the integral. So, move the 1/T inside the integral.

```
» syms T D Ip Im x
```

Defines T, d, Ip, Im, as variables. x is independent.

```
» int((Ip-Im)/(D*T^2)*x+Im/T,0,D*T)
```

$$\int_{0}^{DT} \left(\frac{I_p - I_m}{DT^2} x + \frac{I_m}{T} \right) dx$$

```
ans =
  1/2*(Ip-Im)*D+Im*D
```

$$\frac{1}{2}(I_p - I_m)D + I_m D$$

Example 3-4

   a. Calculate the average value of a trapezoid with the following parameters:

       frequency = 100 kHz
       duty cycle = 35%
       $I_p = 3$ A$_p$
       $I_{ripple} = 400$ mA$_{pp}$

   b. Verify your manual calculation with a simulation.

Solution

   a. Manual calculation

$$I_{min} = I_p - I_{ripple} = 3\,\text{A}_p - 0.4\,\text{A}_{pp}$$

$$I_{min} = 2.6\text{A}$$

$$T = \frac{1}{f} = \frac{1}{100\,\text{kHz}}$$

$$T = 10\mu\text{s}$$

Now that you have all of the parameters that you need, it is time to calculate the average value.

$$I_{ave} = D\left(\frac{I_p - I_{min}}{2} + I_{min}\right)$$

$$I_{ave} = 0.35\left(\frac{3A - 2.6A}{2} + 2.6A\right)$$

$$I_{ave} = 0.98\,A_{dc}$$

b. The PSpice simulation is shown in Figure 3-16. To create the trapezoid current wave shape, a piece-wise linear current source (IPWL) is needed. There are only eight points allowed, permitting two cycles of the wave.

   The transient analysis dialog box shows that the analysis is run over those two cycles, an end time of 20 μs. A step time and a step ceiling of 20 ns are indicated.

   When the analysis is run, the current meter, called IPROBE, indicates 0 A! This is the current produced by the dc sources and initial conditions only. PSpice sets the time variable sources to zero. The dc calculations are initial bias points. They do *not* indicate the average value.

   You must use the Probe feature to obtain average value. When using a piece-wise linear source, always display the source as one of the signals in Probe. This assures that you did indeed specify the signal correctly. To determine the average value of $I(R1)$, add the trace

   AVE(I(R1))

   Notice that the trace varies with time. Each point is the average of all points up to that time. Good results require several cycles for the average value to reach a steady level. Ten cycles gives a reasonably smooth result. But only two cycles of this trapezoid could be entered. The cursor is placed at the very end of the trace. It indicates

   I ave = 982 mA

   This is in good agreement with theory.

Practice: Determine the average value of a ramp with $I_p = 3$ A and $f = 100$ kHz, manually and using PSpice.

Answers: manual $I_{ave} = 1.5\,A_{dc}$,   PSpice $I_{ave} \cong 1.45\,A_{dc}$

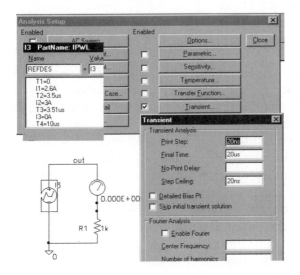

Figure 3-16 (a) PSpice simulation of Example 3-4

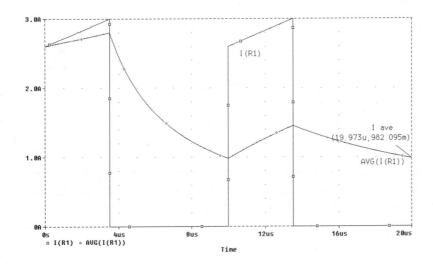

Figure 3-16 (b) Simulation of Example 3-4

3.3 Root-Mean-Squared Value

The average value is indicated by dc measuring instruments. It gives a good measure of the performance you can expect from dc rated loads, such as a dc motor. It is a very poor way, though, to compare the effectiveness of differently shaped signals. The more or less pure sinusoid delivered by the utility company to most residences has an average value of $0 \ V_{dc}$. However, it is capable of lighting your lamps, running your refrigerator and your computer, and heating your microwave oven. It will also kill you, even though a dc meter still indicates zero.

The **root-mean-squared** value (**rms**) of a signal is a much better indicator of the power a signal can deliver, regardless of wave shape. Signals of the same rms value deliver the same power to a resistive load.

Remember, for simple dc and sinusoids,

$$P = \frac{V^2}{R}$$

Therefore, to begin the calculation of rms, the voltage function must be squared.

$$V_{rms} \propto v^2(t)$$

Remember when a sinusoid is applied to a permanent magnet dc motor, for one half-cycle the shaft tries to turn one direction, and for the other half-cycle, it tries to turn the opposite direction. On the average, no work is done. Similarly, it is the average value of the power, not its instantaneous level, that is of interest.

$$V_{rms} \propto \frac{1}{T} \int_0^T v^2(t) \, dt$$

Now, however, the units in the relationship are wrong. There are volts on the left, and v^2 on the right. To correct this and produce an equality, you must take the square root of the right side.

Definition of root-mean-squared

$$V_{rms} = \sqrt{\frac{1}{T} \int_0^T v^2(t) \, dt}$$

DC Value

To serve as a baseline, and to see how to apply the rms equation in its simplest form, calculate the rms value of a steady, dc voltage.

$$v(t) = V_p$$

$$V_{rms} = \sqrt{\frac{1}{T} \int_0^T v^2(t)\, dt}$$

$$V_{rms\ of\ dc} = \sqrt{\frac{1}{T} \int_0^T V_p^2\, dt}$$

Since V_p is a constant, it can be brought out of the integral.

$$V_{rms\ of\ dc} = \sqrt{\frac{V_p^2}{T} \int_0^T dt}$$

The integral of the derivative of t is t, evaluated at the limits.

$$V_{rms\ of\ dc} = \sqrt{\frac{V_p^2}{T} \left(t \Big|_0^T \right)}$$

$$V_{rms\ of\ dc} = \sqrt{\frac{V_p^2}{T} (T - 0)}$$

$$V_{rms\ of\ dc} = V_p$$

The rms of steady dc

This is just what you should have expected.

Rectangular Waves

It is almost as easy to calculate the rms value of the rectangular wave in Figure 3-17.

$$v(t) = \begin{cases} V_p & 0 \le t < DT \\ 0 & DT \le t < T \end{cases} \quad 0 < D < 1$$

$$V_{rms} = \sqrt{\frac{1}{T} \int_0^T v^2(t)\, dt}$$

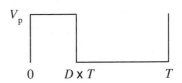

Figure 3-17 Rectangular wave

Since the equation for the wave has two parts, the integral also has two parts.

$$V_{\text{rms rectangle}} = \sqrt{\frac{1}{T}\left(\int_0^{DT} V_p^2\, dt + \int_{DT}^T 0^2\, dt\right)}$$

The integral of zero is zero, so the second term drops out.

$$V_{\text{rms rectangle}} = \sqrt{\frac{1}{T}\int_0^{DT} V_p^2\, dt}$$

This looks almost exactly like the rms calculation for the dc signal. Only, for the rectangular wave, the upper limit is DT.

$$V_{\text{rms rectangle}} = \sqrt{\frac{V_p^2}{T}\left(t\Big|_0^{DT}\right)}$$

$$V_{\text{rms rectangle}} = \sqrt{\frac{V_p^2}{T}(DT)}$$

The rms of a rectangular wave

$$V_{\text{rms rectangle}} = V_p\sqrt{D}$$

---

multiSIM

Example 3-5

a. Calculate the rms value of a rectangular wave with the following parameters:

   frequency = 100 kHz
   duty cycle = 35%
   $V_p = 15$ V

b. Verify your manual calculation with a simulation.

Solution

a. $$V_{\text{rms rectangle}} = V_p\sqrt{D} = 15\,V_p\sqrt{0.35}$$

   $$V_{\text{rms rectangle}} = 8.87\,V_{\text{rms}}$$

b. The Multisim simulation is shown in Figure 3-18. The ac value is wrong! Be very careful to verify that the simulator you are using produces correct results on a simple circuit before you trust its answers on more practical, complicated systems. Also, be cautious when you are told V_{ac}. Is that rms, peak, or peak-to-peak?

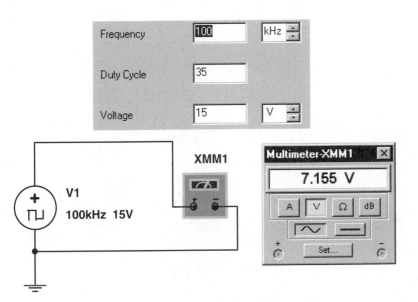

Figure 3-18 Simulation for Example 3-5

Triangular Waves

When you calculate the rms value of the triangular and the trapezoidal waves, things get just a little more complicated. The ramp is shown again in Figure 3-19. Its equation is

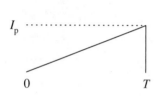

Figure 3-19 Ramp

$$i(t) = \frac{I_\text{p}}{T} t$$

$$I_\text{rms} = \sqrt{\frac{1}{T} \int_0^T i^2(t)\, dt}$$

$$I_\text{rms ramp} = \sqrt{\frac{1}{T} \int_0^T \left(\frac{I_\text{p}}{T} t \right)^2 dt}$$

$$I_\text{rms ramp} = \sqrt{\frac{1}{T} \int_0^T \frac{I_\text{p}^2}{T^2} t^2\, dt}$$

Move the constants out of the integral.

$$I_\text{rms ramp} = \sqrt{\frac{1}{T} \frac{I_\text{p}^2}{T^2} \int_0^T t^2\, dt}$$

$$I_\text{rms ramp} = \sqrt{\frac{I_\text{p}^2}{T^3} \int_0^T t^2\, dt}$$

The integral of t^2 is $^1/_3\, t^3$.

$$I_\text{rms ramp} = \sqrt{\frac{I_\text{p}^2}{T^3} \left(\frac{1}{3} t^3 \Big|_0^T \right)}$$

Evaluate the result at the two limits.

$$I_\text{rms ramp} = \sqrt{\frac{I_\text{p}^2}{T^3} \frac{1}{3} \left(T^3 - 0^3 \right)}$$

$$I_\text{rms ramp} = \sqrt{\frac{I_\text{p}^2}{3T^3} T^3}$$

$$I_\text{rms ramp} = \sqrt{\frac{I_\text{p}^2}{3}}$$

The rms of a ramp

$$I_\text{rms ramp} = \frac{I_\text{p}}{\sqrt{3}}$$

Example 3-6

OrCAD

Determine the rms value of a ramp with $I_p = 3$ A, $f = 100$ kHz, manually and using PSpice.

Solution

$$I_{\text{rms ramp}} = \frac{I_p}{\sqrt{3}} = \frac{3\,A_p}{\sqrt{3}}$$

$$I_{\text{rms ramp}} = 1.73\,A_{\text{rms}}$$

The circuit setup is shown in Figure 3-20(a). The current source must be created piece-wise linearly. The source, IPWL, provides eight points, allowing four full cycles. So, the transient analysis is run for a total time of 40 μs, with the step time and the step ceiling set at 40 ns each.

Figure 3-20(b) shows the resulting Probe waveforms . Be sure to plot the source signal, to verify that you created the IPWL correctly. In Probe, when you add a trace, one of the options that is available in the dialog box is RMS(). This function, on a point-by-point basis, calculates the square root of the average value of the square of the specified function. This is then displayed as another signal. The cursor identifies the value at the end of that trace as

$$\text{RMS(I(R1))} = 1.73\ A_{\text{rms}}$$

This matches the value predicted by manual calculation.

Practice: Use PSpice Probe to simulate the rms value of the rectangular wave from Example 3-5.

Answer: PSpice Probe indicates that RMS(V(out)) = 8.88 V_{rms}. This matches the manual calculations.

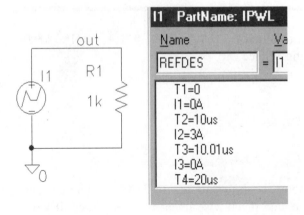

Figure 3-20 (a) PSpice setup for Example 3-6

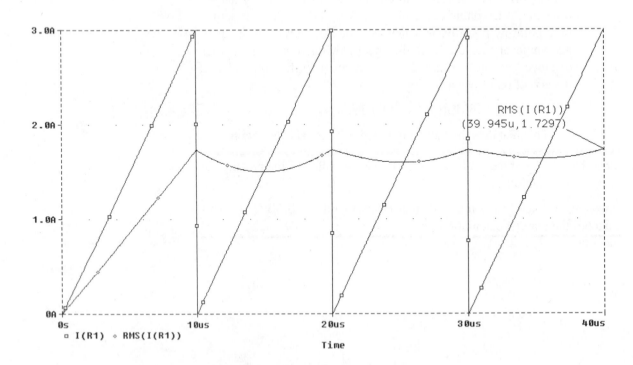

Figure 3-20 (b) Probe results for Example 3-6

The trapezoid is shown in Figure 3-21. Its equation is

$$i(t) = \begin{cases} \dfrac{I_p - I_{min}}{DT}t + I_{min} & 0 \le t < DT \\ 0 & DT \le t < T \end{cases}$$

The algebra becomes a little involved when you square $i(t)$. So, it is important to remember that I_p, I_{min}, D, and T are all constants.

$$I_{rms} = \sqrt{\frac{1}{T}\int_0^T i^2(t)\,dt}$$

$$I_{rms\ trapezoid} = \sqrt{\frac{1}{T}\int_0^{DT}\left(\frac{I_p - I_{min}}{DT}t + I_{min}\right)^2 dt}$$

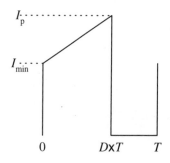

Figure 3-21 Trapezoid

Complete the square, to remove the parentheses.

$$I_{rms\ trapezoid} = \sqrt{\frac{1}{T}\int_0^{DT}\left[\left(\frac{I_p - I_{min}}{DT}t\right)^2 + 2\frac{I_p - I_{min}}{DT}t \times I_{min} + I_{min}^2\right] dt}$$

Each term can be integrated individually.

$$I_{rms\ trapezoid} = \sqrt{\frac{1}{T}\int_0^{DT}\left(\frac{I_p - I_{min}}{DT}t + I_{min}\right)^2 dt}$$

Move the constants out of each of the integrals.

$$I_{rms\ trapezoid} = \sqrt{\frac{1}{T}\left[\left(\frac{I_p - I_{min}}{DT}\right)^2 \int_0^{DT}t^2\,dt + \left(2I_{min}\frac{I_p - I_{min}}{DT}\right)\int_0^{DT}t\,dt + I_{min}^2\int_0^{DT}dt\right]}$$

Now, perform each integration.

$$I_{rms\ trapezoid} = \sqrt{\frac{1}{T}\left[\left(\frac{I_p - I_{min}}{DT}\right)^2 \frac{1}{3}\left(t^3\big|_0^{DT}\right) + \left(2I_{min}\frac{I_p - I_{min}}{DT}\right)\frac{1}{2}\left(t^2\big|_0^{DT}\right) + I_{min}^2\left(t\big|_0^{DT}\right)\right]}$$

Evaluate each element at its limits.

$$I_{\text{rms trapezoid}} = \sqrt{\frac{1}{T}\left[\left(\frac{I_p - I_{\min}}{DT}\right)^2 \frac{1}{3}D^3T^3 + \left(2I_{\min}\frac{I_p - I_{\min}}{DT}\right)\frac{1}{2}D^2T^2 + I_{\min}^2 DT\right]}$$

Now simplify the algebra.

$$I_{\text{rms trapezoid}} = \sqrt{\frac{1}{T}\left[\frac{1}{3}\left(I_p - I_{\min}\right)^2 DT + I_{\min}\left(I_p - I_{\min}\right)DT + I_{\min}^2 DT\right]}$$

$$I_{\text{rms trapezoid}} = \sqrt{\frac{DT}{T}\left[\frac{1}{3}\left(I_p - I_{\min}\right)^2 + I_{\min}\left(I_p - I_{\min}\right) + I_{\min}^2\right]}$$

$$I_{\text{rms trapezoid}} = \sqrt{D\left[\frac{1}{3}\left(I_p - I_{\min}\right)^2 + I_{\min}\left(I_p - I_{\min}\right) + I_{\min}^2\right]}$$

$$I_{\text{rms trapezoid}} = \sqrt{D\left[\frac{1}{3}\left(I_p - I_{\min}\right)^2 + I_{\min}\left(I_p - I_{\min}\right) + I_{\min}^2\right]}$$

$$I_{\text{rms trapezoid}} = \sqrt{D\left[\frac{1}{3}\left(I_p^2 - 2I_p I_{\min} + I_{\min}^2\right) + \frac{3}{3}I_{\min}I_p\right]}$$

$$I_{\text{rms}} = \sqrt{\frac{1}{T}\int_0^T i^2(t)\,dt}$$

The rms of a trapezoid

$$I_{\text{rms trapezoid}} = \sqrt{\frac{D}{3}\left[I_p^2 + I_p I_{\min} + I_{\min}^2\right]}$$

MATLAB verifies this result. The $1/T$ is moved inside the integral to improve the simplification.

```
» syms T D Ip Im x

» sqrt(int(1/T*((Ip-Im)/(D*T)*x+Im)^2,0,D*T))

ans =

1/3*(3*Ip^2*D+3*Im*D*Ip+3*Im^2*D)^(1/2)
```

$$I_{\text{rms trapezoid}} = \sqrt{\int_0^{DT}\frac{1}{T}\left(\frac{I_p - I_{\min}}{DT}t + I_{\min}\right)^2 dt}$$

$$\frac{1}{3}\sqrt{3I_p^2 D + 3I_{\min}I_p D + 3I_{\min}^2 D}$$

It takes a step or two of algebra to further simplify this result to match the equation derived manually.

Sinusoids

You have been told in ac circuits that

$$V_{\text{rms sine}} = \frac{V_p}{\sqrt{2}}$$

Before going on to the half-wave and full-wave rectified versions, confirm that this equation is correct. The general equation for a simple sine wave is

$$v(\theta) = V_p \sin\theta$$

The rms value is defined as

$$V_{\text{rms}} = \sqrt{\frac{1}{T}\int_0^T v^2(t)\,dt}$$

Substitute, and adjust the limits.

$$V_{\text{rms sine}} = \sqrt{\frac{1}{2\pi}\int_0^{2\pi}(V_p\sin\theta)^2\,d\theta}$$

Complete the square.

$$V_{\text{rms sine}} = \sqrt{\frac{1}{2\pi}\int_0^{2\pi}V_p^2\sin^2\theta\,d\theta}$$

Move the constant out of the integral.

$$V_{\text{rms sine}} = \sqrt{\frac{V_p^2}{2\pi}\int_0^{2\pi}\sin^2\theta\,d\theta}$$

There are several forms for the integral of $\sin^2\theta$. A convenient one is

$$\int \sin^2\theta\,d\theta = \frac{1}{2}\theta - \frac{1}{4}\sin 2\theta$$

Apply this integral solution to the rms calculation.

$$V_{\text{rms sine}} = \sqrt{\frac{V_P^2}{2\pi}\left(\frac{1}{2}\theta - \frac{1}{4}\sin 2\theta\right)\Big|_0^{2\pi}}$$

Evaluate this result at the two limits.

$$V_{\text{rms sine}} = \sqrt{\frac{V_p^2}{2\pi}\left[\left(\frac{1}{2}2\pi - \frac{1}{4}\sin 4\pi\right) - \left(\frac{1}{2}0 - \frac{1}{4}\sin 0\right)\right]}$$

$$V_{\text{rms sine}} = \sqrt{\frac{V_p^2}{2\pi}\left[\left(\pi - \frac{1}{4}0\right) - (0-0)\right]}$$

This is getting simpler. Now, just finish the algebra.

$$V_{\text{rms sine}} = \sqrt{\frac{V_p^2}{2\pi}\pi}$$

The rms of a full sine wave

$$V_{\text{rms sine}} = \frac{V_p}{\sqrt{2}}$$

The half-wave rectified sinusoid is shown in Figure 3-22. Its equation is

$$v(\theta) = \begin{cases} V_p \sin\theta & 0 \le \theta < \pi \\ 0 & \pi \le \theta < 2\pi \end{cases}$$

The rms value is defined as

$$V_{\text{rms}} = \sqrt{\frac{1}{T}\int_0^T v^2(t)\,dt}$$

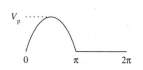

Figure 3-22 Half-wave rectified sine

Substitute and adjust the limits.

$$V_{\text{rms half sine}} = \sqrt{\frac{1}{2\pi}\int_0^\pi (V_p \sin\theta)^2\,d\theta}$$

Since the second part of the waveform and its equation are zero, that integral has been omitted. Notice that this looks *just* like the work done for the full sine wave, except for the upper limit. Now, complete the square.

$$V_{\text{rms half sine}} = \sqrt{\frac{1}{2\pi}\int_0^\pi V_p^2 \sin^2\theta\,d\theta}$$

Move the constant out of the integral.

$$V_{\text{rms half sine}} = \sqrt{\frac{V_p^2}{2\pi} \int_0^\pi \sin^2\theta \, d\theta}$$

$$\int \sin^2\theta \, d\theta = \frac{1}{2}\theta - \frac{1}{4}\sin 2\theta$$

Apply this integral solution to the rms calculation.

$$V_{\text{rms half sine}} = \sqrt{\frac{V_p^2}{2\pi} \left(\frac{1}{2}\theta - \frac{1}{4}\sin 2\theta\right)\Big|_0^{2\pi}}$$

Evaluate this result at the two limits.

$$V_{\text{rms half sine}} = \sqrt{\frac{V_p^2}{2\pi} \left[\left(\frac{1}{2}\pi - \frac{1}{4}\sin 2\pi\right) - \left(\frac{1}{2}0 - \frac{1}{4}\sin 0\right)\right]}$$

$$V_{\text{rms half sine}} = \sqrt{\frac{V_p^2}{2\pi} \left[\left(\frac{\pi}{2} - \frac{1}{4}0\right) - (0 - 0)\right]}$$

Finish the algebra.

$$V_{\text{rms half sine}} = \sqrt{\frac{V_p^2}{2\pi}\frac{\pi}{2}}$$

$$V_{\text{rms half sine}} = \frac{V_p}{2} \qquad\qquad V_{\text{rms full sine}} = \frac{V_p}{\sqrt{2}}$$

The rms of a half-wave rectifed and a full-wave rectified sine wave

Be very careful with this result. It seems a little strange that you can eliminate half of a signal, and its rms value is *not* cut in half.

Example 3-7

  a. Calculate the rms value of a half-wave rectified sine wave with $f = 60$ Hz and $V_p = 170$ V$_p$.

  b. Does Multisim confirm the result?

Solution

  a.
$$V_{\text{rms half sine}} = \frac{170\,V_p}{2} = 85\,V_{\text{rms}}$$

multiSIM

b. The results of the simulation are shown in Figure 3-23. The two meters have been set to ac. The meter before the diode indicates 120 V_{ac}. The other meter shows 65 V_{ac}. It should indicate $85 V_{rms}$. Be careful to check the results of a simulation.

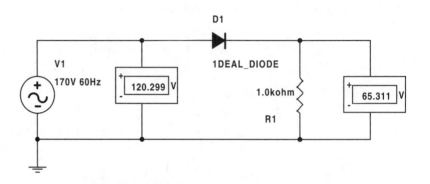

Figure 3-23 Multisim simulation of Example 3-7

Practice: Repeat the simulation using PSpice and the RMS function in Probe. Run the transient analysis for 167 ms. Set the diode model's breakdown voltage above 170 V.

Answer: At the end of the analysis, the RMS is varying between 84.4 V and 86.5 V. This gives an average of the RMS function as 85 V_{rms}.

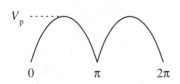

Figure 3-24 Full-wave rectified sine

The full-wave rectified sine wave is shown, again, in Figure 3-24. Its equation is

$$v(\theta) = \begin{cases} V_p \sin\theta & 0 \le \theta < \pi \\ -V_p \sin\theta & \pi \le \theta < 2\pi \end{cases}$$

The rms value is

$$V_{rms} = \sqrt{\frac{1}{T}\int_0^T v^2(t)\,dt}$$

Because the signal is defined in two parts, the integral can be evaluated in two pieces.

$$V_{\text{rms full-wave rect}} = \sqrt{\frac{1}{2\pi}\left[\int_0^\pi \left(V_p\sin\theta\right)^2 d\theta + \int_\pi^{2\pi}\left(-V_p\sin\theta\right)^2 d\theta\right]}$$

Complete each square.

$$V_{\text{rms full-wave rect}} = \sqrt{\frac{1}{2\pi}\left[\int_0^\pi V_p^2\sin^2\theta\, d\theta + \int_\pi^{2\pi} V_p^2\sin^2\theta\, d\theta\right]}$$

The square changes the negative sign in the second integral to a positive. Except for the limits, the two integrals are the same. Next, move the constants out of the integrals.

$$V_{\text{rms full-wave rect}} = \sqrt{\frac{V_p^2}{2\pi}\left[\int_0^\pi \sin^2\theta\, d\theta + \int_\pi^{2\pi}\sin^2\theta\, d\theta\right]}$$

$$\int\sin^2\theta\, d\theta = \frac{1}{2}\theta - \frac{1}{4}\sin 2\theta$$

Apply this integral solution to the rms calculation.

$$V_{\text{rms full-wave rect}} = \sqrt{\frac{V_p^2}{2\pi}\left[\left(\frac{1}{2}\theta - \frac{1}{4}\sin 2\theta\right)\Big|_0^\pi + \left(\frac{1}{2}\theta - \frac{1}{4}\sin 2\theta\right)\Big|_\pi^{2\pi}\right]}$$

Evaluate this result at the two limits.

$$V_{\text{rms full-wave rect}} = \sqrt{\frac{V_p^2}{2\pi}\left[\left(\frac{\pi}{2} - \frac{1}{4}\sin 2\pi\right) - \left(\frac{1}{2}0 - \frac{1}{4}\sin 0\right) + \left(\frac{1}{2}2\pi - \frac{1}{4}\sin 4\pi\right) - \left(\frac{1}{2}2\pi - \frac{1}{4}\sin 2\pi\right)\right]}$$

$$V_{\text{rms full-wave rect}} = \sqrt{\frac{V_p^2}{2\pi}\left[\left(\frac{\pi}{2} - \frac{1}{4}0\right) - (0-0) + \left(\pi - \frac{1}{4}0\right) - \left(\frac{\pi}{2} - \frac{1}{4}0\right)\right]}$$

Finish the algebra. $$V_{\text{rms full-wave rect}} = \sqrt{\frac{V_p^2}{2\pi}\left(\frac{\pi}{2} + \pi - \frac{\pi}{2}\right)}$$

$$V_{\text{rms full-wave rect}} = \sqrt{\frac{V_p^2}{2}}$$

$$V_{\text{rms full-wave rect}} = \frac{V_p}{\sqrt{2}}$$

The rms of a full-wave rectified sine

This is also the same value as for a pure sine wave. In terms of its ability to deliver power, such as heating a resistor, it does not matter if the voltage is positive or negative. When both half-cycles are present, whether they are plus and minus or both positive, the same power is delivered to a resistor. So the two signals have the same rms relationship. Table 3-1, on the following page, summarizes the wave shapes, their equations, and their average and rms values.

3.4 Power

It is now time to review a little physics. Work is a measure of the effort expended, or the energy used (or gained). If a 150 lb person climbs to the top of a 5 story building (50 ft), then

$$W = F \times y$$

$$W = 150\,\text{lb} \times 50\,\text{ft} = 7500\,\text{ft} \cdot \text{lb}$$

In the MKS (meter-kilogram-seconds) system of measure, that 150 lb person has a mass of 68 kg. Each kg of mass exerts 9.8 newtons of force at the Earth's surface. The 50 ft building is 15.2 m tall. This gives

$$W = 68\,\text{kg} \times \frac{9.8\,\text{N}}{\text{kg}} \times 15.2\,\text{m} = 10.13\,\text{kNm}$$

A newton·meter is a Joule

$$W = 10.13\,\text{kJ}$$

Walking leisurely, it may take five minutes to climb the five stories. At a full run, though, that same 150 lb person may be able to climb the 50 ft in 45 s. In either case, the same work is done, 150 lb is raised 50 ft. But it certainly is harder to do it in 45 s at a dead run, than in five minutes at a walk. This is expressed by the power expended. **Power** is the *rate* of doing work.

$$p = \frac{W}{t}$$

During the slow walk up the stairs,

$$p = \frac{150\,\text{lb} \times 50\,\text{ft}}{5\,\text{min} \times 60\,\dfrac{\text{s}}{\text{min}}} = 25\,\frac{\text{ft} \cdot \text{lb}}{\text{s}}$$

Table 3-1 Wave shapes and common values

	function	average	rms
	$v(t) = V_p$	V_p	V_p
	$v(\theta) = V_p \sin\theta$	0	$\dfrac{V_p}{\sqrt{2}}$
	$v(\theta) = \begin{cases} V_p \sin\theta & 0 < \theta < \pi \\ 0 & \pi < \theta < 2\pi \end{cases}$	$\dfrac{V_p}{\pi}$	$\dfrac{V_p}{2}$
	$v(\theta) = \begin{cases} V_p \sin\theta & 0 < \theta < \pi \\ -V_p \sin\theta & \pi < \theta < 2\pi \end{cases}$	$\dfrac{2V_p}{\pi}$	$\dfrac{V_p}{\sqrt{2}}$
	$v(t) = \begin{cases} V_p & 0 < t < DT \\ 0 & DT < t < T \end{cases}$ $0 < D < 1$	DV_p	$\sqrt{D} \times V_p$
	$v(t) = \dfrac{V_p}{T} t$	$\dfrac{V_p}{2}$	$\dfrac{V_p}{\sqrt{3}}$
	$i(t) = \begin{cases} \dfrac{I_P - I_{min}}{DT} t + I_{min} & 0 < t < DT \\ 0 & DT < t < T \end{cases}$	$I_{\text{ave trapezoid}} = D\left(\dfrac{I_P - I_{min}}{2} + I_{min} \right)$ $I_{\text{rms trapezoid}} = \sqrt{\dfrac{D}{3}\left[I_p^2 + I_p I_m + I_m^2\right]}$	

During the run, however, the power being generated is

$$p = \frac{150\,\text{lb} \times 50\,\text{ft}}{45\,\text{s}} = 167\,\frac{\text{ft} \cdot \text{lb}}{\text{s}}$$

You are probably more familiar with power specified in horsepower (hp).

$$1\,\text{hp} = 550\,\frac{\text{ft} \cdot \text{lb}}{\text{s}}$$

So

$$p = 167\,\frac{\text{ft} \cdot \text{lb}}{\text{s}} \times \frac{1\,\text{hp}}{550\,\dfrac{\text{ft} \cdot \text{lb}}{\text{s}}} = 0.304\,\text{hp}$$

This is about the same power that a washing machine delivers.

In the MKS system, this is

$$p = \frac{10.13\,\text{kJ}}{45\,\text{s}} = 225\,\frac{\text{J}}{\text{s}}$$

A J/s is also called a **watt**. $P = 225$ W.

Even though the person at the top of the stairs has gained (or expended) the same amount of energy whether the stairs were climbed at a slow walk or a dead run, the power dissipated during the climb is much greater for the run. Power gives a measure of the rate at which energy must be delivered to a load, and the heat that might be generated.

When you first studied electricity, you saw that voltage is defined as the amount of energy given to each coulomb of charge as it is pumped from the negative to the positive terminal of the source by the generator (magnetic energy) or by the battery (chemical energy). In equation form,

$$v = \frac{W}{Q}$$

The unit is J/C or volts. This tells you how much work each coulomb of charge can do as it flows through your circuit.

It is the rate of moving that charge that makes the difference. How quickly you can deliver charge is defined as current.

$$i = \frac{Q}{t}$$

Power (the rate of doing work) is

$$p = \frac{W}{t}$$

Now, look at the three equations: for v, i, and p.

$$p = \frac{W}{t} = \frac{W}{Q} \times \frac{Q}{t}$$

$$p = vi$$

The power delivered to a load, *at any instant in time*, is the voltage across the load at that instant, times the current through the load at that instant. To emphasize that all three of these quantities are at an instant in time, rewrite the equation as

$$p(t) = v(t) \times i(t)$$

Instantaneous power definition

This is the only correct, general equation for power. It is *not* true to say that average power is average voltage times average current, or that rms power (whatever that is) is rms voltage times rms current. Power is a function that varies with time, as the voltage applied to the load and the resulting current through the load vary.

Example 3-8

Its necessary to raise 1000 lb a height of 40 ft in 30 s. Assuming that the motor is 90% efficient and the mechanics in the system are 50% efficient, calculate the following:

a. the size of the motor needed.
b. the current needed if the dc motor is rated at 220 V_{dc}.

Solution

a. Calculate the work that must be done to get the load to the top.

$$W = 1000\,\text{lb} \times 40\,\text{ft} = 40,000\,\text{ft} \cdot \text{lb}$$

To accomplish this much work in 30 s requires a power of

$$p = \frac{W}{t} = \frac{40,000\,\text{ft} \cdot \text{lb}}{30\,\text{s}}$$

$$p = 1333\,\frac{ft \cdot lb}{s}$$

$$p = 1333\,\frac{ft \cdot lb}{s} \times \frac{1\,hp}{550\,\dfrac{ft \cdot lb}{s}} = 2.42\,hp$$

Since the mechanics in the system are only 50% efficient, half of the power is lost in the gears, pulleys, cables, and the like. Therefore, you will need a motor rated at twice this power. Pick a 5 hp motor.

b. How much electrical power must be provided to the motor?

$$p_{shaft} = 90\%\,p_{electrical}$$

$$p_{electrical} = \frac{p_{shaft}}{0.9} = \frac{5\,hp}{0.9}$$

$$p_{electrical} = 5.56\,hp$$

$$1\,hp = 746\,W$$

$$p = 5.56\,hp \times \frac{746\,W}{hp} = 4.14\,kW$$

$$i = \frac{p}{v} = \frac{4.14\,kW}{\cdot 220\,V}$$

$$i = 18.8\,A$$

Positive power: energy from source ⇒ load

Instantaneous power is the product of the voltage at that instant with the current at that instant. When that power is positive, energy is being delivered from the source to the load. The load may then convert it to some other form of energy, such as heat, light, sound, or motion. Alternately, the load may store that energy in an electrostatic field (capacitance) or an electromagnetic field (inductance).

Negative power: energy from load ⇒ source

The product of voltage times current may result in negative instantaneous power. During the time that power is negative, energy is being returned to the source from the load. The fields within the capacitance or inductance are collapsing, sending the stored energy back to the source. No heat, light, sound, or motion is created by the capacitance or inductance. The source has to absorb that energy, which is something it was not intended to do.

To determine how much useful work (heat, light, sound, motion) the load can provide, you must calculate the **average power**.

$$P_{ave} = \frac{1}{T} \int_0^T v(t) \times i(t)\, dt$$

Average power

Notice that the capital letter **P** is used to designate average power, while the lowercase **p** indicates instantaneous power.

There are a host of voltage and current waveforms that combine in a variety of ways in power electronics circuits. In the following sections, you will calculate the average power for several representative samples of the more common combinations of currents and voltages. However, you are not going to be dragged through every single possible set of signals. Once you have calculated the average power for a sample of different cases, you will be able to complete the integration for others as the opportunity arises.

DC Voltage and Current

The simplest case, and one you have dealt with since your first electricity course, is for a steady voltage applied across the load, resulting in a steady current through the load. These two signals are shown in Figure 3-25. The equations of the voltage and current are

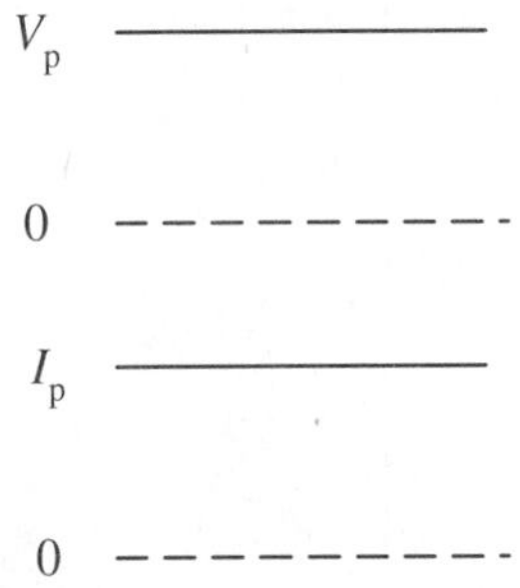

$$v(t) = V_p$$

$$i(t) = I_p$$

To calculate the resulting average power, apply

$$P_{ave} = \frac{1}{T} \int_0^T v(t) \times i(t) dt$$

$$P_{ave} = \frac{1}{T} \int_0^T V_p I_p\, dt$$

Move the two constants out of the integral.

$$P_{ave} = \frac{V_p I_p}{T} \int_0^T dt$$

The integral of the derivative of t is just t, evaluated at the limits.

Figure 3-25 DC waveforms

$$P_{ave} = \frac{V_p I_p}{T} \left(t \Big|_0^T \right)$$

Evaluate the equation at the limits.

$$P_{ave} = \frac{V_p I_p}{T} (T - 0)$$

Average power for steady dc voltage and current

$$P_{ave} = V_p I_p$$

This is the result you have used for years. However, be careful. It only applies to a steady dc voltage and a steady dc current. Any other wave shapes produce a different amount of average power.

Sinusoids

The other waveforms you have seen in previous courses are sinusoids. In the simplest version, the sine wave voltage is applied to a resistor, resulting in a sinusoidal current that is in phase with the voltage. This is shown in Figure 3-26. The two equations are

$$v(\theta) = V_p \sin\theta$$

$$i(\theta) = I_p \sin\theta$$

The average power dissipated by the load is

$$P_{ave} = \frac{1}{T} \int_0^T v(t) \times i(t)\, dt$$

$$P_{ave} = \frac{1}{2\pi} \int_0^{2\pi} \left(V_p \sin\theta \times I_p \sin\theta \right) d\theta$$

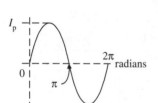

Figure 3-26 In phase sine voltage and current

Move the constants out of the integral and combine the sine terms.

$$P_{ave} = \frac{V_p I_p}{2\pi} \int_0^{2\pi} \sin^2\theta\, d\theta$$

In the previous section you saw the integral of $\sin^2\theta$.

$$P_{ave} = \frac{V_p I_p}{2\pi} \left(\frac{1}{2}\theta - \frac{1}{4}\sin 2\theta \right) \Big|_0^{2\pi}$$

Substitute the values at both limits.

$$P_{ave} = \frac{V_p I_p}{2\pi}\left[\left(\frac{1}{2}2\pi - \frac{1}{4}\sin 4\pi\right) - \left(\frac{1}{2}0 - \frac{1}{4}\sin 0\right)\right]$$

Simplify.

$$P_{ave} = \frac{V_p I_p}{2\pi}\left[\left(\pi - \frac{1}{4}0\right) - (0-0)\right]$$

$$P_{ave} = \frac{V_p I_p}{2\pi}\pi$$

$$P_{ave} = \frac{V_p I_p}{2}$$

Average power from in phase sine voltage and current.

This is not what you saw when you studied ac electronics. Then, you were told that for voltage and current signals that are in phase

$$P_{ave} = V_{rms} I_{rms}$$

Converting these sinusoid rms values to peak,

$$P_{ave} = \frac{V_p}{\sqrt{2}} \frac{I_p}{\sqrt{2}}$$

Combine the radicals.

$$P_{ave} = \frac{V_p I_p}{2}$$

So, the two are indeed the same.

Power for linear amplifiers is usually provided from two opposite dc power supplies, $\pm V_{supply}$. These voltages are steady dc. However, the standard test signal sent to the load is a sine wave, the positive half provided from the positive supply, and the negative half from the negative supply. With a resistive load, this output waveform draws a half-sine current from each supply. These two waveforms are shown in Figure 3-27. Their equations are

$$v(\theta) = V_{sup}$$

$$i(\theta) = \begin{cases} I_p \sin\theta & 0 \le \theta < \pi \\ 0 & \pi \le \theta < 2\pi \end{cases}$$

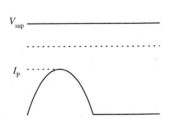

Figure 3-27 Supply voltage and half-wave rectified sine current

The load dissipates power according to the sine voltage and current equation just derived. But each power supply provides power from a steady voltage and a half cycle sine current. The average power supplied from each is

$$P_{ave} = \frac{1}{T} \int_0^T v(t) \times i(t)\, dt$$

Substitute the equations for the voltage and the current.

$$P_{ave} = \frac{1}{2\pi} \left(\int_0^\pi V_{sup} \times I_p \sin\theta\, d\theta + \int_\pi^{2\pi} V_{sup} \times 0\, d\theta \right)$$

The second integral is zero. Also, bring the constants out of the first integral.

$$P_{ave} = \frac{V_{sup} I_p}{2\pi} \int_0^\pi \sin\theta\, d\theta$$

The integral of $\sin\theta$ is $-\cos\theta$.

$$P_{ave} = \frac{V_{sup} I_p}{2\pi} \left(-\cos\theta \right) \Big|_0^\pi$$

Evaluate the equation at the two limits. Be very careful with the negative signs.

$$P_{ave} = \frac{V_{sup} I_p}{2\pi} \left[(-\cos\pi) - (-\cos 0) \right]$$

$$P_{ave} = \frac{V_{sup} I_p}{2\pi} \left[(--1) - (-1) \right]$$

Average power from dc supply and half sine current

$$P_{ave} = V_{sup} \frac{I_p}{\pi}$$

Look carefully at this result. The average power provided from a dc power supply is the dc voltage of the supply times the average value of the current, *not* the rms value of current. Therefore, it is *not* true that power is rms voltage times rms current! It depends on the wave shapes. You have to do the integral.

Example 3-9

a. Calculate the power delivered to the load in Figure 3-28, and the power provided by each supply.

b. Verify your calculations with a simulation.

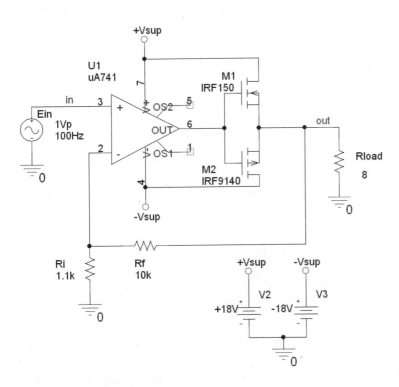

Figure 3-28 Schematic for Example 3-9

Solution

a. The power delivered to the load is in the form of a sine voltage into a resistor, producing a sinusoidal current.

$$V_{Ri} = E_{in} = 1\,V_p$$

$$I_{Rf} = I_{Ri} = \frac{1\,V_p}{1.1\,k\Omega} = 909\,\mu A_p$$

$$V_{load\,p} = I_{Rf\&i}\left(R_f + R_i\right) = 10.1\,V_p$$

The current through the resistor is

$$I_\text{p} = \frac{V_\text{p}}{\text{R}} = \frac{10.1\,\text{V}_\text{p}}{8\,\Omega}$$

$$I_\text{p} = 1.26\,\text{A}_\text{p}$$

The power is delivered to the load by a sinusoidal voltage and an in-phase full sine wave. It dissipates

$$P_\text{load ave} = \frac{V_\text{p}I_\text{p}}{2} = \frac{10.1\,\text{V}_\text{p} \times 1.26\,\text{A}_\text{p}}{2}$$

$$P_\text{load ave} = 6.36\,\text{W}$$

The voltage from each power supply is steady. The current from each supply is a half sine wave, $+V_\text{sup}$ providing the positive half cycle, and $-V_\text{sup}$ providing the negative current.

$$P_\text{supply ave} = V_\text{sup}\frac{I_\text{p}}{\pi} = 18\,\text{V}_\text{dc}\frac{1.26\,\text{A}_\text{p}}{\pi}$$

$$P_\text{supply ave} = 7.22\,\text{W}$$

This is the power provided by each of the two power supplies. So the total power provided from the supplies is twice this, or 14.4 W.

b. The simulation Probe display is shown in Figure 3-29. The transient analysis was run for ten cycles of the input, 100 ms, enough to let the average function create a steady value. The step time and step ceiling were both set to 100 μs. This gave 1000 data points, plenty to produce a good representation of the signals.

Before calculating the power, be sure that the simulation is working correctly. This is done by looking at V(out). The manual calculations indicate a sine wave of 10 V$_\text{p}$ at the output. The bottom probe shows the correct output.

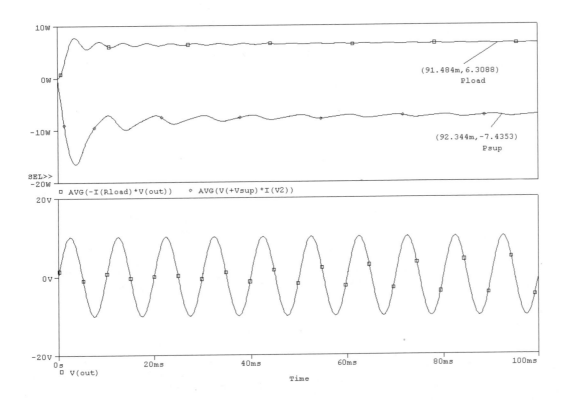

Figure 3-29 Simulation result for Example 3-9

The average power dissipated by the load is

$$AVG(I(Rload)*V(out))$$

When you enter this equation, Probe may display a negative power. Negative power is interpreted as power being returned by the load (to the amplifier and supplies), not dissipated by the load. However, the load resistor dissipates power. The problem is that PSpice assumes a direction for the current through each resistor. If you get a negative power when power is being dissipated, either flip the part, or change the calculation to

$$AVG(-I(Rload)*V(out))$$

Power provided by the positive power supply is

$$\text{AVG(V(+Vsup)*I(V2))}$$

Since this is supplied power, a negative result is appropriate. Both P_{load} and P_{sup} from the simulation match the manually calculated values.

Practice: Calculate the power dissipated by each transistor, M1 and M2.

Answer: Each transistor must dissipate 4.0 W.

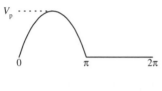

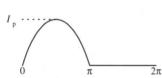

Figure 3-30 Half-wave sine voltage and current

A simple diode converts the full sinusoid into a wave with only a positive (or only a negative) half cycle. This is an easy way to drive a dc load, such as a motor or a power supply. The positive half-wave voltage, when applied to a resistive load, creates a positive half-wave current. These are shown in Figure 3-30. The equations for the voltage and the current are

$$v(t) = \begin{cases} V_p \sin\theta & 0 \le \theta < \pi \\ 0 & \pi \le \theta < 2\pi \end{cases}$$

$$i(t) = \begin{cases} I_p \sin\theta & 0 \le \theta < \pi \\ 0 & \pi \le \theta < 2\pi \end{cases}$$

The power that this voltage and this current dissipates (or generates) is

$$P_{ave} = \frac{1}{T} \int_0^T v(t) \times i(t)\, dt$$

Since each wave is defined in two parts, there should be two integrals.

$$P_{ave} = \frac{1}{2\pi}\left(\int_0^\pi V_p \sin\theta \times I_p \sin\theta\, d\theta + \int_\pi^{2\pi} 0\times 0\, d\theta \right)$$

The second integral is zero. Bring the constants out of the first integral.

$$P_{ave} = \frac{V_p I_p}{2\pi} \int_0^\pi \sin^2\theta\, d\theta$$

You have integrated $\sin^2\theta$ several times before.

$$P_{ave} = \frac{V_p I_p}{2\pi}\left(\frac{1}{2}\theta - \frac{1}{4}\sin 2\theta \right)\Big|_0^\pi$$

Apply the limits.

$$P_{ave} = \frac{V_p I_p}{2\pi}\left[\left(\frac{1}{2}\pi - \frac{1}{4}\sin 2\pi\right) - \left(\frac{1}{2}0 - \frac{1}{4}\sin 0\right)\right]$$

$$P_{ave} = \frac{V_p I_p}{2\pi}\frac{1}{2}\pi$$

$$P_{ave} = \frac{V_p I_p}{4}$$

Average power from a half sine voltage and half sine current

Example 3-10

a. Calculate the power delivered from a 170 V_p half-wave sine voltage to a 51 Ω resistor.

b. Verify your manual calculations with a simulation.

Solution

a. The current to the resistor is

$$I = \frac{V}{R} = \frac{170\,V_p}{51\,\Omega}$$

$$I_p = 3.33\,A_p$$

$$P_{ave} = \frac{V_p I_p}{4} = \frac{170\,V_p \times 3.33\,A_p}{4}$$

$$P_{ave} = 142\,W$$

b. The simulator Multisim contains a wattmeter. Connect the voltage terminals across the part whose power you are measuring, and arrange the wiring so that the current through that part also flows through the current terminals (just as with an ammeter). The results are shown in Figure 3-31.

Practice: Derive the equation for the power delivered to a resistive load from a full-wave rectified sine.

Answer: $P_{ave} = \frac{V_p I_p}{2}$

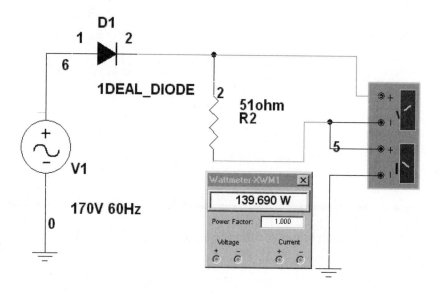

Figure 3-31 Multisim simulation for Example 3-10

Rectangular Waves

When a rectangular pulse is applied to a resistive load, it produces an identically shaped rectangular current through the load. The equation for the voltage and for the current are

$$v(t) = \begin{cases} V_p & 0 \le t < DT \\ 0 & DT \le t < T \end{cases}$$

$$i(t) = \begin{cases} I_p & 0 \le t < DT \\ 0 & DT \le t < T \end{cases}$$

The power dissipated by the load is

$$P_{ave} = \frac{1}{T} \int_0^T v(t) \times i(t)\, dt$$

$$P_{ave} = \frac{1}{T} \left(\int_0^{DT} V_p I_p\, dt + \int_{DT}^T 0 \times 0\, dt \right)$$

The second integral is zero. Move the constants in the first integral out front.

$$P_{ave} = \frac{V_p I_p}{T} \int_0^{DT} dt$$

The integral of the derivative of t is just t.

$$P_{ave} = \frac{V_p I_p}{T} (t)\Big|_0^{DT}$$

Apply the limits.

$$P_{ave} = \frac{V_p I_p}{T} (DT - 0)$$

$$P_{ave} = DV_p I_p$$

Average power from a pulse into a resistor

Example 3-11

For the circuit in Figure 3-32, calculate the power dissipated by the load and by the transistor. When the pulse is high, the transistor's resistance is 0.2 Ω.

Solution

When the pulse is high, the transistor's resistor is 0.2 Ω. So the current in the series circuit is

$$I_p = \frac{E_{supply}}{R_Q + R_L} = \frac{7.2V}{0.2\Omega + 4\Omega}$$

$$I_p = 1.71A_p$$

$$V_{load} = I_p R_{load} = 1.71A_p \times 4\Omega$$

$$V_{load} = 6.86V_p$$

$$P_{load\ ave} = DV_p I_p = 0.6 \times 6.86V_p \times 1.71A_p$$

$$P_{load\ ave} = 7.04W$$

When the transistor is on, it is a 0.2 Ω resistance. So, the same relationship developed for the resistor also works for this transistor.

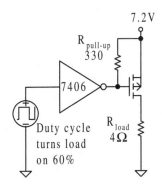

Figure 3-32 Schematic for Example 3-11

$$V_Q = I_p R_{Q\,on} = 1.71\,\text{A}_p \times 0.2\,\Omega$$

$$V_Q = 0.34\,\text{V}_p$$

$$P_{Q\,ave} = DV_p I_p = 0.6 \times 0.34\,\text{V}_p \times 1.71\,\text{A}_p$$

$$P_{Q\,ave} = 0.35\ \text{W}$$

MATLAB can be used to calculate the average power. A full-wave rectified sine wave voltage applied to a resistive load results in the same shaped current.

$$v(t) = \begin{cases} V_p \sin\theta & 0 < \theta < \pi \\ -V_p \sin\theta & \pi < \theta < 2\pi \end{cases}$$

$$i(t) = \begin{cases} I_p \sin\theta & 0 < \theta < \pi \\ -I_p \sin\theta & \pi < \theta < 2\pi \end{cases}$$

The MATLAB solution is

```
» syms Vp Ip x
»
int(1/(2*pi)*Vp*sin(x)*Ip*sin(x),0,pi)+int(
1/(2*pi)*(-Vp)*sin(x)*(-Ip)*sin(x),pi,2*pi)

ans =

5734161139222659/36028797018963968*pi*Vp*Ip
```

When you actually calculate the result of that very long number multiplied by π, the result is $V_p I_p/2$. Table 3-2, on the following page, summarizes the more common waveform combinations found in power electronics circuits.

Summary

Electricity is usually delivered to your circuits as either steady dc or as a sinusoid. The efficient control of this power as it is passed to the load, however, requires that it be converted to half-wave, full-wave rectified

Table 3-2 Common power waveforms and equations

$v(t)$	$i(t)$	P_{ave}
V_p ——————— $- - - - - - -$	I_p ——————— $- - - - - - -$	V_pI_p
V_p (sine wave)	I_p (sine wave)	$\dfrac{V_pI_p}{2}$
V_{dc} ——————— $- - - - - - -$	I_p (half sine pulse)	$V_{dc}\dfrac{I_p}{\pi}$
V_p (half sine pulse)	I_p (half sine pulse)	$\dfrac{V_pI_p}{4}$
V_p (full-wave rectified)	I_p (full-wave rectified)	$\dfrac{V_pI_p}{2}$
V_p (pulse, $D \times T$, T)	I_p (pulse, $D \times T$, T)	DV_pI_p

sinusoids, rectangles, triangles, or trapezoids. The average value, the rms, the instantaneous, and the average power provided by these wave shapes are the subject of this chapter.

Before any of these values can be calculated, you must determine the signal's equation as a function of time or phase. Most of the signals presented in this chapter a `re separated into two parts, each part with its own equation that is valid over a limited range of time or phase.

DC loads, such as motors or power supplies, respond to the average value of the voltage provided. Average value is obtained by finding the area under the curve created by the signal. Then divide that area by its length of time. For all but the simplest signals, finding the area under the curve involves evaluating an integral. When the equation for the signal has several parts, write an integral for each part. The limits for each integral define the region over which the equation is valid. The area under the curve, then, is the sum of these integrals.

A wave that has negative as well as positive sections has its area reduced by these negative parts, even though the negative part still delivers energy to the load. In the extreme case of a perfect sinusoid, the average value is zero, although the signal may be thousands of volts high, and do a great deal of work. The root-mean-squared value resolves this conflict. All signals of the same rms value deliver the same power to a resistive load, regardless of shape. The rms value is obtained by squaring the function of the signal, finding the average value of that square, then taking the square root of the average.

In this section of the chapter you found the rms value for a variety of rectangular, triangular, and sinusoidal signals. You also saw the results of simulations.

Power is the rate of doing work or delivering energy. As such it is a time varying function and is the instantaneous voltage times the instantaneous current. Positive power is power delivered to a load. Negative power is being returned to the source by the load. The net result is obtained by calculating the average. To obtain the correct average power you must calculate the average of the product of the voltage across the element times the current through it. Average power is *not* just average voltage times average current. If the load is a pure resistance, average power is the rms voltage across the load times the rms current through it. But if the load is not purely resistive, this simplification is also wrong. To get the correct average power every time, calculate the average (that involves an integral) of the instantaneous voltage across the part times the instantaneous current though the part.

The first three chapters laid a foundation for your studies of power electronics: reviewing op amps, discovering how to build the circuit to actually work as it was designed, and calculating the voltages and power you can expect. It is now time to look at the power circuits. The next chapter presents the simplest and most commonly used power circuits. The power op amp brings the power stage inside the IC. This makes the part appear to be as simple to use as any other op amp. But issues of power dissipation, power delivered to the load, efficiency, heat, heat sinking, safe operating areas of voltage and current, thermal shutdown, and current limiting are all addressed. The audio power amplifier IC is more popular than the power op amp. This part is optimized to drive a speaker with signals that you can hear. These restrictions alter several of your concerns and applications.

Problems

Waveforms

3-1 Determine the equation for the waveform in Figure 3-33. The point at which the sine is turned on is ϕ, and may vary. So, your answer should be a function of ϕ. *Hint:* the answer should have 3 parts.

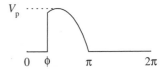

Figure 3-33 Waveform for Problem 3-1

3-2 Determine the equation for the waveform in Figure 3-34. This is a symmetric triangle wave. The equation has two parts. Each segment is of the form $y = mx + b$. The second segment has a negative slope and a y intercept of $2I_p$.

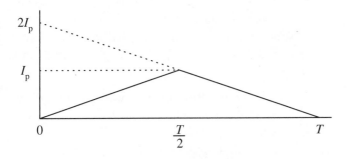

Figure 3-34 Waveform for Problem 3-2

Average Value

3-3 **a.** Calculate the equation for the average value of the partial sine wave from Figure 3-33.

b. Given an amplitude of 170 V_p and a firing angle, $\phi = 60°$, calculate the average value.

c. Devise a circuit to create this wave. Verify the waveform and your calculation of part b with a simulation.

3-4 **a.** Calculate the equation for the average value of the triangle wave from Figure 3-34.

b. Given a peak amplitude of 2 A_p and a period $T = 10$ μs, calculate the average value.

c. Verify your calculation of part b with a simulation.

Root-Mean-Squared Value

3-5 **a.** Calculate the equation for the rms value of the partial sine wave from Figure 3-33.

b. Given an amplitude of 170 V_p and a firing angle, $\phi = 60°$, calculate the rms value.

c. Devise a circuit to create this wave. Verify the waveform and your calculation of part b with a simulation.

3-6 **a.** Calculate the equation for the rms value of the triangle wave from Figure 3-34.

b. Given a peak amplitude of 2 A_p and a period $T = 10$ μs, calculate the rms value.

c. Verify your calculation of part b with a simulation.

Power

3-7 **a.** Calculate the energy (in joules) needed to lift a 12 lb bowling ball from the floor to the top of a 6 ft rack.

b. Calculate the power (in watts) needed to lift the bowling ball of part a to the rack in 2 s.

c. Assuming that the motor in the ball-return system is 40% efficient and is powered by 120 V, calculate its rated hp and current.

3-8 **a.** Calculate the equation for the power delivered by a current with the shape of the partial sine wave from Figure 3-33, and the voltage as a steady dc value of V_{on}.

b. Given a peak amplitude of 2.5 A_p, a firing angle, $\phi = 60°$, and an on voltage of $V_{on} = 1.8$ V_{dc}, calculate the power dissipated.

3-9   **a.** Calculate the equation for the power delivered by a current in the shape of the triangle wave from Figure 3-34, and a voltage that is a rectangle, with a peak voltage of V_p and a 50% duty cycle.

      b. Given a peak currnt of 2 A_p, a peak voltage of 18 V_p, and a period $T = 10 \ \mu s$, calculate the power dissipated.

4

Linear Power Amplifier Integrated Circuits

Introduction

The most convenient way to control power to a load is with a power op amp. These op amps have characteristics that are very similar to the low power ICs. However, they may be powered by over 50 V at 5 A.

Working with this much voltage and current requires that you accurately anticipate the power delivered to the load, the power provided by the dc supplies and the power that the op amp itself must dissipate. The power that the op amp dissipates is converted to heat. To move that heat from the silicon wafer requires effective heat sinking. Should a failure occur (by breakdown of a component, or because you botched the heat sink design), thermal shutdown allows the op amp to turn itself off. On-chip current limiting also allows you to set the maximum current the IC delivers to the load. This protects the load during a failure.

Audio power ICs have several special problems. Often you must run them from a single battery. Noise and total harmonic distortion are critical issues, as are circuit features such as bass or treble boost. You will see a low power tabletop amplifier IC and a 56 W boomer IC.

Objectives

By the end of this chapter, you will be able to:

- Determine the gain setting resistors and decoupling capacitors needed by a high voltage, high current, op amp based amplifier.

- Calculate the power delivered to the load, provided by the supply and dissipated by the IC.

- Calculate the IC and the case temperature and select the heat sink.

- Set the short circuit current limiting.

- Define special audio amplifier concerns and parameters.

- Properly apply a desktop amp, and a high power audio amplifier IC.

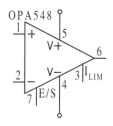

Figure 4-1 OPA548
(*courtesy of Burr-Brown*)

Figure 4-2 OPA548
(*courtesy of Burr-Brown*)

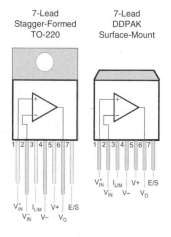

Figure 4-3 Footprints
(*courtesy of Burr-Brown*)

4.1 OPA548 Operational Amplifier IC

The OPA548 is a high voltage/high current operational amplifier made by Burr-Brown. Its schematic symbol is shown in Figure 4-1. It is available in two packages, a staggered lead, 7 pin TO-220 package, and a surface mount DDPAK. These are given in Figure 4-2. The circuit board footprints are shown in Figure 4-3.

If you want only to output positive voltage to the load, this op amp may be powered from a single supply, as high as +60 V. Connect V− to the circuit common. Unlike older designs, when powered from a single supply the OPA548 operates correctly with the input signal as much as a half of a volt *below* ground. The minimum single supply voltage is 8 V. So you will have to provide a separate, higher, more powerful supply if the rest of your system is running from +5 V.

To drive the load either positive or negative, reversing the direction of a motor, or to send audio to a loudspeaker, you can run the OPA548 from split supplies. Typically this means the dual supplies may range from ±4 V to ±30 V. But as long as the total difference between the power pins is 60 V or less, any combination of supply voltages is acceptable. Just remember to assure that the voltage on each input also always lies between V+ and V−.

Only CMOS implemented op amps are able to drive their output voltage all the way to the supplies. Typically the maximum output is several volts smaller than the supplies. This difference between supply voltage and maximum output voltage is called the **saturation voltage**. For the OPA548, the saturation voltage may be as large as 4 V. With supplies of ±28 V, the output voltage may be no more than 24 V_p.

The OPA548 can source or sink up to 5 A to the load. This is an instantaneous (peak) value. So, be sure to use the correct peak to root-mean-squared relationship and peak to average equations. These are summarized in Table 3-1. There are two other factors that limit the instantaneous current. You can connect a resistor between the I_{lim} pin (pin 3) and V− (pin 4). This resistor limits the maximum current to a value below 5A. This allows you to protect the load. The IC also has an on-wafer temperature sensor. Should the chip get too hot (>160°F), the op amp turns its output transistors off, until they cool down. This protects the IC. Later in this section, you will see how to set the current limit and to provide heat sinking to take advantage of the thermal protection feature.

There are two nonideal dc characteristics that affect how you use this IC. With proper negative feedback, an ideal op amp has no difference in potential between its inputs. However, in reality, the two input transistors cannot be made identically. So there is some small voltage between these two pins. This is the **input offset voltage**. For the OPA548 its worst case value is ±10 mV.

$V_{ios} < 10\,mV$

To figure out the effect this has on the circuit, you can model the input offset voltage as a small dc supply in series with the noninverting input. A voltage follower is shown in Figure 4-4. The input offset voltage drives V_{out} to equal V_{ios}, instead of 0 V.

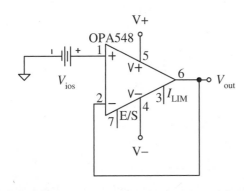

Figure 4-4 Voltage follower with input offset voltage

In some critical applications this ±10 mV output dc voltage could cause a problem. However, when you add gain to the circuit, the problem gets worse. Look at Figure 4-5. Ideally, with 0 V in, the output should also be 0 V. But since the input offset voltage is actually between the op amp's input and common, that voltage looks like an input signal. It is multiplied by the noninverting gain of the circuit.

$$V_{out} = \left(1 + \frac{R_f}{R_i}\right) V_{ios}$$

So, for reasonable gains, the output dc voltage could very easily be shifted significant parts of a volt, or more. If this shift in the output dc level causes a problem, you can add a coupling capacitor *immediately* before the load. Be sure to connect the feedback resistor, R_f, directly to the op amp's output. You could also add a voltage divider and potentiometer to the input, to allow you to tweak the output dc. Or, an integrator could be used to automatically sense any dc voltage at the output and send a small correction signal back into the OPA548. This is shown in Chapter 1, Figure 1-40.

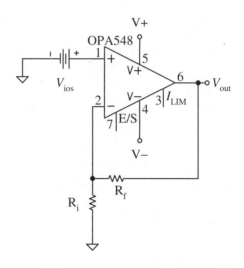

Figure 4-5 Amplifier with input offset voltage

$I_{bias} \leq 500nA$

The second nonideal dc characteristic that may cause you problems is the input current. You have assumed that there is no current flowing into the input of an op amp. But for the OPA548 both the input bias current and the input offset current may be as large as 500 nA. The results of this large an input current are shown in Example 4-1

Example 4-1

Calculate the output voltage for the circuit in Figure 4-5 with

$$R_f = 1 \text{ M}\Omega, \quad R_i = 10 \text{ k}\Omega, \quad V_{ios} = 0 \text{ V}, \quad I_{into \ pin \ 2} = 500 \text{ nA}$$

Solution

With $V_{ios} = 0$ V, the voltage at the noninverting input pin (pin 3) is

$$V_{NI} = 0 \text{ V}$$

Since negative feedback makes the inverting input track the non-inverting (ignoring the input offset voltage),

multi**SIM**

$$V_{INV} = V_{NI} = 0\,V$$

This means that there is no voltage difference across R_i. So no current flows through it.

$$I_{Ri} = 0\,A$$

But, there is 500 nA of bias current flowing into the inverting input pin. It must come from somewhere. Since none of it can come through R_i, it must all come from the output of the op amp, flowing left to right through R_f. This current creates a voltage drop across R_f of

$$V_{Rf} = 500\,nA \times 1\,M\Omega = 500\,mV$$

Since the left end of R_f is held at 0 V, the output of the op amp must go up by 500 mV.

The simulation in Figure 4-6 confirms these calculations. Notice that it is necessary to edit the model of the op amp.

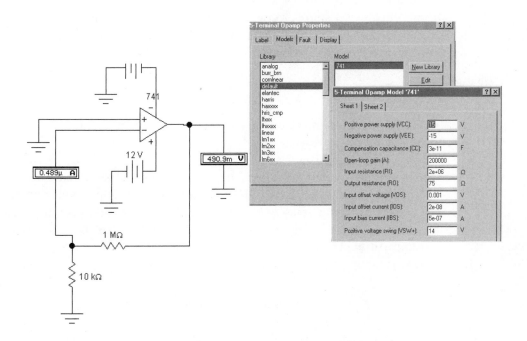

Figure 4-6 Multisim simulation of Example 4-1

Practice: What is the effect on the V_{out} of lowering the resistors to

$$R_i = 100 \ \Omega, \quad R_f = 10 \ k\Omega$$

Answer: $V_{out} = 5 \ mV$

$R_f < 10 \ k\Omega$

To minimize the effect of this unusually large bias current, keep the feedback resistor for the OPA548 as small as practical, typically a few kΩ.

The **gain bandwidth product** of the OPA548 is typically 1 MHz. This is the same as the 741 small signal amplifier. The gain bandwidth product is defined for *small* signals (1 V_p). It is

GBW = 1 MHz

$$GBW = A_o \times f_H$$

where

A_o = the amplifier's closed loop gain

f_H = the amplifier's high frequency cutoff, the frequency at which the gain has dropped to 0.707 A_o.

You may certainly input small signals into the OPA548. So you must be sure to consider the gain bandwidth product. But the main purpose of this op amp is to output large signals. For signals above 1 V_p, the op amp's speed is limited by its **slew rate**. The slew rate defines how rapidly the output may change.

SR = 10 V/μs

$$SR = \frac{dv_{out}}{dt}$$

For the OPA548 the slew rate is typically 10 V/μs when driving an 8 Ω load with a 50 V_{pp} signal.

If that output is a triangle wave, the slew rate defines the maximum slope of the output. For a rectangular output, the slew rate sets the smallest rise and fall times. But, for a sinusoidal output, you actually have to take the derivative of the signal, and look at that function at its maximum value. The result for a sine wave is called the **full power bandwidth**. This is the maximum frequency for a large undistorted sine wave output.

$$f_{max} = \frac{SR}{2\pi V_p}$$

Example 4-2

An audio amplifier has an input of 300 mV$_{rms}$ and is to deliver 35 W to an 8 Ω speaker. Can the OPA548 be used?

Solution

Assuming that the speaker is resistive, and that the signal is a sine wave,

$$P_{load} = \frac{V_{load\ rms}^{\ \ \ 2}}{R_{load}}$$

$$V_{load\ rms} = \sqrt{P_{load}R_{load}} = \sqrt{35\,W \times 8\,\Omega}$$

$$V_{load} = 16.7\,V_{rms} = 23.7\,V_{p}$$

$$I_{load} = \frac{23.7\,V_{p}}{8\,\Omega} = 3\,A_{p}$$

These levels are within the OPA548's ability.

$$f_{max} = \frac{SR}{2\pi V_{p}}$$

$$f_{max} = \frac{10\dfrac{V}{\mu s}}{2\pi \times 23.7\ V_{p}} = 67.2\,kHz$$

Be careful with the units and the powers of ten. The slew rate is specified in V/μs. So the computed answer is 0.0672 MHz. Since the top of the audio band is only 20 kHz, the OPA548 has more than enough slew rate to pass an audio sine wave to the load.

$$A_{o} = \frac{V_{out}}{E_{in}}$$

$$A_{o} = \frac{16.7\,V_{rms}}{300\,mV_{rms}} = 55.7$$

$$GBW = A_o \times f_H$$

$$GBW = 55.7 \times 20\,\text{kHz} = 1.1\,\text{MHz}$$

The OPA548 has high enough voltage, current, and slew rate to serve for this 35 W audio amplifier. But it cannot amplify the high frequency signals enough. That is, its gain bandwidth product is too small.

Practice: Prove that the amplifier can be built with a small signal 741 op amp amplifier with a gain of 5, followed by the OPA548.

Answer: $V_{\text{out 741}} = 2.12\,\text{V}_p$, $SR_{\text{741 needed}} = 0.27\,\text{V/}\mu\text{s}$, $GBW_{\text{741 needed}} = 100\,\text{kHz}$, $GBW_{\text{OPA548 needed}} = 223\,\text{kHz}$

4.2 Power Calculations

In considering the power that a linear power IC delivers to a load, it helps to consider the simplified model shown in Figure 4-7. Of course, the actual circuit is *much* more complex. But this highlights the key points. The traditional, low power op amp drives two power transistors. The *npn* transistor, Q1, turns on and **sources current** from V+ out of the output terminal, through the load, to circuit common. This produces a positive load voltage. On the negative load cycle, the *pnp* power transistor turns on. It **sinks current**, from circuit common, up through the load, through Q2 and to the negative supply, V−. All of the load current passes through either Q1 and/or Q2.

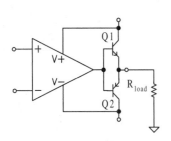

Figure 4-7 Simplified model of a power op amp

The power a signal can deliver to the load depends on several factors: the load's resistance, the op amp's power supply voltage, and the wave shape. These same factors determine the power that the supply must provide, and the power that the IC must dissipate. In Chapter 3, you saw how to calculate the power resulting from a wide variety of currents and voltages. These are summarized in Table 3-2. The two most prevalent wave shapes, dc and the sinusoid, will be applied to the power amplifier of Figure 4-7. But you can use these same techniques for any signal shape.

DC Signal to the Load

When a positive voltage is applied to the load, current flows from V+, through Q1, and to the load. The power that the load dissipates is

$$P_{\text{load}} = V_{\text{load dc}} \times I_{\text{load dc}}$$

Assuming that the load is purely resistive, this power equation can be combined with Ohm's law to give two alternate versions.

$$P_{\text{load}} = I_{\text{load dc}}{}^{2} \times R_{\text{load}}$$

Load power

$$P_{\text{load}} = \frac{V_{\text{load dc}}{}^{2}}{R_{\text{load}}}$$

This is the power that the load uses, converting it to heat, light, sound, or motion. Generally, that is the end product and the entire reason for the existence of the electronics.

This power is provided by the dc supply. Both V+ and V− are available to allow either positive or negative load voltages. If the load voltage only goes one direction, then the opposite supply just plays a small, biasing role. It does not deliver any significant power. The power provided by the supply is

$$P_{\text{supply}} = V_{\text{supply}} \times I_{\text{supply}}$$

Since the power supply, Q1, and the load form a series circuit,

$$I_{\text{supply}} = I_{\text{load dc}}$$

Combining these gives

$$P_{\text{supply}} = V_{\text{supply}} \times I_{\text{load dc}}$$

Supply power

Since the supply voltage is greater than the load voltage, the power supply's power is always greater than the power delivered to the load. The power that is provided by the supply but is not delivered to the load must be dissipated by Q1 (or Q2 if the output is negative).

$$P_{\text{IC}} = P_{\text{supply}} - P_{\text{load}}$$

IC power

This power is dissipated as heat in the IC output transistor, raising the temperature of the silicon. You will see in the next section how to provide the correct heat sink to protect the IC.

___

Example 4-3

Calculate the power delivered to the load, provided by the supply, and dissipated by the op amp, for a circuit running from $\pm 12\ V_{\text{dc}}$, delivering $+5\ V_{\text{dc}}$ to a 10 Ω resistive load.

Solution

The power delivered to the load is

$$P_{load} = \frac{V_{load}^2}{R_{load}}$$

$$P_{load} = \frac{(5\,V)^2}{10\,\Omega} = 2.5\,W$$

The current through the load, and therefore from the power supply is

$$I = \frac{5\,V_{dc}}{10\,\Omega} = 0.5\,A_{dc}$$

So, the power supply must provide

$$P_{supply} = V_{supply} \times I_{load\ dc}$$

$$P_{supply} = 12\,V_{dc} \times 0.5\,A_{dc} = 6\,W$$

The power supply is providing 6 W, but the load is using only 2.5 W. The IC dissipates the power that the load does not use.

$$P_{IC} = 6\,W - 2.5\,W = 3.5\,W$$

In this particular configuration, more power is going up as waste heat than is being delivered to the load. The simulation is shown in Figure 4-8.

Practice: What is the effect of decreasing the power supply to 9 V?

Answer: $P_{load} = 2.5$ W, $P_{supply} = 4.5$ W, $P_{IC} = 2$ W

Changing the supply voltage did *not* change the voltage or power delivered to the load. As long as the op amp has adequate headroom, the voltage passed to the load is set by the input. Changing the input voltage changes the load voltage, and that changes the power delivered to the load and the power provided by the supply. In turn, these changes alter the power that the IC must dissipate, how hot the IC becomes, and the heat sink that you must provide. So in designing a power amplifier, you must consider two points of operation. The first is at the top end, at the maximum load voltage. Here, be sure to verify that the circuit can deliver adequate voltage and current to the load.

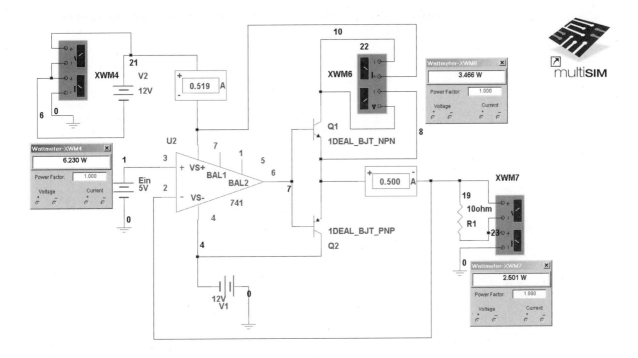

Figure 4-8 Simulation for Example 4-3

The second operational point to consider is the worst case power dissipation for the amplifier. At low output voltage, there is very little current flowing from the power supply through the op amp to the load. So the op amp is required to dissipate little power.

At the other extreme, where maximum voltage is being delivered to the load from the fixed supply voltage, the op amp's output voltage is driven very close to its supply voltage. This leaves only a small *difference* in potential between the collector and emitter of the op amp's power transistor. So, even though a lot of current is passing through the op amp, there is little voltage across it, so the op amp dissipates little power at the top end.

At the low end, there is little current through the op amp, so it does not have to dissipate much power. At the top end, there is little voltage *across* the op amp. So, again the IC does not dissipate much power. Precisely where is the worst case for the IC? Example 4-4 calculates the op amp's worst case operating point.

Example 4-4

Given a supply voltage of 12 V and a load voltage that varies from 0 V to 12 V, plot the IC power (y axis) versus load voltage (x axis), for a 10 Ω load.

Solution

Create a spreadsheet, with load voltage in the first column, load current in the second, load power in the third, supply power in the fourth, and IC power in the fifth.

Step the load voltage in each row from 0 V to 12 V, in 0.2 V increments.

"Pull down" all of the equations to fill the table. The result is shown in Table 4-1.

Table 4-1 Spreadsheet to determine the worst case IC power dissipation

V_{load} volts	I_{load} amps	P_{load} watts	P_{supply} watts	P_{IC} watts	
0.0	0.0	0.00	0.00	0.00	
0.2	0.0	0.00	0.24	0.24	
0.4	0.0	0.02	0.48	0.46	
0.6	0.1	0.04	0.72	0.68	
0.8	0.1	0.06	0.96	0.90	
4.8	0.5	2.30	5.76	3.46	matches
5.0	0.5	2.50	6.00	3.50	◄— Example 4-3
5.2	0.5	2.70	6.24	3.54	
5.4	0.5	2.92	6.48	3.56	
5.6	0.6	3.14	6.72	3.58	
5.8	0.6	3.36	6.96	3.60	P_{IC} peaks &
6.0	0.6	3.60	7.20	3.60	then drops
6.2	0.6	3.84	7.44	3.60	
6.4	0.6	4.10	7.68	3.58	
11.4	1.1	13.00	13.68	0.68	
11.6	1.2	13.46	13.92	0.46	
11.8	1.2	13.92	14.16	0.24	
12.0	1.2	14.40	14.40	0.00	

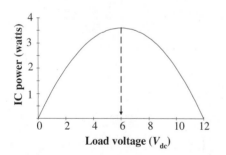

Figure 4-9 Maximum IC power

There are two points to notice from Table 4-1. The row that starts with $V_{in} = 5$ V duplicates the calculations in Example 4-3. This should give you confidence that the calculations in the table are correct.

Now, scan down the P_{IC} column. It starts low, grows as I_{load} increases, peaks, then decreases as the difference between V_{supply} and V_{load} falls. The maximum power that the IC must dissipate occurs when

$$V_{load} = \frac{1}{2}V_{supply}$$

Worst case IC power for a dc signal occurs when:

Figure 4-9 is a plot of the data from Table 4-1. It, too, shows a parabola with the IC power peaking when the load voltage is half of the supply voltage.

Practice: Duplicate Table 4-1 and Figure 4-9, with $V_{supply} = 15$ V, and $R_{load} = 5\ \Omega$.

Answer: $P_{IC\ max} = 11.3$ W at $V_{load} = 7.5$ V

Sinusoidal Signal to the Load

The steps you used to determine the effects of a dc signal on the load, the supply, and the power IC can be applied to any wave shape. But the results are different for each different signal.

From the calculations in Chapter 3, you saw that the power delivered by a sine wave to a resistive load is

$$P_{load} = \frac{V_p I_p}{2}$$

Load power

In more familiar rms terms, for a sine wave,

$$P_{load} = \frac{V_p I_p}{\sqrt{2}\sqrt{2}} = \frac{V_p}{\sqrt{2}} \times \frac{I_p}{\sqrt{2}}$$

$$P_{load} = V_{rms} I_{rms}$$

During the positive half cycle, the upper *npn* transistor, Q1, inside the op amp turns on, sourcing a half-cycle sine wave current to the load. During the negative half-cycle, Q1 turns off. It passes no current. But Q2 turns on, sinking a half cycle current from the load. So, the positive power supply provides current to the load during the positive output half-cycle, and rests during the negative part of the cycle. Similarly, the negative power supply rests during the positive output half-cycle and sinks current during the negative part of the cycle.

Again, look back at Table 3-1. The two power supplies must be able to provide I_p. But, a dc ammeter indicates the average value.

$$I_{p\,supply} = I_{p\,load} = \frac{V_{p\,load}}{R_{load}}$$

Supply current

$$I_{dc\,supply} = \frac{I_{p\,supply}}{\pi}$$

To determine the power that the supply is delivering, you must consider the wave shapes of both the voltage from the supply and the current it provides. The power supply outputs a steady, dc voltage. But the current is a half sine wave, positive sourcing on one half-cycle and negative sinking on the other half-cycle. For the positive supply this is shown in Figure 4-10. To determine the power provided by this combination of voltage and current, you must complete the integral

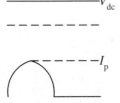

Figure 4-10 Power supply waveforms

$$P_{each\,supply} = \frac{1}{2\pi}\left(\int_0^\pi V_{dc} \times I_p \sin\theta\,d\theta + \int_\pi^{2\pi} V_{dc} \times 0\,d\theta \right)$$

This was done in Chapter 3, and the result is in Table 3-2.

$$P_{each\,supply} = V_{dc} \frac{I_p}{\pi}$$

Since there are two supplies, the *total* power supplied is

Supply power

$$P_{total\,supply} = 2V_{dc} \frac{I_p}{\pi}$$

Finally, only some of $P_{total\,supply}$ is delivered to the load. The rest must be dissipated by the amplifier.

IC power

$$P_{IC} = P_{total\,supply} - P_{load}$$

Remember, this power is converted to heat at the IC's silicon wafer. The heat must be removed from the IC and passed on to the environment. The greater the heat, the larger the heat sink, and the hotter the inside of the equipment. Forced air (fans) may even be required. These thermal management components add cost and bulk to the equipment.

Example 4-5

Calculate the power delivered to the load, provided by the supply, and dissipated by the op amp, for a circuit running from ± 12 V_{dc}, delivering a 5 V_p sine wave to a 10 Ω resistive load. Verify your calculations with a simulation.

Solution

The current through the load is

$$I_{load} = \frac{V_{load}}{R_{load}}$$

$$I_{load} = \frac{5\,V_p}{10\Omega} = 0.5\,A_p = 0.35\,A_{rms}$$

The power delivered to the load is

$$P_{load} = \frac{V_p I_p}{2}$$

$$P_{load} = \frac{5\,V_p \times 0.5\,A_p}{2} = 1.25\,W$$

Or, in terms of rms voltage and current,

$$P_{load} = \frac{5\,V_p}{\sqrt{2}} \times 0.35\,A_{rms} = 3.5\,V_{rms} \times 0.35\,A_{rms} = 1.23\,W$$

Each power supply must be able to provide 0.5 A_p. But this is in the shape of a half sine. So, the dc ammeter indicates

$$I_{dc\ supply} = \frac{I_p}{\pi} = \frac{0.5\,A_p}{\pi} = 160\,mA_{dc}$$

The power delivered by the two supplies is

$$P_{total\ supply} = 2V_{dc}\frac{I_p}{\pi}$$

$$P_{total\ supply} = 2 \times 12\,V_{dc}\frac{0.5\,A_p}{\pi} = 3.8\,W$$

The IC must dissipate the power that is provided by the supplies but not delivered to the load.

$$P_{IC} = P_{total\ supply} - P_{load}$$

$$P_{IC} = 3.8\,W - 1.3\,W = 2.5\,W$$

The simulation is shown in Figure 4-11. Look carefully at the values. The load's ammeter has been set to ac, and shows 0.354 A$_{rms}$. Load power also matches theory at 1.25 W. The dc current from the +12 V supply is shown as 158 mA$_{dc}$. Theory indicates 160 mA$_{dc}$. The wattmeter of the top supply indicates 1.93 W, for a total supply power of 3.8 W. This also matches theory. Finally the top transistor is dissipating 1.27 W, giving a total IC power of 2.5 W.

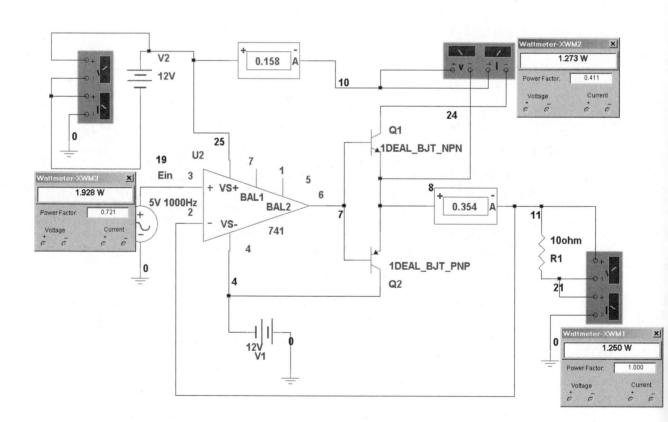

Figure 4-11 Simulation for Example 4-5

Practice: What is the effect of increasing the load voltage to 10 V_p?

Answer: $I_{load} = 1 \text{ A}_p,$ $P_{load} = 5 \text{ W},$ $P_{supply} = 7.6 \text{ W},$ $P_{IC} = 2.6 \text{ W}$

As with a dc signal, the worst case power dissipation for the IC comes at moderate levels of output, not at the top end. However, since the wave shapes have changed, the maximum IC power point has also shifted. In Table 4-2, the appropriate equations for a sine wave have been entered into the spreadsheet and pulled down the column.

Table 4-2 Worst case IC power for a sinusoid

V_{load}	I_{load}	P_{load}	P_{supply}	P_{IC}
volts	amps	watts	watts	watts
0.0	0.0	0.00	0.00	0.00
0.2	0.0	0.00	0.15	0.15
0.4	0.0	0.01	0.31	0.30
0.6	0.1	0.02	0.46	0.44
4.6	0.5	1.06	3.51	2.46
4.8	0.5	1.15	3.67	2.51
5.0	0.5	1.25	3.82	2.57 ← matches Example 4-5
5.2	0.5	1.35	3.97	2.62
5.4	0.5	1.46	4.13	2.67
5.6	0.6	1.57	4.28	2.71
7.2	0.7	2.59	5.50	2.91
7.4	0.7	2.74	5.65	2.92
7.6	0.8	2.89	5.81	2.92 P_{IC} peaks & then drops
7.8	0.8	3.04	5.96	2.92
8.0	0.8	3.20	6.11	2.91
11.4	1.1	6.50	8.71	2.21
11.6	1.2	6.73	8.86	2.13
11.8	1.2	6.96	9.01	2.05
12.0	1.2	7.20	9.17	1.97

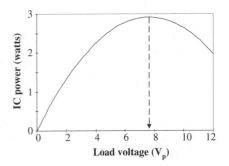

Figure 4-12 Plot of load voltage vs IC power for a sine wave

When the signal applied to the load is steady dc, the IC dissipates maximum power at

$$V_{\text{load p @ IC worst case}-\text{DC}} = 0.5 V_{\text{supply}}$$

But, when the load voltage is a sine wave, the worst case for the IC occurs at a higher point, because the signal only spends a small part of each cycle near the peak.

$$V_{\text{load p @ IC worst case}-\text{sine}} = \frac{2}{\pi} V_{\text{supply}}$$

For other wave shapes, the worst case occurs at a different place. You have to complete the entire analysis for that particular shape. It will even involve some integrals, since the waveforms you are using are probably not in Table 3-2. Calculate

$$V_{\text{load p}}, I_{\text{load p}}, P_{\text{load}}, P_{\text{supply}}, P_{\text{IC}}$$

Then use a spreadsheet to investigate the effect of $V_{\text{load p}}$ on P_{IC}.

4.3 Heat Sinks

So what is to become of the power that the IC must dissipate? It is converted to heat by the resistance in the main transistors, Q1 and Q2. The hotter the IC gets, the poorer its performance, and the sooner it will fail. It is critical that you provide an adequate way to remove the heat from the silicon. Selecting the proper heat sink is central to the success of any electronic power device.

The key parameters are illustrated in Figure 4-13. Power dissipated at the silicon level produces heat that must flow through the IC packaging material, to the case, through the thermal interface between the case and the heat sink, to the heat sink, and finally into the surrounding air. These parameters combine to define the junction temperature.

$$T_J = T_A + P\left(\Theta_{JC} + \Theta_{CS} + \Theta_{SA}\right)$$

where

T_J = silicon junction temperature of the IC–a specification
T_A = ambient temperature (of the air *immediately* surrounding the package)
P = power that the IC must dissipate
Θ_{JC} = thermal resistance between the *junction* and the *case*. This too is a specification of the IC.
Θ_{CS} = thermal resistance between the *case* and the heat *sink*.

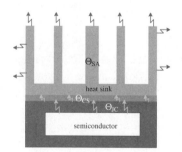

Figure 4-13 IC and heat sink layout and parameters

Θ_{SA} = thermal resistance between the heat *sink* and the *air*.
This is the key heat sink parameter.

The cooler you can keep the silicon, the longer the IC will run without a failure. But for even the worse case conditions, the manufacturer recommends that you must keep

$$T_J \leq 125°C$$

Thermal resistance tells you how much hotter the IC gets as heat flows through the packaging material. It is measured in °C of temperature rise for each watt of power dissipated. The thermal resistance, junction-to-case, Θ_{JC}, is set by how the manufacturer packages the silicon wafer. So, it is a specification. For the OPA548 TO220 package, typically

$$\Theta_{JC} = 2.5 \frac{°C}{W}$$

For every watt of power the IC must dissipate in passing a signal to the load, its junction's temperature goes up 2.5°C as the heat flows out to the case.

The interface between the IC's case and the heat sink is critical. You must not ignore it. The simplest technique is to apply a liberal coat of heat sink **grease** or **compound** between the case of the IC and the heat sink. This fills in the surface imperfections and helps the heat transition to the heat sink. This is simple, inexpensive, and effective. But it is messy. Properly applied, a typical value is

$$\Theta_{CS} \approx 0.2 \frac{°C}{W} \qquad \text{grease only}$$

Other forms of thermal adhesives and interface pads are also available. Be sure to verify their thermal resistance before selecting them.

In most power ICs the metal package is electrically connected to the most negative potential used by the IC. For the OPA548 the metal tab is tied to V−. If you connect the case directly to the heat sink, then the heat sink, too, is tied to V−. This most certainly can present a shock hazard, and a short circuit with destructive consequences if that heat sink ever touched anything else. So it is often a very good idea to *electrically* insulate the IC package from the heat sink, while keeping as good a thermal connection as you can. You can do this by inserting a wafer (often made of mica) between the IC and the heat sink. Be sure to also use an electrically *nonconductive* thermal grease as well. To provide a solid mechanical structure, either bond the sandwich together with an electrically *nonconductive* glue, or use a *nylon* bolt, nut, and washers. This provides the electrical insulation you need. But it raises the thermal resistance to, typically,

$$\Theta_{CS} \approx 2\frac{{}^\circ C}{W} \text{ mica wafer, electrically insulated}$$

The most common way to determine the junction temperature is to measure the case temperature with a temperature sensor. You may do this during prototype testing to assure that the design meets its specifications. Or you may actually build a temperature monitor that measures the case temperature and takes some action to protect the electronics if things get too hot. With the sensor mounted on the heat sink,

$$T_J = T_C + P(\Theta_{JC} + \Theta_{CS})$$

So, measuring the temperature of the case, and knowing the specifications of the IC and the power that it is handling lets you determine the junction's temperature.

This leaves only the heat sink itself. The larger the heat sink, the more fins it contains, and the easier it is for heat to flow through it to the surrounding air. The thermal resistance is inversely proportional to the heat sink's surface area. A large heat sink has a small thermal resistance and heat flows through it easily, keeping the IC junction cool. This is the only part of the thermal elements over which you have much influence. If you want a different thermal resistance, pick a different heat sink. But be sure it fits into the space you have available. Rearranging the basic thermal equation to solve for the maximum allowable heat sink thermal resistance gives

$$\Theta_{SA\,max} = \frac{T_{J\,max} - T_{A\,max}}{P_{IC\,worst\,case}} - \Theta_{JC} - \Theta_{CS}$$

In still air, the temperature of the layer of air immediately surrounding the heat sink may be considerably higher than the air only a few centimeters away. This raises T_A, forcing you to obtain a larger heat sink with a lower thermal resistance. An alternative is to force air past the heat sink with a fan. This moves the heat away from the surface of the heat sink, rather than just relying on convection. The faster the air moves, the more heat is pulled from the sink. The heat sink's manufacturer gives the numerical effect of forced air cooling in a plot of the thermal resistance of the heat sink on the *y* axis, versus the speed of the

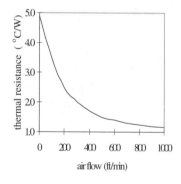

Figure 4-14 Air flow's effect on thermal resistance (*courtesy of Avid*)

air on the *x* axis. Figure 4-14 shows this effect for a TO220 heat sink by Avid.

You may see large areas of a printed circuit board with the copper left in place. The power semiconductor is then bent over and bolted to this expanse of metal. If you have large undedicated areas of your PCB, this may seem like a cheap and easy way to provide a heat sink. However, most printed circuit board material expands seven times more in thickness than it does in the plane of the surface, as it heats up. The co-efficient of expansion for the board is also significantly different from that of the leads, the solder, and any nearby vias. So, as the heat transfers to the board, a great deal of force is exerted on the solder joint. After a surprisingly few cycles of heating and cooling, the leads break free from their solder connections, down inside the board (where you cannot see the break). And worse, the breaks separate when the parts heat up, but when you first turn on the equipment, to try to find the problem, everything is cool, and the broken ends are touching again. Do *not* use the printed circuit board as a heat sink unless you have PCB material specially designed for that purpose.

Example 4-6

Determine the heat sink's thermal resistance for the op amp from Example 4-5. Assume an ambient temperature of 40°C.

Solution

When picking a heat sink, you must consider the IC's worst case condition. From Table 4-2 this occurs at

$$P_{\text{IC worst case}} = 2.92\,\text{W}$$

Solving the fundamental heat sink equation for Θ_{SA},

$$\Theta_{\text{SA max}} = \frac{T_{\text{J max}} - T_{\text{A max}}}{P} - \Theta_{\text{JC}} - \Theta_{\text{CS}}$$

From the specifications for the OPA548, $T_{\text{J max}} = 125°\text{C}$

$$\Theta_{\text{JC}} = 2.5\frac{°\text{C}}{\text{W}}$$

Since the metal package of the OPA548 is connected to V−, it is wise to isolate the case from the heat sink with a mica wafer.

$$\Theta_{\text{CS}} = 2\frac{°\text{C}}{\text{W}}$$

Make the substitutions.

$$\Theta_{SA\,max} = \frac{125°C - 40°C}{2.92W} - 2.5\frac{°C}{W} - 2\frac{°C}{W} = 24.6\frac{°C}{W}$$

This is the *largest* thermal resistance that can be used without the IC overheating.

Example 4-7

If a heat sink with a thermal resistance of $4\dfrac{°C}{W}$ is used on the OPA548, with ±24 V supplies, what is the largest sine wave that can be delivered to a 10 Ω load?

Solution

The basic thermal relationship is

$$T_J = T_A + P(\Theta_{JC} + \Theta_{CS} + \Theta_{SA})$$

Solve this for P.

$$P_{IC\,worstcase} = \frac{T_{J\,max} - T_{A\,max}}{\Theta_{JC} + \Theta_{CS} + \Theta_{SA}}$$

$$P_{IC\,worstcase} = \frac{125°C - 40°C}{2.5\dfrac{°C}{W} + 2\dfrac{°C}{W} + 4\dfrac{°C}{W}} = 10\,W$$

Working the power equations backwards to determine the operating conditions that result in the IC dissipating 10 W becomes rather involved. A simpler solution is to alter the spreadsheet in Table 4-2, then look at the row that results in P_{IC} = 10 W. That row is shown in Table 4-3.

Table 4-3 Solution to Example 4-7

V_{load} volts	I_{load} amps	P_{load} watts	P_{supply} watts	P_{IC} watts
9.4	0.94	4.42	14.36	9.94
9.6	0.96	4.61	14.67	10.06

Practice: How much power can the IC dissipate and how much power can be delivered to the load if the heat sink's thermal resistance is raised to $8\dfrac{°C}{W}$ and the power supplies are lowered to ±18 V?

Answer: $P_{\text{IC worst case}} = 6.8\text{ W}, \quad P_{\text{load}} = 6.5\text{ W}$

4.4 Protecting the OPA548

The OPA548 power op amp has internal **thermal protection**. When the semiconductor's temperature reaches about 160°C, the output transistors are automatically disabled. They are kept off until the wafer's temperature falls below approximately 140°C. This feature is intended to protect the IC from unexpected failures, enabling it to survive *occasional* problems. Allowing the IC to repeatedly operate above 125°C degrades its performance. This feature was *not* intended to replace proper heat sinking. Without the correct heat sink, as calculated in the previous section, the IC can never reliably deliver the required power to the load. This internal thermal protection is only a fail-safe.

Thermal protection only activates when the power dissipated by the IC has caused its temperature to build up over 160°C. This may take several cycles of the load current. If the load demands over 5 A from the OPA548, the IC, welds, and bonds may be damaged before the IC heats up enough to cause the thermal protection to activate. In addition, this excessive current flows from the power supply and through the printed circuit board traces, connectors, wires, and the load itself. Excessive current may damage all of these elements.

Adding a single, low power resistor to the OPA548 allows it to limit load current. The connection is shown in Figure 4-15. Select R_{CL} as

$$R_{\text{CL}} = \frac{15\text{k}\Omega \times 4.75\text{A}}{I_{\text{LIM}}} - 13.75\text{k}\Omega$$

Current limiting is illustrated in Figure 4-16. When the load resistance is low, the output current is the value dictated by the output voltage and load resistance, as set by Ohm's law. But once I_{LIM} is reached, the output current is fixed. Since Ohm's law cannot be violated, once current limiting is entered, further reductions of the load resistance causes the load *voltage* to drop proportionally.

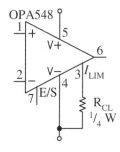

Figure 4-15 Current limiting with the OPA548 (*courtesy of Burr-Brown*)

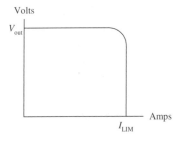

Figure 4-16 Current limiting characteristic curve

Current limiting is instantaneous. It affects only those parts of a waveform that result in currents above I_{LIM}. All load currents above I_{LIM} are held at I_{LIM}. This effectively shaves off the peaks.

Current limiting only limits the current. It does *not* prevent the amplifier from overheating. In fact, should the load be shorted, the load voltage goes to 0 V. This means that the load dissipates no power. The amplifier holds the current to I_{LIM}. The amplifier must now dissipate all of the power being delivered by the supply. If the heat sink is sized for normal operation, it cannot handle this excessive power. The IC's temperature rises. The OPA548 quickly goes into thermal limiting. Without thermal limiting, the IC would burn up. Current limiting does *not* protect the amplifier from overheating.

Do *not* try to test current limiting by shorting the load. You will just force the amplifier into thermal shutdown. Instead, raise the output voltage to its maximum. Then, gradually lower the load resistance until the peaks just begin to flatten. During that flattened part of the output signal, current limiting is being activated. The current limit is

$$I_{LIM\ measured} = \frac{V_{flattened\ peak}}{R_{load}}$$

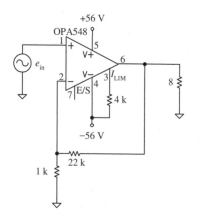

Figure 4-17 Schematic for Example 4-8

Example 4-8

  a. Calculate the gain and the current limit for the circuit in Figure 4-17.

  b. Accurately draw the output voltage waveform for an input of 2 V_p.

Solution

  a.

$$A_v = 1 + \frac{R_f}{R_i} = 1 + \frac{22\,k\Omega}{1\,k\Omega} = 23$$

$$R_{CL} = \frac{15\,k\Omega \times 4.75\,A}{I_{LIM}} - 13.75\,k\Omega$$

$$I_{LIM} = \frac{15\,k\Omega \times 4.75\,A}{R_{CL} + 13.75\,k\Omega} = 4.0\,A$$

b. For an input of 2 V_p, the output should be

$$V_{out} = 2\,V_p \times 23 = 46\,V_p$$

But the maximum output current is 4 A_p. With an 8 Ω load, this current limit sets the maximum output voltage to

$$V_{out\ I\,lim} = 4\,A_p \times 8\,\Omega = 32\,V_{max}$$

The output voltage is shown in Figure 4-18.

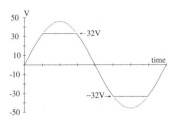

Figure 4-18 Clipped output for Example 4-8

Practice: How must you change the circuit in Figure 4-16 to deliver a 50 W sine wave to the 8 Ω load?

Answer: Change $R_{CL} < 6.4$ kΩ

4.5 Audio Power Parameters

Have you ever been told (or told some one) "Turn the volume down. It's too loud!"? Just how loud is loud enough? If the loudspeakers are on the stage, what happens to the volume as you move toward them? Will a more expensive loudspeaker make the music louder? How much power do you need from your amplifier? Music is not a sine wave. So what do you do about the peak-to-rms relationship? How do you rate the amplifier's power with a music signal? Before you can build an audio amplifier that actually produces the sound you want at the listener, you have to answer all of these questions.

Sound exerts a pressure on the eardrum. The higher that pressure, the louder the sound's volume. The softest sound that a young, undamaged ear can detect comes from a pressure of 20 μPascals. This level has been selected as a reference.

$$p_{ref} = 20\mu P$$

Softest detectable sound

But, our ear responds logarithmically to amplitude. So loudness is measured in **sound pressure level (spl)**. It is defined as

$$dB_{spl} = 20\log\frac{p_{ear}}{p_{ref}}$$

The softest sound generally detectable produces $p_{ear} = 20 \ \mu P$. The threshold of hearing, then, is at $dB_{spl \ threshold} = 0 \ dB_{spl}$. A change of 1 dB_{spl} is barely perceivable, although it represents a 26% change in sound pressure. Your television remote control alters the volume by about 1.5 dB_{spl} each time you press the volume button. Most people can detect a 3dB_{spl} increase in volume. This requires *twice* the power from the amplifier.

Common volume levels are given in Table 4-4.

Table 4-4 Typical sound pressures of various sources (*courtesy of Yamaha's <u>Sound Reinforcement Handbook</u>*)

dB_{spl}	sound
140	.45 ACP Colt pistol at 25 feet
130	Siren at 100 feet
120	Threshold of pain
	Rock music at 10 feet
110	
100	Film musical score at 20 feet
	Loud classical music
90	Heavy street traffic at 5 feet
80	Cabin of a jet aircraft
70	
	Average conversation at 3 feet
60	
50	Average suburban home at night
40	Quiet auditorium
30	
	Quiet whisper at 5 feet
20	
	Rustling leaves
10	
0	Threshold of hearing

Since a sound wave usually spreads out both vertically and horizontally as it travels away from the source, sound pressure level decays as the square of distance. As compared with the level at 1 meter, the sound *x* meters away has changed by

$$\Delta dB_{spl} = 10 \log \left(\frac{1m}{x} \right)^2$$

The square of the argument of a log is the same as twice the log.

$$\Delta dB_{spl} = 20\log\left(\frac{1\,m}{x}\right)$$

Effect of distance on sound

Example 4-9

A singer wants to be "loud enough" 8 m from the loudspeakers. The sound from a loudspeaker is rated at 1 m. How loud must the music be 1 m in front of the loudspeaker?

Solution

"Loud enough" is interpreted as 90 dB$_{spl}$. Softer would not be "commanding." Much louder could be painful to the people closer to the stage.

$$\Delta dB_{spl\,@\,15m} = 20\log\frac{1\,m}{8\,m} = -18\,dB_{spl}$$

The sound drops −18 dB$_{spl}$ as it travels from the loudspeaker to the listener, 8 m away. So, at 1 m from the loudspeaker the sound pressure must be

$$dB_{spl\,@\,1m} = 90\,dB_{spl} + 18\,dB_{spl} = 108\,dB_{spl}$$

This at the top end of acceptable. Much louder would too loud for the audience close to the loudspeaker.

Practice: A teacher can project a sound of 85 dB$_{spl}$ in a class room. Given that the listener in the back of the room must hear the sound at 70 dB$_{spl}$, how large can the room be without needing a sound reinforcement system?

Answer: The room may be 5.6 m = 18.5 ft deep.

Loudspeakers convert electrical energy into sound pressure. How effectively they do this is indicated by their rating:

$$dB_{spl} \text{ @ 1 m @ 1 W}$$

The power delivered to the speaker can also be rated in dBW:

$$dBW = 10\log\frac{P}{1\,W}$$

This allows the direct conversion between sound pressure level and power delivered to the loudspeaker.

Example 4-10

In Example 4-9, it was decided that 108 dB$_{spl}$ is needed at 1 m in front of the loudspeaker. How much power must be provided by the amplifier if the loudspeaker is rated at 95 dB$_{spl}$ @ 1 m @ 1 W?

Solution

When 1 W of power is applied to the loudspeaker, it outputs a 98 dB$_{spl}$ sound. But you want the sound of 108 dB$_{spl}$.

$$\Delta dB = 108\,\text{dB} - 95\,\text{dB} = 13\,\text{dB}$$

The power to the loudspeaker must be increased by 13 dBW above 1 W.

$$dBW = 10\log\frac{P}{1\,\text{W}} = 13\,\text{dBW}$$

Solve this equation for P.

$$P = 1\,\text{W} \times 10^{\frac{13}{10}} = 20\,\text{W}$$

Practice: How much power is needed if it is decided to lower the volume at the listener by 6 dB$_{spl}$?

Answer: The needed power drops to 5 W, a fourfold decrease.

Most testing is done with a sine wave. It is well defined, its shape is not altered by linear components, distortion is easy to detect, and it is commonly available. However, audio program content is *not* sinusoidal. It is **random**. That is exactly what makes it interesting to listen to. You have seen repeatedly that the peak of a sine wave is 1.414 times its root mean squared value. And it is this rms value that is used in determining the power delivered to a resistive load (the loudspeaker). The ratio of peak to rms is called the **crest factor**.

$$crest\,factor = \frac{V_p}{V_{rms}}$$

Table 4-5 Crest factors

sound	crest factor voltage	power
sine	1.414	
speech	1.73	x 2
music	3.2	x 10

The rms value is used to determine the power that the signal delivers to the load. This in turn sets the average sound pressure level that the loudspeaker produces and that the audience hears. However, the amplifier must be able to provide a signal whose random, instantaneous, peak voltage is up to three times larger than the rms voltage. This peak is not repetitive. It happens occasionally and unpredictably during the program. The durations of the peaks also vary widely, but the peaks are a very small percentage of the entire program.

The amplifier's power rating is based on a sine wave. Knowing the gain and maximum power rating of the amplifier, you can input a sinusoidal tone, deliver the amplifier's rated maximum power to the loudspeaker, and provide a known sound pressure level at 1 m and at any given spot in the audience. The tone sounds loud and clear. However, as soon as you connect the same rms speech or music source, the peaks are far above that of the sine wave. The amplifier is already at its maximum. So the peaks from the audio source are clipped off. During the clipped peak, the amplifier's output goes to a high, constant level. The loudspeaker's cone moves out and stays there during the clipping. The sound is distorted. In addition, since the cone is not moving it does not cool itself. There is a danger that the loudspeaker may be damaged. The loudspeaker is damaged because the amplifier is *under*powered!

The solution is to buy a bigger amplifier than needed to provide the sound pressure level calculated in Examples 4-9 and 4-10. Generally, sound professionals recommend that the amplifier's power be increased by the signal's crest factor. Therefore, if the amplifier from Example 4-10 is to be used in a classroom, select a 2x20 W = 40 W amplifier. This assures that the continuous volume meets the levels set in Example 4-9, but that there is plenty of **head room** when the teacher's peaks occur. If you want to play music through the same system, producing the same sound levels, then the amplifier must be 10x20 = 200 W to assure that there are crisp, clear, undistorted peaks.

The amplitude from the audio signal processor (mixer) is usually measured in **dBu**.

$$dBu = 20\log\frac{V_{rms}}{0.775\,V_{rms}}$$

The reference voltage of 0.775 V_{rms} is the amplitude needed to deliver 1 mW into 600 Ω. Signals coming from professional sound reinforcement equipment are typically set at about 4 dBu. This translates into 1.23 V_{rms}. The **line out** from consumer electronics is usually −10 dBu, 0.25 V_{rms}.

Example 4-11

What output voltage and what gain are needed by an amplifier that delivers 20 W to an 8 Ω loudspeaker?

Solution

To deliver 20 W to 8 Ω:

$$P = \frac{V^2}{R}$$

$$V = \sqrt{P \times R} = \sqrt{20\,W \times 8\Omega} = 12.7\,V_{rms}$$

From a professional mixer, the line level is 1.23 V_{rms}, so

$$gain = \frac{V_{out}}{V_{in}}$$

$$gain = \frac{12.7\,V_{rms}}{1.23\,V_{rms}} = 10.3$$

Practice: What range of gain must you provide if this amplifier is also to work with consumer electronics.

Answer: 10–51

4.6 Low Power Audio Amplifier IC

The LM386 is a low voltage audio power amplifier. It has been optimized to require a minimum of components while delivering as much as 1 W to a loudspeaker. Its schematic is shown in Figure 4-19.

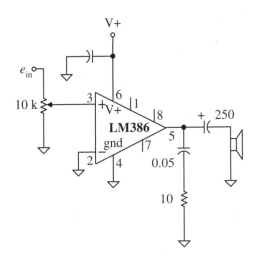

Figure 4-19 Low power audio amplifier
(*courtesy of National Semiconductor*)

Unlike the op amps that you have seen, the LM386 has been designed to operate from a *single* power supply voltage, rather than ±V. This voltage may range from 4 V to 12 V for most versions. The -4 version allows a supply voltage as large as 18 V. This supply requirement allows you to power the amplifier from three 1.5 V batteries, from the +5 V of your computer, from a 9 V battery, or from your car battery (which may rise to 18 V).

However, without a negative supply voltage, what happens when the output signal needs to swing down? The LM386 biases its output pin to half of whatever voltage is provided as the supply. So, when the input is 0 V, the output is at half of the supply. The output can then swing up toward the supply, or down toward circuit common. For a 5 V supply driving an 8 Ω load, the output voltage can swing up to within 1 V of the supply, and down to within 1 V of circuit common. As the supply voltage and resulting load current increase, this saturation level also increases. With a 12 V supply and 8 Ω load, the output can swing up to 9.5 V and down to 2.5 V. Refer to the manufacturer's specifications for a detailed graph of supply voltage versus output peak to peak swing for various loads.

The bias voltage at the output must be blocked from the load, while the signal should be passed. That is the role of the 250 μF capacitor just before the loudspeaker. These two form a high pass filter, blocking the dc bias voltage (0 Hz) and passing the signal. The low frequency cut-off is

Low frequency cut-off

$$f_l = \frac{1}{2\pi R_{load} C}$$

For the components in Figure 4-18, signals below 80 Hz are attenuated. If you want to pass more bass, increase this output capacitor. If you are working with speech rather than music, set f_l = 300 Hz. This allows you to reduce the size and cost of that capacitor.

Another major difference between the LM386 and the general purpose op amp is that negative feedback is provided internally by the LM386. Apply the input signal directly to the noninverting pin and tie the inverting input to circuit common. This provides a gain of 20. To invert the signal, apply the signal to the inverting pin and connect the noninverting input to circuit common. This sets the gain to −20.

Internal gain = 20

With this gain and an output swing of less than 10 V_{pp}, you must restrict the input amplitude. The manufacturer rates the maximum input as ±0.4 V. That is the purpose of the input potentiometer. Even though the IC is powered only with a positive voltage, the input may swing both positive and negative 400 mV, without an *RC* input coupler. Either input may be referenced directly to analog common. Each input provides a 50 kΩ resistor to common. Unlike the op amp, the input impedance of the LM386 is not extremely large. It is 50 kΩ. The input potentiometer is set to 10 kΩ to reduce the loading effect of this 50 kΩ input impedance.

Input impedance = 50 kΩ

The internal negative feedback is provided by a 15 kΩ resistor between the output (pin 5) and pin 1. The lower part of the negative feedback voltage divider is provided internally by a 1.35 kΩ resistor (between pins 1 and 8) and a 150 Ω resistor. The gain is

$$A = 2\frac{15\,k\Omega}{R_{pin\,1-8} + 150\,\Omega}$$

Increase the gain above 20 by placing a resistor and a dc blocking capacitor between pins 1 and 8. This resistor parallels the 1.35 kΩ internal resistor. If you place only a capacitor between pins 1 and 8, the gain rises to 200. Remember to account for the effect of the internal 1.35 kΩ resistor. Select the capacitor's value so that at the lowest frequency of interest, its reactance is small (1/7) compared to the resistor it affects.

$$C > \frac{1}{2\pi f_1 \dfrac{R}{7}}$$

Example 4-12

Determine the resistor and capacitor you must add between pins 1 and 8 to set the gain at 40, with $f_1 = 80$ Hz.

Solution

The gain is

$$A = 2\frac{15\,k\Omega}{R_{pin\,1-8} + 150\,\Omega}$$

Solve this for $R_{pin\,1-8}$.

$$R_{pin\,1-8} = 2\frac{15\,k\Omega}{A} - 150\,\Omega$$

$$R_{pin\,1-8} = 2\frac{15\,k\Omega}{40} - 150\,\Omega = 600\,\Omega$$

Pick a 620 Ω resistor.

The capacitor in series with the 620 Ω resistor should be

$$C > \frac{1}{2\pi \times 80\,Hz \times \dfrac{620\,\Omega}{7}} = 22\mu F$$

Pick a 27 µF capacitor.

Practice: How much gain error would using a 620 Ω resistor produce?

Answer: $A = 39$ This is an error of 2.5 %.

It is often desirable to boost the low and the high frequency signals. This may be used to compensate for an inexpensive loudspeaker, or to implement the *loudness* feature. It is usually enough to give these bass and treble signals twice the gain that the middle frequency signals have.

You can boost the bass by placing an *RC* series pair between pins 1 and 5. At low frequencies, this capacitor looks like an open, and the gain is 20. As the frequency goes up, the capacitor begins to look like a short. This places the external resistor between pins 1 and 5 in parallel with the 15 kΩ internal resistor. Setting this external resistor to 15 kΩ then reduces the gain to

$$A_{mid} = 2\frac{15\,\text{k}\Omega // 15\,\text{k}\Omega}{1.35\,\text{k}\Omega + 150\,\Omega} = 10$$

This is the lowest stable gain that the amplifier produces.

The low frequency boost point is set by the capacitor in series with the 15 kΩ resistor between pins 1 and 5. The precise equation is rather *complex* since it involves a series-parallel combination of two resistors and a capacitor. Select its initial value near

$$C_{\text{bass boost}} \approx \frac{1}{2\pi \times 15\,\text{k}\Omega \times f_{\text{bass boost}}}$$

Then test select the final value by substitution at the bench.

This *RC* pair between pins 1 and 5 has dropped the gain above $f_{\text{bass boost}}$ to 10. To raise the gain back up to 20 for the treble signals, you must add the *RC* pair between pins 1 and 8. But because the resistor between pins 1 and 5 is now paralleling the internal 15 kΩ resistor, the equation changes to

$$R_{\text{pin 1-8}} = 2\frac{15\,\text{k}\Omega // 15\,\text{k}\Omega}{20} - 150\,\Omega = 600\,\Omega$$

But instead of selecting the capacitor to look like a short at 80 Hz, pick it to pass the signal to $R_{\text{pin1-8}}$ at the treble boost frequency. Below that frequency, the capacitor looks like an open, and the gain is 10. Above that frequency the capacitor looks like a short and the gain is 20.

$$C_{\text{treble boost}} \approx \frac{1}{2\pi R_{\text{treble boost}} f_{\text{treble boost}}}$$

The schematic, with both bass and treble boost, and the resulting frequency response are given in Figure 4-20.

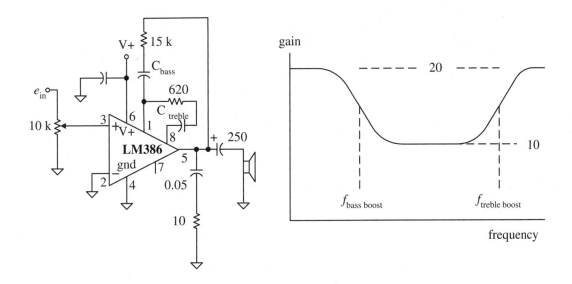

Figure 4-20 LM386 with bass and treble boost

All of the power and heat sinking relationships that you used for the power op amp of the previous section apply equally to the LM386. They are demonstrated in the following example.

Example 4-13

Calculate the maximum power you can deliver to an 8 Ω loudspeaker, the worst case power that the LM386 must dissipate, and the heat sink's thermal resistance using a +5 V supply.

$$T_{J\,max} = 150°C \qquad \Theta_{JC} = 37°C/W \qquad \Theta_{JA\,no\,sink} = 107°C/W$$

How loud is the sound produced?

Solution

The output dc voltage centers at half of the supply.

$$V_{out\,dc} = 2.5\ V$$

This output voltage can rise to within 1 V of the supply.

$$V_{out\,max} = 4\ V$$

The ac peak swing, then, is

$$V_{p\,out} = 4\,V - 2.5\,V = 1.5\,V_p$$

Assuming that this signal is a sine wave means that

$$V_{rms\,out} = \frac{1.5\,V_p}{\sqrt{2}} = 1.06\,V_{rms}$$

The loudspeaker dissipates

$$P_{load} = \frac{\left(V_{rms}\right)^2}{R_{load}}$$

$$P_{load} = \frac{\left(1.06\,V_{rms}\right)^2}{8\,\Omega} = 140\,mW$$

In terms of dBW, this is

$$dBW = 10\log\frac{140\,mW}{1\,W} = -8.5\,dBW$$

How loud this sounds depends on the efficiency of the loudspeaker and how close to it you are. Having a +5 V supply suggests that this circuit is to be used for a desktop computer monitor. So, a distance of 1 m is reasonable.

The loudspeaker from earlier examples was rated at

$$98\ dB_{spl}\ @\ 1W\ @\ 1\ m$$

The sound heard, then, is

$$dB_{spl} = -8.5\,dBW + 98\,dB_{spl} = 89.5\,dB_{spl}$$

Table 4-4 suggests that this is not quite as loud as heavy street traffic at 5 ft. That certainly should be loud enough for a computer monitor on your desk.

The worst case output level for the IC (assuming a sine wave) is when the signal is at 63% of the supply voltage. Since the signal swings up and down from 2.5 V, the worst case for the IC is when

$$V_{p\,worst\,case} = 0.63 \times 2.5\,V = 1.6\,V_p$$

But the maximum signal is only 1.5 V_p. So the worst case for the IC is at maximum signal. The current from the supply is

$$I_{\text{p supply}} = \frac{1.5\,\text{V}_\text{p}}{8\,\Omega} = 188\,\text{mA}_\text{p}$$

This current flows from the supply, charges the capacitor, and then flows to the load, when the output voltage swings positive. On the negative half-cycle, the IC's sourcing power transistor turns off, and its sinking transistor is turned on. Current flows from the capacitor, through the IC's lower transistor to ground, and then up from ground through the load to the negative side of the capacitor. This is how current reverses direction through the load without a negative supply voltage.

The key point is that current flows from the supply only during the positive half-cycle. The power delivered comes from a dc voltage and a half sine current. From Table 3-2 that power is

$$P_{\text{supply}} = V_{\text{dc}} \frac{I_\text{p}}{\pi}$$

$$P_{\text{supply}} = 5\,\text{V}_{\text{dc}} \frac{188\,\text{mA}_\text{p}}{\pi} = 299\,\text{mW}$$

The IC must dissipate the power that is provided by the power supply but is not passed to the load.

$$P_{\text{IC}} = P_{\text{supply}} - P_{\text{load}}$$

$$P_{\text{IC}} = 299\,\text{mW} - 140\,\text{mW} = 159\,\text{mW}$$

The maximum allowable thermal resistance is

$$\Theta_{\text{JA max}} = \frac{T_{\text{J max}} - T_{\text{A}}}{P}$$

For an amplifier that sits on a desk, without forced cooling, it is reasonable to expect that the air next to the enclosed IC gets no hotter than about 50°C.

$$\Theta_{\text{JA max}} = \frac{150°\text{C} - 50°\text{C}}{0.159\,\text{W}} = 629\,\frac{°\text{C}}{\text{W}}$$

This is the *maximum* allowable thermal resistance. Any value below this will keep the junction temperature below 150°C. The specifications of the thermal resistance for the IC without a heat sink is 107°C/W for the DIP package. So the IC needs no heat sink.

Practice: Calculate the maximum power you can deliver to an 8 Ω loudspeaker, the worst case power that the LM386 must dissipate, and the heat sink's thermal resistance if this amplifier is to be used in a car (+12 V supply).

Answer: $V_{\text{load p}} = 3.5\ V_p$, $P_{\text{load}} = 0.77\ W$,   97 dB$_{\text{spl}}$ in the front seat, 91 dB$_{\text{spl}}$ in the back seat, $I_{\text{supply p}} = 438\ mA_p$, $P_{\text{supply}} = 1.7\ W$, $P_{\text{IC}} = 0.9\ W$, $\Theta_{\text{JA}} = 111°C/W$.

With *good* loudspeakers, the LM386 delivers reasonable sound levels if you are close, *and* if you ignore the fact that music produces peaks that are 10 times its rms value. To keep from clipping these peaks, distorting the sound and damaging the loudspeakers, the amplifier must be able to provide more power. The **bridge amplifier** of Figure 4-21 provides up to four times the power without a larger power supply voltage.

The loudspeaker is *not* tied to circuit common. Both ends are driven, one by U1, and the other by U2. The second IC, U2, is configured as an inverting amplifier. So, when U1 drives the left side of the load to its maximum level, the right amplifier, U2, drives that end down

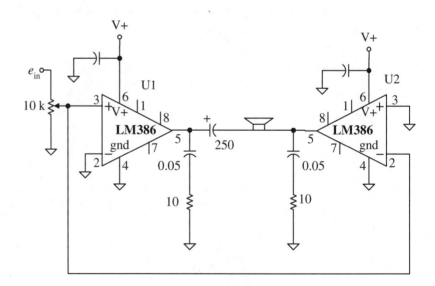

Figure 4-21 Bridge amplifier

just above circuit common. So the total positive peak across the load is almost the supply voltage, not half of the supply as it is when the loud-speaker is tied to common. On the negative half-cycle the left end is driven down near common, while the right end goes close to the supply. However, the current is flowing in the opposite direction. The current has reversed direction without having a negative power supply.

Example 4-14

Calculate the maximum power delivered to an 8 Ω load by a bridged amplifier using two LM386s running from a 5 V supply.

Solution

On the positive input peak, U1's output goes to

$$V_{\text{left max}} = 5\,\text{V} - 1\,\text{V} = 4\,\text{V}$$

U2 drives the right end of the load close to common.

$$V_{\text{right min}} = 1\,\text{V}$$

The *peak* across the load is

$$V_{\text{p}} = 4\,\text{V} - 1\,\text{V} = 3\,\text{V}_{\text{p}}$$

Assuming a sine wave,

$$V_{\text{rms}} = \frac{3\,\text{V}_{\text{p}}}{\sqrt{2}} = 2.1\,\text{V}_{\text{rms}}$$

The power delivered to the load is

$$P_{\text{load}} = \frac{\left(2.1\,\text{V}_{\text{rms}}\right)^2}{8\,\Omega} = 551\,\text{mW}$$

Practice: Calculate the maximum power delivered to an 8 Ω load by a bridged amplifier using two LM386s running from a 12 V supply. At this supply level, there is a 2.5 V saturation.

Answer: 3 W

4.7 High Power Audio Amplifier IC

The few watts available from the LM386 are fine for personal listening. But to be heard more than a meter or two from the loudspeaker, a sound must be delivered at much higher power levels. The LM3875 can deliver much more power.

Up to 56 W of continuous power can be provided to an 8 Ω speaker. At that power level, the gain is flat from 20 Hz to 20 kHz and there is typically no more than 0.1% of distortion and noise. The IC also provides a variety of protection circuits. It monitors its internal temperature, turning itself off if the junction temperature exceeds 165°C, allowing operation to begin again once it has cooled to 155°C. Output current is limited to a peak of 4 A. Output protection is also provided for under voltage and over voltage transients.

A typical schematic is shown in Figure 4-22. At first glance, it looks just like an op amp. However, the IC has been optimized to deliver a lot of power, over a narrow frequency range, through long leads, at low noise and distortion levels. It handles these jobs much better than a general purpose power op amp.

The pin out diagram is given in Figure 4-23. It is a staggered lead power package, with the case tied to V−. The leads are spaced less than 0.1″ apart, so be careful if you plan to use a traditional universal breadboard with solderless connections.

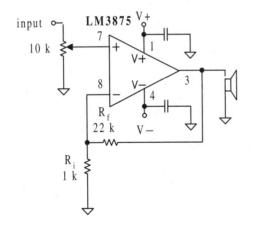

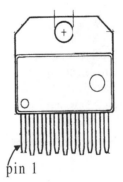

Figure 4-22 High power audio amplifier
(*courtesy of National Semiconductor*)

Figure 4-23 LM3875 pin out diagram
(*courtesy of National Semiconductor*)

There are several key specifications that affect how you use the amplifier. These are shown in Table 4-6

Table 4-6 LM3875 key parameters

Parameter	Value	Unit
Power supply voltage	20–84	V
Output drop out voltage	5	V
Output current limit	4	A
Junction temperature	150	°C
Thermal resistance Θ_{JC}	1	°C/W
Gain bandwidth product	2	MHz

The power supply voltage is the difference between V+ and V−. You can run the IC from either split supplies (±10 V to ±42 V) or from a single supply (usually +20 V to +84 V with V− connected to common). If you choose to use a single supply be sure to provide appropriate biasing and *RC* coupling at the input and output. The minimum supply of 20 V means that you cannot power this IC *directly* from a car's 12 V battery, or the four D cells in a boom box. The maximum voltage (84 V or ±42 V) sets the upper limit on the largest voltage you can deliver to the load.

The output drop out voltage indicates how close to the power supplies the output can be driven before the transistors saturate. Although under certain conditions it may be less, it is wise to count on being able to drive the output no closer to the supplies than 5 V. So, if you are using the maximum ±42 V supplies, the largest output peak is ±37 V_p.

The output current typically is limited internally to 6 A. But the guaranteed limit is 4 A (or more). So you can count on being able to deliver at least 4 A to the load, assuming the part does not overheat.

The highest temperature on the silicon wafer that the IC can reliably run is 150°C. Even though there is internal temperature monitoring and shutdown, repeatedly pushing the chip over 150°C degrades its performance and may eventually cause its failure. So provide the appropriate heat sink to assure that under worst case, normal operation (i.e., not a failure in the system), the wafer temperature does not exceed 150°C. To determine this heat sink, you need to know the thermal resistance between the junction and the case, Θ_{JC}. It is no larger than 1°C/W.

Finally, the maximum closed loop gain is determined by the highest frequency to be amplified and the IC's gain bandwidth product. Although the GBW is typically 8 MHz, the worst case guaranteed value is 2 MHz. So, for a sine wave at the upper end of the audio range, 20 kHz, you can have a closed loop gain of 100. But, remember, at that frequency, the *actual* gain has dropped by 0.707 (−3 dB).

In addition to the limitations dictated by these specifications, there are several precautions you must observe to assure that your amplifier works well. Review Chapter 2 on the proper methods to breadboard and to lay out a printed circuit board for a power amplifier. *All* of the techniques and warnings presented there apply when you are using the LM3875. Be sure to:

- Keep the input and the output connections well separated.

- Rigorously decouple each power supply pin, as close to that pin as possible with a 10 µF electrolytic and a 0.1 µF film capacitor.

- Connect the speaker return *directly* to the analog common in its own separate lead, not back to the amplifier board's common.

The leads that run from the output of the amplifier to the loudspeaker may be quite long. Running two conductors, side by side, creates a capacitor. The longer the distance, the greater is the capacitance to common. At high frequencies, this capacitance may cause the amplifier to oscillate. The parallel *RL* circuit shown in Figure 4-24 looks like a short at audio frequencies, but forms a low pass filter, blocking the high frequency oscillations.

In many applications, long leads may also be connected to the input, as the customer runs a cable to the CD player on another shelf. Noise coupled into this cable from the large amplitude output may cause the circuit to oscillate. The capacitor *directly* across the input leads prevents this, without degrading the audio signal.

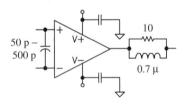

Figure 4-24 Input and output compensation (*courtesy of National Semiconductor*)

Example 4-15

Using the LM3875, with a sinusoidal signal, and assuming an ambient temperature of 35°C, calculate the maximum power you can deliver to an 8 Ω load and the heat sink needed for worst case operation.

Solution

The supply voltages limit the power to the load. The maximum voltages for the LM3875 are

$$V_{max\ supply} = \pm 42V$$

The maximum output voltage from the LM3875 is the drop out voltage below its supply.

$$V_{max\ out} = 42\,V - 5\,V = 37\,V_p$$

This is the peak voltage for the output wave. Assuming a sinusoid,

$$V_{p\ load} = 37\,V_p$$

$$V_{rms\ load} = \frac{37\,V_p}{\sqrt{2}} = 26.2\,V_{rms}$$

The 8 Ω load dissipates

$$P_{max\ load} = \frac{\left(26.2\,V_{rms}\right)^2}{8\Omega} = 85.5\,W$$

This is a very impressive amount of power from an integrated circuit. What must be done to assure that the IC does not overheat? From Table 4-2 and Figure 4-11 you saw that the worst case for the IC, assuming a sine wave, is *not* at the maximum load voltage. The worst case for the IC is at

$$V_{p\ load\ @\ IC\ worst\ case-sine} = 0.63 \times V_{supply} = 26.5\,V_p$$

Repeat the sequence of calculations above. This leads to

$$P_{load\ @\ IC\ worstcase} = 44\,W$$

At this level, the current that the power supply provides is

$$I_{supply\ @\ IC\ worst\ case} = \frac{26.5\,V_p}{8\Omega} = 3.3\,A_p$$

The power delivered by these two supplies is

$$P_{supplies} = 2V_{dc}\frac{I_p}{\pi}$$

$$P_{\text{supplies @ IC worstcase}} = 2 \times 42 \, \text{V}_{\text{dc}} \frac{3.3 \, \text{A}_{\text{p}}}{\pi} = 89 \, \text{W}$$

The IC must dissipate the power that is not delivered to the load.

$$P_{\text{IC @ worstcase}} = 89 \, \text{W} - 44 \, \text{W} = 45 \, \text{W}$$

The basic temperature relationship is

$$T_{\text{J}} = T_{\text{A}} + P\left(\Theta_{\text{JC}} + \Theta_{\text{CS}} + \Theta_{\text{SA}}\right)$$

Solve this for the thermal resistance of the heat sink, Θ_{SA}.

$$\Theta_{\text{SA}} = \frac{T_{\text{J}} - T_{\text{A}}}{P} - \Theta_{\text{JC}} - \Theta_{\text{CS}}$$

$$\Theta_{\text{SA}} = \frac{150°\text{C} - 35°\text{C}}{45\text{W}} - 1\frac{°\text{C}}{\text{W}} - 0.2\frac{°\text{C}}{\text{W}} = 1.4\frac{°\text{C}}{\text{W}}$$

Although this is a low thermal resistance, with proper forced air flow, even the simple TO220 heat sink from Figure 4-13 can reach this level. A larger heat sink may not require a fan.

Practice: Calculate the maximum power you can deliver to an 8 Ω load with the LM3875 powered from ±28 V supplies. What heat sink is needed?

Answer: $P_{\text{load max}} = 33.1 \, \text{W},\quad \Theta_{\text{SA}} = 4.6\frac{°\text{C}}{\text{W}}$

Summary

The OPA548 power op amp allows a supply voltage up to 60 V. You can set the load current limit with a single, low wattage, high resistance resistor. Should there be a failure, the IC turns itself off when its temperature exceeds 160°C. The input offset voltage, offset current, gain bandwidth, and slew rate all can have a significant effect on the output, so do not forget to consider each.

Power depends on the signal's shape. The two most common shapes are dc and sinusoidal. The power dissipated by the load is

$$P_{\text{load}} = \frac{\left(V_{\text{dc load}}\right)^2}{R_{\text{load}}} \text{ or } \frac{\left(V_{\text{rms load}}\right)^2}{R_{\text{load}}}$$

The power supply to the amplifier provides

$$P_{\text{supply}} = V_{\text{dc}} \times I_{\text{dc}} \text{ or } 2V_{\text{dc}} \frac{I_{\text{p}}}{\pi}$$

The amplifier must dissipate the power that is provided by the supply but not delivered to the load.

$$P_{\text{amp IC}} = P_{\text{supply}} - P_{\text{load}}$$

Calculate these power levels at the maximum load power to assure that the supply can provide enough voltage and current, and that the load can indeed dissipate the power that reaches it. But the worst case for the amplifier IC is not at maximum load power. So, you must repeat the analysis again at IC worst case:

$$V_{\text{dc load}} = 0.5V_{\text{supply}} \text{ or } V_{\text{p load}} = 0.63V_{\text{supply}}$$

The more power that the IC must dissipate, the hotter it becomes. Its precise temperature depends on the ambient temperature, how much power it is dissipating, the thermal resistance between the wafer and the case, how the heat sink is mounted to the IC, and the characteristics of the heat sink. The air flowing over the heat sink lowers its thermal resistance.

One of the major uses of power ICs is as an audio amplifier. The loudness of a sound is measured in dB_{spl}, ranging from 25 dB_{spl} for a whisper to 140 dB_{spl} for a pistol shot. Sound decreases logarithmically as you move beyond the 1 m specification of the loudspeaker. Power may also be rated logarithmically in dBW. Audio signals are random, not sinusoidal. The crest factor varies from 1.7 for speech to 3.2 for music. Typically, signal levels from professional consoles are centered around 4 dBu, 1.23 V_{rms}. Consumer electronics output signals of about 310 mV_{rms}.

Tabletop applications require little more than a watt. This can be accomplished by the LM386. It is optimized for single supply operation from 4 V to 12 V. It has a fixed internal gain of 20. But this gain can be altered with external resistors, or its frequency response can be shaped with external *RC* networks. Bridging two LM386s allows you to deliver four times the power to the load.

For more power (over 50 W), consider the LM3875. It is configured very much like an op amp. It may be run from a single supply or from split supplies (20 V to 80 V). The load current is internally limited to at least 4 A, and the junction temperature is monitored and automatically limited.

The power ICs that you have seen in this chapter are convenient. But convenience is purchased at the expense of flexibility. Different voltage and power levels, increased frequency performance, custom current foldback, current output (to drive particularly low load resistances), and load power in the hundreds of watts require that you design the amplifier yourself. In Chapter 5 you will build on the fundamentals introduced here to construct custom MOSFET linear power amplifiers.

Problems

OPA548 Power Op Amp

4-1   For the circuit in Figure 4-25, calculate the dc and rms values of each of the four indicated voltages.

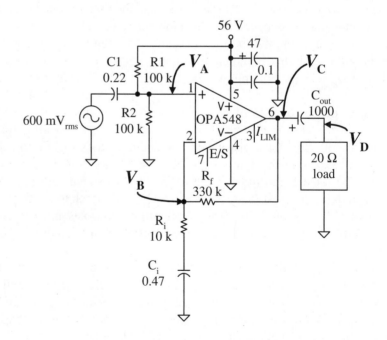

Figure 4-25 Schematic for Problem 4-1

4-2 Explain the purpose of each of the following components in Figure 4-25: C1, R1, R2, C_i, and C_{out}

4-3 The nonideal input offset voltage, bias currents, and offset currents are shown in Figure 4-26. Calculate the indicated voltages and currents. Hint: do the calculations in the indicated order.

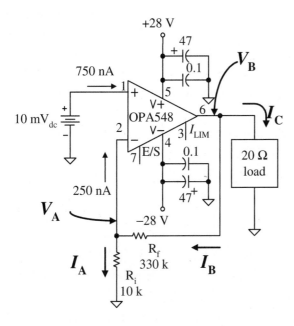

Figure 4-26 Schematic for Problem 4-3

4-4 For the circuit in Figure 4-26, if the 10 mV_{dc} offset voltage source is replaced with a 500 mV_{rms} sinusoid,

a. calculate the high frequency cut-off, f_H.

b. What is the output amplitude at f_H?

c. Is the output signal distorted by the op amp's slew rate limit? Prove your answer with a calculation.

d. Can a larger input be applied without distorting the output? Explain your answer with calculations.

Power Calculations

4-5   An OPA548 is powered from ±28 V, and is driving a 20 Ω resistive load. For a dc input signal, calculate the following at the maximum *load* power: P_{load}, P_{supply}, P_{IC}.

4-6   Repeat Problem 4-5 at the operating point that causes the IC to dissipate maximum power.

4-7   An OPA548 is powered from ±28 V, and is driving a 20 Ω resistive load. For a sinusoidal input signal, calculate the following at the maximum *load* power: P_{load}, $P_{both\ supplies}$, P_{IC}.

4-8   Repeat Problem 4-7 at the operating point that causes the IC to dissipate maximum power.

Heat Sinks

4-9   For the following conditions, determine the OPA548's junction temperature:

$$T_A = 35°C, \ \ P_{IC} = 10\ W, \ \ \Theta_{SA} = 4\frac{°C}{W}, \text{mica wafer}$$

4-10   For the following conditions, calculate the largest acceptable heat sink thermal resistance for a OPA548.

$$T_A = 35°C, \ \ P_{IC} = 10\ W, \text{mica wafer}$$

4-11   Locate the specifications for a heat sink that fulfills the requirements from Problem 4-10, in still air.

4-12   For the following conditions, calculate the maximum power that can be delivered by a sinusoidal signal: to a 20 Ω load, by a OPA548 and a sinusoidal signal from ±28 V supplies.

$$T_A = 35°C, \text{ mica wafer}, \ R_{load} = 20\ \Omega, \ \text{OPA548}, \ V_{supply} = \pm 28\ V$$

Audio Power Parameters

4-13   A sound level meter indicates 105 dB_{spl}.
   a. How loud is that in qualitative terms?
   b. How much pressure is being exerted on the eardrum?
   c. If you wanted to lower the volume to that of average conversation, how many TV remote clicks does it take?

4-14   If the 105 dB_{spl} is measured 20 feet from the loudspeaker, how far must you move from the loudspeaker for the sound to drop to 70 dB_{spl}?

4-15   The loudspeaker producing the 105 dB$_{spl}$ at 20 feet is rated at 90 dB$_{spl}$ @ 1 W @ 1m. How much power must you provide to the loudspeaker to produce the 105 dB$_{spl}$ at 20 feet?

4-16   An audio amplifier is capable of outputting 24 V$_p$. How much music power can be delivered to an 8 Ω loudspeaker without distorting the sound?

4-17   To deliver 10 W to a 4 Ω loudspeaker, how much gain must the audio amplifier have if it is to be driven from a typical consumer electronics signal?

Low Power Audio Amplifier IC

4-18   Design an amplifier circuit using the LM386 to run from a 5 V supply, driving a 4 Ω loudspeaker. Select all components. Determine the heat sink's thermal resistance assuming an ambient temperature of 35°C.

4-19   Repeat Problem 4-18 using a 12 V power supply.

4-20   Alter your design from problem 4-18 to allow a 20 mV$_{rms}$ input to produce 0.5 W into the 4 Ω loudspeaker.

High Power Audio Amplifier IC

4-21   Design an amplifier circuit using the LM3875 to run from ±28 V supply, driving an 8 Ω loudspeaker. Assume that an input signal of 500 mV$_{rms}$ produces the maximum output power. Select all components. Determine the heat sink's thermal resistance assuming an ambient temperature of 40°C.

4-22   Complete the design of the bridge amplifier shown in Figure 4-27. Set the gain to allow a 500 mV$_{rms}$ signal to produce the maximum output power. Calculate the power delivered to the load and the heat sinks' thermal resistance assuming an ambient temperature of 40°C.

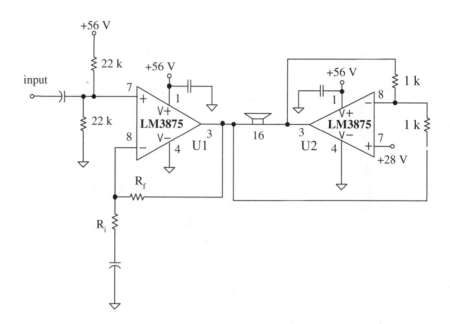

Figure 4-27 Bridge amplifier for Problem 4-22

Power Op Amp Lab Exercise

Prototyping

The pins of the OPA548 are *not* spaced on 0.1″ centers. They also do not tolerate repeated bending. Power supply and load currents of 3 A are required in this exercise. For these reasons it is recommended that a printed circuit board for the OP548 be developed, rather than using the traditional universal breadboard with 0.1″ solderless connections. The layout in Figure 4-28 works well. Notice that the power and load are brought onto the board through screw terminals, and traces are sized to handle the current. But 0.1″ spaced pins are provided to allow you to breadboard the other connections. Decoupling capacitors are provided on the PCB because it is important to keep them as close as possible to the IC's power pins. The diode protects the IC if you reverse the power supply polarities.

This exercise *can* be run without the PCB. Be sure to handle the op amp's leads with great care, and keep all connections between the op amp, power, and the load as short as possible.

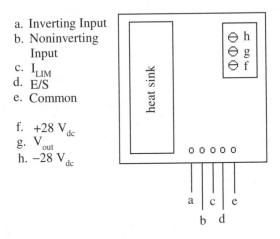

a. Inverting Input
b. Noninverting Input
c. I_{LIM}
d. E/S
e. Common

f. $+28\ V_{dc}$
g. V_{out}
h. $-28\ V_{dc}$

Figure 4-28 OP548 board connections

A. Voltage Follower

1. Build a voltage follower using the OPA548. Be sure to *rigorously* follow the procedures for prototyping power electronic circuits explained in Chapter 2.

2. Replace e_{in}, the input signal generator, with a short to common.

3. Assure the load is set to at least 100 Ω.

4. Position the board as close as practical to the power supply. Use simple 22 AWG wires from the power supply *directly* to the IC. Keep the leads as short as practical. Apply the ±18 V power.

5. Measure the input and the output voltage with your digital multimeter. Record *all* stable digits. The output voltage, with the input tied to common, is the **input offset voltage**. It should be a few millivolts. If your output voltage is significantly larger, there is a problem. Do *not* continue until the circuit is working correctly.

6. Replace the short to common at the input with a sine wave of 5 V_{rms}, 100 Hz as measured on the digital multimeter.

7. Display the input and the output on the oscilloscope.

8. When the output is an *undistorted* sine wave, with little offset, measure the input and the output with your digital multimeter. Record all stable digits.

9. Set the input to 0 V_{rms}.

10. Switch the power connections to a ±28 V, high current supply.

11. Turn the power supply on while monitoring the supply current.

12. Set the input to 5 V_{rms}. Gradually lower the load's resistance until the supply current is 1 A_{dc}.

13. When the output is an *undistorted* sine wave, with little offset, measure the input and the output with your digital multimeter. Record all stable digits.

14. Return the load to its maximum resistance and turn the power supply off.

15. Calculate the gain and compare it to theory.

B. Noninverting Amplifier

1. Build a noninverting amplifier with the OPA548. Set $R_f = 33$ kΩ, and $R_i = 1$ kΩ. Carefully review your layout to assure that it closely follows the guidelines from Chapter 2.

2. Measure the value of R_f and R_i. Record all stable digits.

3. Using these *measured* resistor values, calculate and record the amplifier's theoretical gain.

4. Replace e_{in}, the input signal, with a short to circuit common. Apply ±18 V.

5. Measure the input and the output V_{dc} with your digital multimeter. Record all stable digits.

6. Compare this offset voltage with that from the voltage follower. Is it the same, or did it go up by 34? Is the input offset voltage affected by the gain?

7. Replace the short to common with a sine wave of 100 mV$_{rms}$, 100 Hz as measured on the digital multimeter. Also display the input and output signals on the oscilloscope.

8. When the output is an undistorted sine wave, with little offset, measure the input and the output V_{rms} with your digital multimeter. Record all stable digits.

9. Calculate the measured gain and compare it to the theoretical gain calculated (with actual resistor values) in step B3.

10. Set the input to 0 V$_{rms}$.

11. Switch the power connections to a ±28 V, high current supply.

12. Turn the power supply on while carefully monitoring the supply current.

13. Set the input to 0.5 V$_{rms}$. Gradually lower the load's resistance to 8 Ω. The supply's current should indicate about 1 A$_{dc}$.

14. When the output is an *undistorted* sine wave, with little offset, measure the input and the output with your digital multimeter. Record all stable digits.

15. Calculate the gain at full load, and compare it to theory (step B3).

16. Calculate the power being delivered to the load.

C. Amplifier High Speed Performance
1. Lower the input amplitude to 20 mV$_{rms}$.

2. Raise the input signal's frequency until the *gain* has fallen to 0.707 of that measured in step B15. Be sure that the input amplitude remains constant.

3. This frequency is the high frequency cut-off, f_H. Calculate the amplifier's gain bandwidth product.

$$GBW = f_H \times A_{low\,frequency}$$

4. Change the input to a square wave with a 600 mV$_p$ amplitude and a 5 kHz frequency.

5. Measure the *slope* of the output. This slope is expressed in V/μs. This is the amplifier's slew rate.

6. Using your amplifier's slew rate, calculate the highest frequency a 24 V$_p$ sine wave output can obtain before distortion begins.

$$f_{large\,signal\,max} = \frac{slew\,rate}{2\pi V_{p\,out}}$$

7. Change the input to a sine wave and adjust its amplitude to 0.5 V_{rms}, at 100 Hz.

8. Raise the frequency slowly until the output begins to distort, changing into a triangle. Record that frequency and compare it to that calculated in step C7.

D. Audio Amplifier

1. Set the input signal generator to 0. Turn the power supplies off.

2. Replace the 8 Ω resistive load with a loudspeaker capable of dissipating at least 50 W.

3. Turn the power supplies on.

4. *Gradually* increase the input amplitude until the output is a clean 100 Hz, 2.83 V_{rms} sine wave. This signal is delivering 1 W to the loudspeaker.

5. Measure the volume in dB_{spl} at 1 m directly in front of the loudspeaker.

6. Increase the input amplitude until the volume is slightly uncomfortably loud.

7. Measure the rms voltage delivered to the loudspeaker. Calculate the power being delivered to the loudspeaker.

8. Measure the volume in dB_{spl} 1 m in front of the loudspeaker. Compare the expected increase in volume with that which you just measured.

9. Move to a point 4 m directly in front of the loudspeaker. Measure the volume. Did the dB_{spl} decrease according to theory?

10. Repeat steps 4 through 8 at 1 kHz and 5 kHz. Discuss the effect of the signal's frequency (tone).

11. Reduce the input signal's input to 0. Replace it with a music source, with the volume all the way down.

12. Gradually increase the music source's volume until the sound is slightly uncomfortable. Repeat steps 7 and 8. Discuss the differences in the power and sound from a sine wave and from a music source.

5

Discrete Linear Power Amplifier

Introduction

In Chapter 4 you saw power integrated circuits that could deliver up to 56 W to a load. Protection of the load and the IC were provided automatically.

To deliver more power you have to build the amplifier from a collection of resistors, capacitors, diodes, transistors, and op amps. Though more involved, this allows you to better match the circuit to its requirements. You can build amplifiers capable of driving several hundred watts of power, provide current limiting and thermal protection, and match performance to a resistive or a reactive load.

During this chapter you will use the central elements of each of the previous chapters. Op amps will be combined with transistors and a suspended supply amplifier in a composite configuration. All of the power calculations you used with power ICs are directly applicable to the power transistors in this chapter. Heat sinking is as important to these discrete transistors as it is to power ICs. Therefore, you have already mastered many of the details. In this chapter you will learn to apply these procedures to arrays of discrete transistors and their supporting electronics.

Objectives

By the end of this chapter, you will be able to:

- Describe the performance of an enhancement mode MOSFET and interpret its data sheet to determine key parameters.

- Explain the problems of a class A MOSFET power amplifier.

- Analyze and design a class B MOSFET power amplifier with an op amp driver and a suspended supply amplifier intermediate stage. In an analysis, determine all currents, voltages, and power. For a design, select all component values, all power ratings, and heat sinks.

- Properly parallel MOSFETs operated in their linear region.

- Analyze and design required current limiting and a transistor case temperature sensing circuit.

- Determine the effect on the amplifier of a load with both resistive and reactive elements.

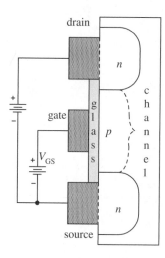

Figure 5-1 Basic MOSFET structure

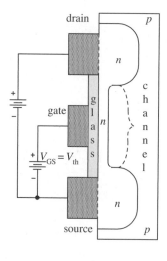

Figure 5-2 MOSFET biased on

5.1 The Enhancement Mode MOSFET

N-Channel

A simplified physical layout of an *n*-channel enhancement mode metal oxide semiconductor field effect transistor (MOSFET) is shown in Figure 5-1. There are three connections. The **source**, at the bottom, is made of **n-type** material and has an abundance of free electrons. The **drain**, at the top, is also made of *n* material, with lots of free electrons. But between these two islands is a **channel** made of *p* material. This area has no free electrons. In fact, there are many **holes** in the crystal structure. Each hole is a missing electron bond. So any electron that might stray into the channel from the source or drain falls into a hole, and becomes fixed in the crystal structure. It disappears from the horde of electrons free to respond to external voltages and produce conduction. All of this is to remind you that the *p* material forms an open between the *n*-type source and drain. Even with an external voltage between the drain and the source, there is no current flow.

The third terminal is the **gate**. It is separated from the rest of the transistor by a layer of silicon dioxide. This is an insulator, preventing current flow into or out of the gate.

With zero volts applied between the gate and the source terminals, there is still a barrier of *p*-type material between the *n* material of the source and the *n* material of the drain. So there are no free electrons in this channel and no current can flow between the source and the drain. The transistor is off.

Applying a positive voltage to the gate with respect to the source attracts electrons into the channel near the gate. As the gate-to-source voltage is made more positive, more electrons are attracted into the channel, and the holes of the *p* material are filled. Eventually, there are enough free electrons near the gate to form a complete channel of *n* type material between the source and the drain. Conduction can now take place in response to the voltage supply connected between the drain and the source terminals. This is shown in Figure 5-2.

The gate-to-source voltage needed to **enhance** the channel enough to allow 250 µA of drain (and source) current is called the **threshold voltage**.

$$V_{th} = V_{GS}\big|_{I_D=250\mu A}$$

This parameter is a specification of the transistor. Before the transistor can begin to respond to the input (gate-to-source) voltage, you must apply this much dc voltage, just to turn the transistor on. Figure 5-3 is a part of the data sheet for the IRF530N.

IRF530N

International
I⌀R Rectifier

Electrical Characteristics @ T_J = 25°C (unless otherwise specified)

	Parameter	Min.	Typ.	Max.	Units	Conditions
$V_{(BR)DSS}$	Drain-to-Source Breakdown Voltage	100	——	——	V	V_{GS} = 0V, I_D = 250µA
$\Delta V_{(BR)DSS}/\Delta T_J$	Breakdown Voltage Temp. Coefficient	——	0.12	——	V/°C	Reference to 25°C, I_D = 1mA
$R_{DS(on)}$	Static Drain-to-Source On-Resistance	——	——	0.11	Ω	V_{GS} = 10V, I_D = 9.0A ④
$V_{GS(th)}$	Gate Threshold Voltage	2.0	——	4.0	V	V_{DS} = V_{GS}, I_D = 250µA
g_{fs}	Forward Transconductance	6.4	——	——	S	V_{DS} = 50V, I_D = 9.0A

Figure 5-3 Threshold voltage specification for the IRF530N (*courtesy of International Rectifier*)

The manufacturer indicates that no IRF530N begins to conduct with less than 2 V applied between the gate and the source. At some level between 2 V and 4 V the threshold is reached and drain current flows. The precise threshold varies from one IRF530N to another. But you can be assured that every IRF530N conducts at least 250 µA when V_{GS} = 4 V. You will see how to handle this wide variation when building an amplifier later.

As the gate-to-source voltage is increased beyond the threshold, the *n*-type channel grows both deeper and wider. This increases the cross-sectional *area* of the channel. As a consequence, once the threshold voltage has been reached, the drain (and source) current is proportional to the *square* of the gate-to-source voltage. It takes very little increase in gate-to-source voltage to have a significant effect on the drain current.

Example 5-1

Use simulation software to determine the relationship of I_D to V_{GS} for an IRF530N with V_{DS} = 56 V_{dc}. Plot the resulting data.

multiSIM

Solution

The Multisim schematic at $V_{GS} = 4.0$ V is shown in Figure 5-4, along with the plot of I_D versus V_{GS}. $V_{th} = 3.6$ V. A few tenths of a volt change beyond 3.6 V causes a major increase in I_D.

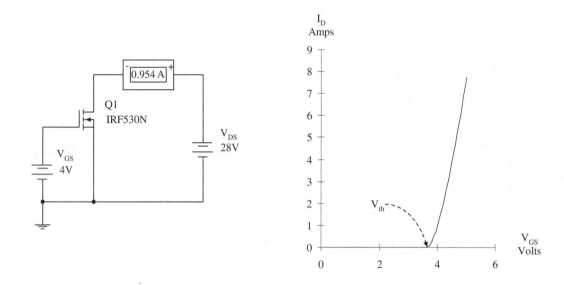

Figure 5-4 Simulation of the IRF530N gate control of drain current

Practice: What effect does changing the drain voltage have?

Answer: It has practically no effect. Once V_{DS} is over 4 V, drain current is relatively independent of drain-to-source voltage.

The manufacture's plot of V_{GS} versus I_D for the IRF530N is given in Figure 5-5. There are several points to notice. First, the shape does not seem to match the shape of the graph from the simulation, shown in Figure 5-4. However, the vertical axis for the specifications in Figure 5-5 is logarithmic, while Figure 5-4 was produced with a linear scale. Secondly, the manufacturer assumes a threshold voltage of 4 V. But you already have seen that this may be as low as 2 V. So, you must offset the horizontal axis, shifting it right or left, to make it apply to your particular MOSFET.

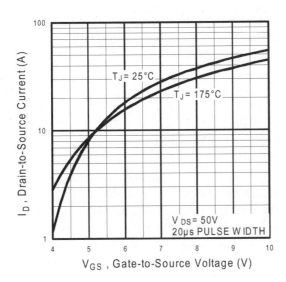

Figure 5-5 Gate control of an IRF530N (*courtesy of International Rectifier*)

Example 5-2

Determine the drain current for an IRF530N with $V_{GS} = 4.5$ V at 25°C.

Solution

The $V_{GS} = 4.5$ V vertical grid intersects the 25°C curve at about $I_D = 3.8$ A, assuming that $V_{th} = 4$ V. For a transistor with a lower threshold, the drain current is much larger.

Practice: What happens to the drain current when the temperature rises to 175°C?

Answer: The drain current *increases* to 5.5 A, as the temperature rises.

From Example 5-2, as temperature rises, the drain current increases, dissipating more power, heating up the transistor more, producing more current, heating up the transistor more, and so on, and so on. In *linear* applications, MOSFET transistors may thermally run away, and burn themselves up. So it is critical that you properly manage the power dissipation, temperatures and heat sinks when building linear amplifiers using MOSFETs.

The transistor discussed so far is an *n-channel enhancement mode* MOSFET. The source and drain are made from *n*-type material. With no gate bias voltage, there is no channel. To establish drain current, the gate bias voltage must be made positive enough to *enhance* the channel. So, on the schematic symbol, the channel is shown as a broken line. The arrow points toward the channel, since semiconductor arrows point toward the *n*-type material.

Once the channel is established, *n*-type current carriers (electrons) move from the source to the drain in response to an external positive V_{DS}. Since the electrons are negatively charged, moving from source to drain, *conventional* current moves from the positive terminal of V_{DS}, into the drain of the MOSFET, to the source of the MOSFET, and then to analog common, and back to the negative terminal of the V_{DS} supply. There is a layer of insulating glass between the gate terminal and the channel, so there is no gate current. This is shown in Figure 5-6.

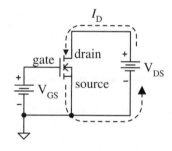

Figure 5-6 Basic *n*-channel enhancement mode schematic

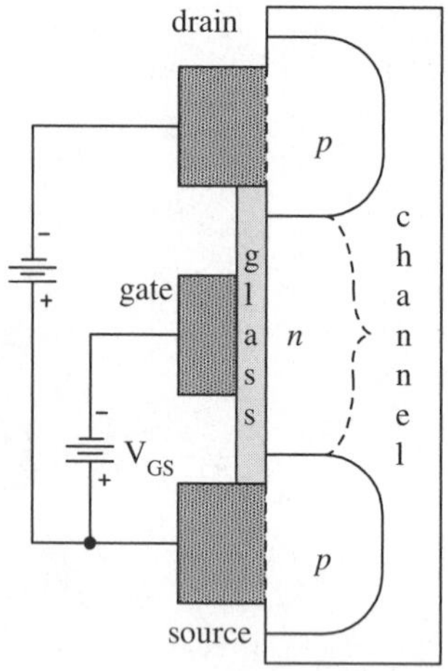

P-Channel

A *p*-channel enhancement mode MOSFET is shown in Figure 5-7. The substrate is *n*-type material, and the source and drain are made from *p* material. The current carriers are the holes in the crystal structure (vacancies that *should* contain an electron). Since holes are the absence of an electron, they act as if they were positive particles. The channel is established (enhanced) by making V_{GS} *negative* enough to repel the free electrons of the *n*-type substrate, and force electrons from their bonds within the crystal structure, leaving holes behind. This changes the region near the gate into *p* material. Once there is a channel from source

Figure 5-7 *P*-channel MOSFET

to drain, holes can flow from the source to the drain in response to the negative V_{DS}.

Holes act as if they are positive charges. They move in the same direction as conventional current. The schematic for a *p*-channel enhancement mode MOSFET with circuit current is shown in Figure 5-8.

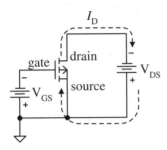

Figure 5-8 Basic *p*-channel enhancement mode schematic

The *p*-channel and the *n*-channel enhancement mode MOSFETs are complementary. Table 5-1 compares their characteristics. In linear applications, *n*-channel MOSFETs are used to handle the positive part of a signal, and *p*-channel devices process the negative voltages.

Table 5-1 *N*- and *p*-type MOSFET comparison

characteristic	*n*-channel	*p*-channel
carriers	electrons	holes
bias voltage	positive	negative
I_D	into drain	out of drain

Example 5-3

The IRF9530N is the *p*-channel complement to the IRF530N. From its specifications find V_{th} and I_D at $V_{GS} = -5V$.

Solution

$$V_{th} = -2 \text{ V to } -4 \text{ V} \qquad I_{D \text{ at VGS} = -5V} = 2.3 \text{ A}$$

5.2 Class A Common Drain Amplifier

A **class A amplifier** conducts during the entire cycle of the input signal. This amplifier most faithfully reproduces the input signal at the output. It produces the lowest distortion. Figure 5-9 is a class A amplifier built with an *n*-channel enhancement mode MOSFET. It is called a common drain amplifier because the drain is connected to the dc supply voltage. But a dc supply is a short to ac signals. So the drain is connected to ac common.

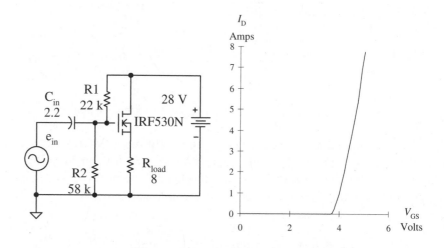

Figure 5-9 Class A common drain amplifier

Bias

The 8 Ω load is connected between the source and circuit common, since power loads are normally tied to common. Resistors R1 and R2 form a voltage divider to set the dc voltage on the gate. They assure that the amplifier is biased well beyond V_{th}, so that the transistor conducts during the entire cycle of the input signal, e_{in}. That is how class A is defined. Capacitor C_{in} blocks the dc bias voltage at the gate from the signal generator, but is sized to pass the lowest signal the amplifier must process.

With a 56 V supply, it is reasonable to set the dc voltage at the source to about 28 V_{dc}. This allows the ac signal to drive the voltage up toward 56 V and down toward common, giving a wide signal swing. An 8 Ω load dropping 28 V_{dc} must pass 3.5 A_{dc}. According to the transistor's V_{GS} versus I_D graph, 3.5 A_{dc} is produced by $V_{GS} \cong 4.4\ V_{dc}$. With 28 V_{dc} on the source and 4.4 V_{dc} between the gate and the source,

$$V_G = V_S + V_{GS} = 3.5\,A_{dc} \times 8\,\Omega + 4.4\,V_{dc}$$

$$V_G = 32\,V_{dc}$$

The voltage divider, R1 and R2 set V_G to

$$V_G = \frac{30\,k\Omega}{22\,k\Omega + 30\,k\Omega} 56\,V_{dc} = 32.3\,V_{dc}$$

So the amplifier in Figure 5-9 is biased in the middle of its supply range, with 3.5 A_{dc} of bias current constantly running through it. This is necessary to assure class A operation.

AC Operation

The ac performance of this amplifier requires a model and a little math. The ac signal causes the gate-to-source voltage to *change*. This change in gate-to-source voltage causes the drain current to *change*. The key ac parameter tells how much the drain current changes in response to changes in gate-to-source voltage. It is called the amplifier's **forward transconductance**.

$$g_{fs} = \frac{\Delta I_D}{\Delta V_{GS}}\bigg|_{V_{DS}\ \text{constant}}$$

It is actually the slope of the I_D versus V_{GS} graph, at the bias point (3.5 A_{dc} in Figure 5-9). The manufacturer specifies that for the IRF530,

$$g_{fs\ \text{IRF530}} > 6.4\frac{A}{V}$$

So, the ac signal does not need to change the gate-to-source voltage very much to have a major effect on the current through the transistor, and through the load.

The ac model for the class A amplifier is shown in Figure 5-10. The capacitor and the dc voltage supply have been replaced with shorts. Resistors R1 and R2 are in parallel for the ac signal, so they have been replaced with a single resistor. The MOSFET is modeled as a voltage controlled current source. The gate is open. Between the drain and the

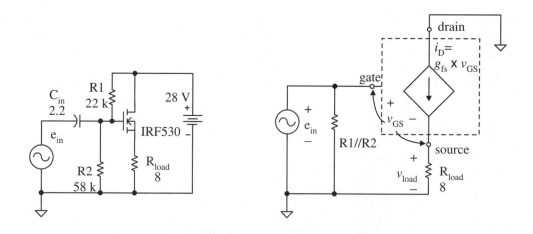

Figure 5-10 AC model of the class A MOSFET amplifier

source is a dependent current source. Its value is determined by the gate-to-source voltage and the transistor's forward transconductance.

To determine the load's ac voltage, apply Kirchoff's voltage law by summing the loop from e_{in}, v_{GS}, and v_{load}.

$$e_{in} - v_{GS} - v_{load} = 0$$

$$e_{in} - v_{GS} - i_D R_{load} = 0$$

But, i_D is set by v_{GS}. $$e_{in} - v_{GS} - (g_{fs} v_{GS}) R_{load} = 0$$

Collecting terms and rearranging gives

$$e_{in} = v_{GS}(1 + g_{fs} R_{load})$$

Or $$v_{GS} = \frac{e_{in}}{1 + g_{fs} R_{load}}$$

Amplifier output voltage   The load voltage is $$v_{load} = i_D R_{load} = g_{fs} v_{GS} R_{load}$$

Substitute for v_{GS}. $$v_{load} = \frac{g_{fs} R_{load}}{1 + g_{fs} R_{load}} e_{in}$$

$A_v < 1$ Since the denominator is always 1 more than the numerator, the load voltage is always a little less than the input signal. For this amplifier, $A_v = 0.98$, but it can output several amps of current to the 8 Ω load.

Example 5-4

Verify the dc and the ac performance of the class A, common drain amplifier of Figure 5-9 with $e_{in} = 20 V_p$.

Solution

The Multisim simulation is shown in Figure 5-11.

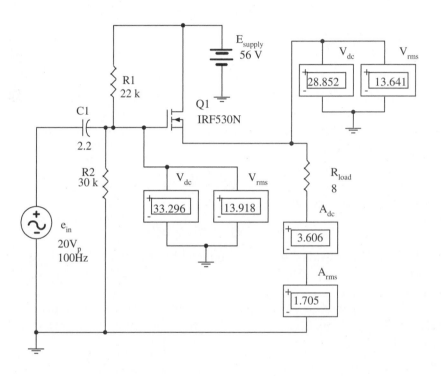

Figure 5-11 Simulation results for Example 5-4

Practice: Using the oscilloscope function, determine how large the output can be before it begins to distort. Explain the distortion.

Answer: The output may go to 28 V_p before it distorts. The distortion is caused by the output running into analog common.

Class A amplifiers may produce very little distortion, when properly biased and not overdriven. However, there are several major problems. In the schematics you have seen so far, the load is connected directly to the source terminal of the transistor. The *bias* current of 3.5 A$_{dc}$ flows continuously through the load, even when no signal is applied. It is necessary to put that much bias current through the transistor and the load to be sure that when the input signal goes negative, the transistor continues to output a smaller signal. It does *not* cut off, outputting zero. Just because of this bias

$$P_{\text{supply}} = 56 \text{ V}_{dc} \times 3.5 \text{ A}_{dc} = 196 \text{ W}$$

The load must dissipate

$$P_{\text{load}} = (3.5 \text{ A}_{dc})^2 \times 8\,\Omega = 98\,\text{W}$$

If the load is an 8 Ω loudspeaker, this dc current causes the cone to deflect once. It does *not* vibrate back and forth as it does with an ac signal. The coil is *not* cooled, and the loudspeaker burns out, without ever making a sound. Even if the load is not a loudspeaker, it still must dissipate the 98 W even before an ac signal is applied.

The transistor must dissipate the power provided by the supply but not dissipated by the load.

$$P_Q = 196\,\text{W} - 98\,\text{W} = 98\,\text{W}$$

This power is wasted. It is not produced by the signal. It is there because the transistor must be biased into the middle of its conduction region.

Class A power amplification is extremely wasteful. It is rarely used. But only a few modifications produce the much more efficient class B amplifier.

5.3 Class B Push-Pull Amplifier

In a **class B** amplifier the transistor is biased *off*, but just on the edge of conduction. That means

$$V_{\text{GS}} = V_{\text{th}}$$

When the input is 0 V, the transistor is off, and that 98 W wasted by the class A amplifier in the examples of the previous section is not used at all.

The Push Amplifier

The class B amplifier is laid out just a little differently from the class A of the previous section. Look at Figure 5-12. There is no input capacitor. The input signal generator drives the lower end of resistor R2. The value of R2 must be changed to shift the bias from the middle of the linear region down to the threshold voltage.

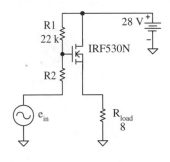

Figure 5-12 The push side of a class B push pull amplifier

Example 5-5

For the circuit in Figure 5-12:

1. Calculate R2's value to establish class B operation.
2. Verify that the transistor is biased at $I_D = 0 \, A_{dc}$.
3. Determine v_{load}'s shape and amplitude with $e_{in} = 40 \, V_p$.
4. Determine P_{load}, P_{supply}, and P_Q for $e_{in} = 40 \, V_p$.

Solution

1. Assuming that the transistor has the same characteristics as seen in the previous sections, with e_{in} replaced with a dc common,

$$V_{GS} = V_{th} = 3.6 \, V_{dc}$$

This means that the voltage across R1 is

$$V_{R1} = 56 \, V_{dc} - 3.6 \, V_{dc} = 52.4 \, V_{dc}$$

So, the current through R1 and R2 is

$$I_{R1\&R2} = \frac{52.4 \, V_{dc}}{22 \, k\Omega} = 2.38 \, mA_{dc}$$

For this current to produce 3.6 V_{dc} when flowing through R2,

$$R2 = \frac{3.6 \, V_{dc}}{2.38 \, mA_{dc}} = 1.5 \, k\Omega$$

2. The bias simulation is shown in Figure 5-13. It indicates 3.57 V_{dc} at the gate. This causes 56 μA_{dc} of drain current. With no input signal, the transistor is off, dissipating no power.

3. The two resistors form a voltage divider, dropping the voltage as it travels from the generator to the gate.

$$v_G = \frac{22 \, k\Omega}{22 \, k\Omega + 1.5 \, k\Omega} 40 \, V_p = 37.5 \, V_p$$

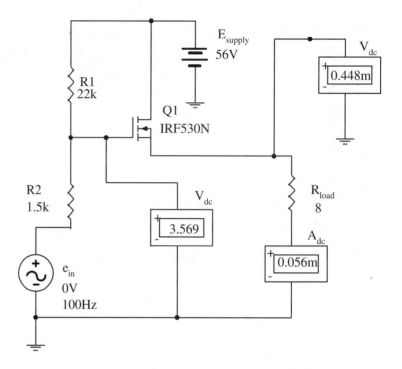

Figure 5-13 Multisim bias simulation results for Example 5-5

You have just seen that the gain for this amplifier is 0.98.

$$v_{load} = 0.98 \times 37.5\,V_p = 36.7\,V_p$$

The probe result from an OrCAD simulation is shown in Figure 5-14. The positive half of the input passes to the load at 36.5 V_p.

4. The voltage wave shape across the load is a half-sine. Through the resistance of the load, this produces a half-sine current.

$$I_{load} = \frac{36.5\,V_p}{8\Omega} = 4.6\,A_p$$

From Table 3-2, the power dissipated by a half-sine voltage and half-sine current is

$$P_{load} = \frac{V_p I_p}{4}$$

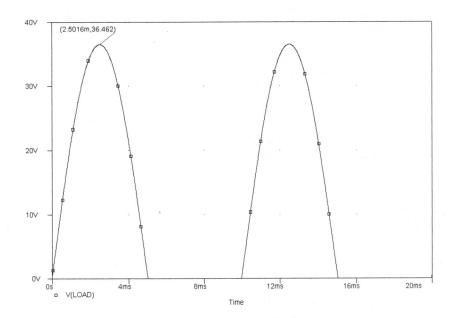

Figure 5-14 OrCAD probe voltage results for Example 5-5

$$P_{\text{load}} = \frac{36.5\,V_p \times 4.6\,A_p}{4} = 41.6\,W$$

The power supply provides a constant 56 V_{dc}, but the current is the half-sine delivered to the load. Again, from Table 3-2

$$P_{\text{supply}} = V_{dc}\frac{I_p}{\pi}$$

$$P_{\text{supply}} = 56\,V_{dc}\frac{4.6\,A_p}{\pi} = 82\,W$$

The transistor must dissipate the power that is provided by the power supply, but not delivered to the load.

$$P_Q = P_{\text{supply}} - P_{\text{load}}$$

$$P_Q = 82\,W - 41.6\,W = 40.4\,W$$

Using OrCAD's probe, each of these powers can be calculated.

$$P_Q = AVG((VD(M1)\text{-}VS(M1))*ID(M1))$$

$$P_{supply} = AVG(V(Esupply\text{:}+)*I(Esupply))$$

$$P_{load} = AVG(\text{-}V(load)*I(Rload))$$

The results of the OrCAD simulation and probe calculations are shown in Figure 5-15.

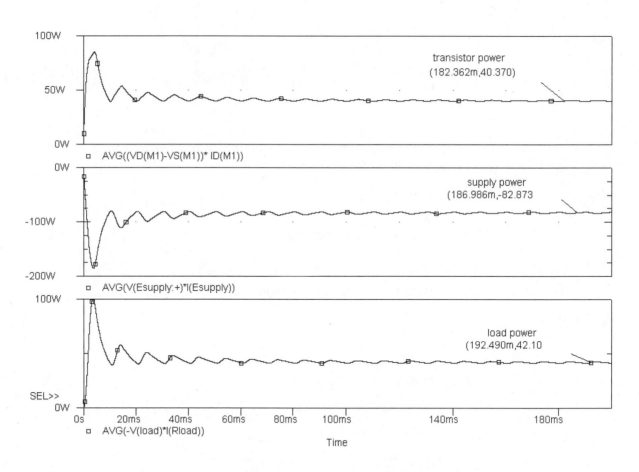

Figure 5-15　OrCAD probe power displays for Example 5-5

The Pull Amplifier

With the *n*-channel, push side of the class B amplifier, you are able to deliver the positive side of the signal to the load with only a small drop in voltage and at close to 4 A. No power has been wasted on biasing the transistor. For the class A amplifier, half of the power was squandered just setting the amplifier up. Manual calculations of bias and power for the positive side (**push**) of the class B amplifier agree well with those from simulation.

But the entire negative half of the signal is missing, because as soon as the input goes negative, the gate voltage falls below the transistor's threshold voltage. The *n*-channel transistor turns off.

A separate (**pull**) stage is needed to process the negative side of the input. The *n*-channel MOSFET is biased on with a positive signal and off when the signal goes negative. The *p*-channel enhancement mode MOSFET handles the negative side, turning on for negative signals, and off when the input goes positive. The *p*-channel pull side of the amplifier and its output wave form are shown in Figure 5-16.

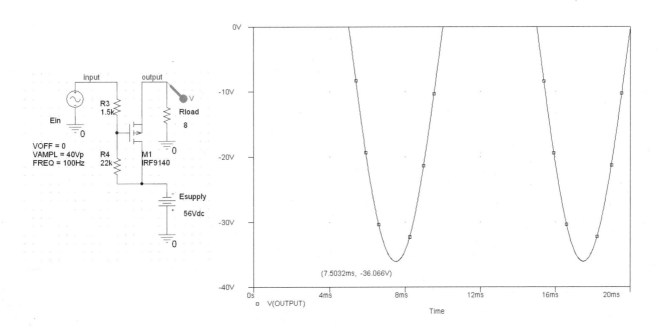

Figure 5-16 *P*-channel pull amplifier and OrCAD probe output wave form

The Push-Pull Amplifier

The combined push-pull (*n*-channel and *p*-channel) amplifier is shown in Figure 5-17. The performance and all of the calculations are just as you saw in Example 5-5. Each transistor is biased to the *edge* of conduction. But neither is on, so no dc current flows through the load, and it does not have to dissipate power just for the transistors to be properly biased.

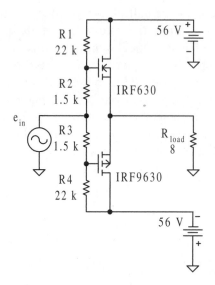

Figure 5-17 Class B push-pull amplifier schematic

From Example 5-5 you saw that when the input goes to positive 40 V_p, the *n*-channel transistor turns *on*, delivering 42 W to the load, and requiring that the transistor dissipate 40 W. The positive power supply must provide 82 W. During that time, the *p*-channel transistor is off, cooling. The negative power supply provides no significant power during this part of the cycle. It too can cool down.

During the negative cycle, the input goes to −40 V_p. As soon as the input drops below common, the *n*-channel transistor's gate is pulled below +3.6 V. That transistor goes off. The negative input now drives the *p*-channel transistor on, outputting as much as −36.5 V_p. This half-cycle also delivers 42 W to the 8 Ω load. The negative supply provides 82 W and the *p*-channel transistor must also dissipate 40 W.

The two half signals combine at the load, producing a full sine wave, at 36.5 V_p, 73 V_{pp}, 25.8 V_{rms}, and 84 W. Each supply delivers 82 W, for a total power from the power supplies of 164 W. *Each* transistor must dissipate 40 W. You have seen all of these calculations in Example 5-5.

The *n*-channel transistor has been changed from the IRF530 to the IRF630. In the push-pull amplifier of Figure 5-17, the *p*-channel may drive the load (and therefore the source of the *n*-channel transistor) to almost −56 V. With +56 V on its drain, the *n*-channel transistor must have a rated maximum V_{DS} of over 112 V. The IRF630 (*n*-channel) /IRF9630 (*p*-channel) transistors each have a maximum drain voltage of 200 V.

The class A amplifier is so inefficient that it is impractical for use in delivering more than a few watts to a load. The class B amplifier solves this problem, allowing you to build a linear amplifier capable of providing hundreds of watts to the load. Even so, the simple class B amplifier of Figure 5-17 has several problems.

The input signal is divided by R1 and R2 or R3 and R4 before it is passed to the transistors. The voltage gain of the common drain amplifier is less than 1.

Changes in drain current (the ac drain current) depend on the *square* of the change in gate-to-source voltage. The load voltage is created by this change in drain current flowing through the load resistance. There is a *nonlinear* relationship between the input (gate) voltage and the load voltage. The output is distorted. The larger the signal, the worse the distortion.

An even bigger contributor to output distortion is the uncertainty in the transistors' threshold voltages. The manufacturer specifies that *none* of the IRF630s produced turn on at a voltage below 2 V. But, *all* turn *on* at some point below 4 V. That is a 100% variation! In the previous examples, it was possible to precisely set the bias to the edge of conduction because the threshold was measured in Example 5-1. The transistor model *always* turns on at 3.6 V. But when you build a push-pull amplifier with real transistors, some may turn on at 2.1 V and others may not go *on* with 3.9 V between the gate and the source. You *could* measure every transistor you put in every amplifier that you build or repair, and then hope that time, temperature, humidity, or the phase of the moon do not cause the transistor to change its characteristics.

You may ask, is a volt or two such a big thing? Look back at the curve in Figure 5-4. A change of 1 V sends the current from 1 A to 8 A.

Even a small error or shift in behavior could have a *huge* effect. The only safe solution is to change the bias so that $V_{GS} = 2$ V. This assures that every transistor is *off*. The positive part of the input signal then pushes the *n*-channel's gate up until eventually it is positive enough to turn *on* and drive current to the load. Similarly, some part of the input negative signal is needed to augment the *p*-channel's negative gate voltage and eventually turn it *on*, creating the negative part of the output.

Example 5-6

1. Calculate new values for R2 and R3 in Figure 5-17 to set the gate voltages at ±2 V, at the bottom of the rated threshold voltage.

2. Simulate the amplifier with the new biasing resistors. Plot the output voltage in response to a 10 V_p input.

Solutions

1. To place 2 V_{dc} across R2 (and R3), resistor R1 (and R4) must drop the rest of the supply voltage. The current through that resistor is

$$I_{R1} = \frac{56\,V_{dc} - 2\,V_{dc}}{22\,k\Omega} = 2.45\,mA_{dc}$$

This current flows through R2, dropping 2 V_{dc}.

$$R2 = \frac{2\,V_{dc}}{2.45\,mA_{dc}} = 816\,\Omega$$

Pick R1 = R4 = 22 kΩ, R2 = R3 = 820 Ω.

2. The composite waveform is shown in Figure 5-18. When the input falls below a few volts, the voltage on the *n*-channel MOSFET's gate drops below V_{th} and the transistor turns off. That's too early. The *p*-channel MOSFET does not turn on until the input is a volt or two negative. That's too late. The result is a flat spot on the output as the input crosses 0 V. This is the result of the 2 V to 4 V threshold variation. Biasing at ±2 V assures that all versions of the transistor are off. But that conservative practice may then require some of the input signal to bring the transistors into full conduction.

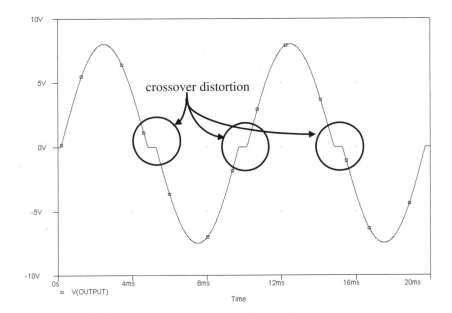

Figure 5-18 Properly biased class B amplifier shows crossover distortion

5.4 Class B Amp with Op Amp Driver

All of the problems with the simple class B push-pull amplifier can be solved by adding an op amp to create a composite amplifier. This was presented in Section 1.3. Look back over that information before you continue. Enclosing the push-pull MOSFET amplifier *within* the negative feedback loop of an op amp based amplifier allows the open loop gain of the op amp to compensate for any nonlinearities, or lack of adequate biasing.

Noninverting Amplifier

The schematic of a noninverting push-pull composite amplifier is given in Figure 5-19.

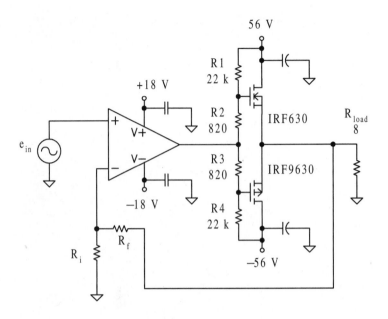

Figure 5-19 Noninverting push-pull composite amplifier

The key feature is that the feedback voltage divider, R_f and R_i, is connected to the *load*. It is *not* connected to the output of the op amp IC. Connecting the feedback directly to the load sends any imperfections at the load (such as nonlinearity or crossover distortion) back to the inverting input of the op amp. Any difference in voltage between that fed back signal and the input on the noninverting pin is amplified by the open loop gain of the op amp (often 100,000 or more).

The output of the op amp jumps significantly, driving the voltage at the load into proper shape. This then assures that the divided version of the load voltage fed back to the op amp's input pin matches the input signal.

Example 5-7

Simulate the performance of the circuit in Figure 5-19 with

e_{in} = 350 mV$_{rms}$, 100 Hz
R_f = 10 kΩ, R_i = 520 Ω

Plot the waveform at the load, and the op amp's output.

Solution

The OrCAD probe waveforms are shown in Figure 5-20. There are several items to notice.

1. The output voltage should be

$$V_{p\,load} = \left(1 + \frac{10\,k\Omega}{520\,\Omega}\right) 350\,mV_{rms} \times \sqrt{2} = 10.0\,V_{p}$$

This is precisely the load voltage. The loss across the bias network and across the transistor have been compensated for by the op amp. The op amp's output is 12 V_{p}. Overall gain has been realized and set by R_f and R_i.

2. The crossover distortion is gone. Look carefully at the output of the op amp. During the crossover time, when both transistors try to turn off, the output of the op amp makes a jump of several volts to bias the opposite transistor on. So the flat spot in the output has been removed by the step in the op amp's output voltage.

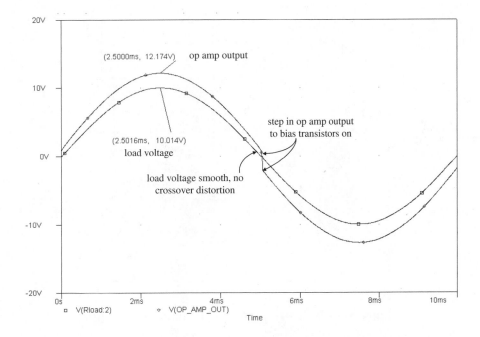

Figure 5-20 Composite amplifier's waveforms for Example 5-7

Practice: How large can the load voltage be driven? What causes this limitation?

Answers: Less than 16 V_p. The op amp's power supply limits its output, and therefore limits the MOSFETs' input and output.

The Suspended Supply Amplifier

To produce a large voltage to the load, the op amp must provide a slightly larger voltage at its output, to drive the MOSFETs' gates. But a conventional op amp's output is limited by its supply voltages, usually ±18 V or less. Delivering a 100 W sinusoidal signal to an 8 Ω load means that the op amp must be able to output over 40 V_p. This cannot happen from an op amp with ±18 V supplies.

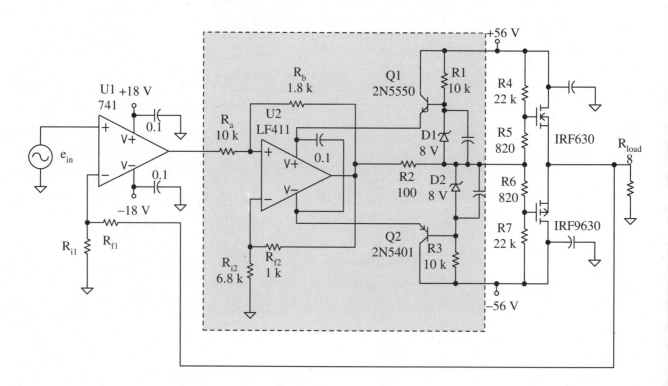

Figure 5-21 Three-stage push-pull composite amplifier

But the suspended supply amplifier presented in Section 1.4 can be powered from the same ± 56 V supplies used for the MOSFETs, and can output 45 V_p. Look back over that material before continuing. The suspended supply amplifier is slipped between the output of the op amp and the input to the MOSFET push-pull amplifier, as shown in Figure 5-21.

Layout Concerns

With all of these high voltages and high currents running around in your circuit, *how* you connect the parts is critical. The principles of good high voltage, high current layout are covered in Chapter 2. You should review these before you begin building an amplifier. Incorrectly positioning a single high current wire can drive the entire amplifier into oscillations, or reduce the gain by 30%.

Run separate power lines to the op amps and the power transistors, but keep each as short as possible. This reduces the possibility that the major variation in current from the power transistor is coupled back into the op amp along a shared power bus. With separate power wires, the massive capacitance of the power supply sits between the transistor power pins and the op amp power pins.

The same is done with the common return lines. The op amp needs the cleanest common line you can provide. So do not contaminate it with current from the power output stage. Any noise appearing on the op amp's common line may be interpreted as part of the signal and amplified. Certainly keep the common return from the power transistors and the load separate from the small signal amplifier common, until they get back to the single point common at the power supply connector.

Figure 5-22 is a repeat of Figure 2-15. It shows the interconnection of an amplifier and test instruments. Look carefully at how the supply leads are kept short and separate, as are the two common returns. The signal enters from the left. The suspended supply amplifier stage has been omitted, but there is plenty of space for it in the center of the protoboard. The load is connected at the extreme right. Finally, current from the load returns *directly* to the supply common in its own lead. This assures that any signal created by the large load return current cannot get back into the input small signal amplifier.

Inverting Amplifier

All of the previous examples are for *noninverting* amplifiers. However, you can just as easily use a push-pull power configuration to build an inverting amplifier, an inverting summer, a difference amplifier, an inte-

Figure 5-22 Proper amplifier and load connection
(*photo by John Zimmerman*)

grator, an oscillator, or any other configuration in which the simple op amp is used. It is as easy as building the op amp circuit, then inserting the suspended supply second stage and push-pull output stage between the output of the op amp and the load. The negative feedback must stay at the load, *not* at the op amp IC's output. It's as simple as that.

Example 5-8

Design an inverting amplifier to meet the following requirements:

e_{in} = +4 dBu at audio frequencies   input impedance 4.7 kΩ
50 W to a 4 Ω load T_A = 40°C, $T_{J\,max}$ = 140°C

power supplies: ±18 V_{dc}, ± 28 V_{dc}, ± 56 V_{dc}, each at 6 A_{dc}

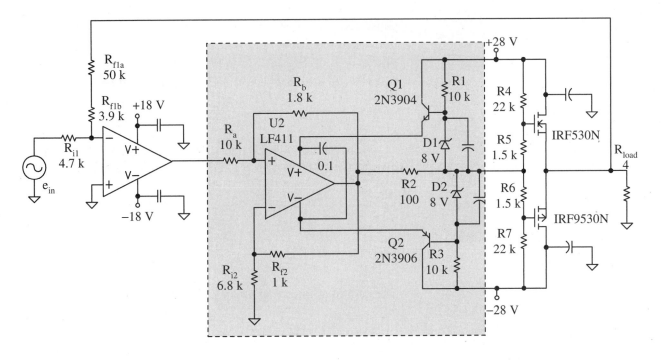

Figure 5-23 Inverting power amplifier

Solution

The power delivered to the load is sinusoidal.

$$P_{\text{load}} = \frac{v_{\text{rms load}}^2}{R_{\text{load}}}$$

Solve this for $v_{\text{rms load}}$

$$v_{\text{rms load}} = \sqrt{P_{\text{load}} \times R_{\text{load}}}$$

$$v_{\text{rms load}} = \sqrt{50\,\text{W} \times 4\,\Omega} = 14.1\,\text{V}_{\text{rms}} = 20.0\,\text{V}_{\text{p}}$$

The op amp can have no more than ±18 V_{dc} supplies. But it needs at least 3 V of head room. So the op amp alone cannot output a 20 V_{p} signal. A suspended supply second stage is needed. It requires about 6 V of head room. For a 20 V_{p} output the suspended supply stage needs about ±28 V_{dc} supplies. The ±28 V_{dc} supplies are adequate. The ±56 V_{dc} supplies would just cause the transistors to have to dissipate more power, running hotter and requiring more heat sinking. Choose the $\pm 28\ V_{dc}$ supplies.

The input impedance of an inverting op amp amplifier is set by R_i.

$$R_i = Z_{in} = 4.7\ k\Omega$$

From Section 4.5 the line level of professional sound equipment is typically +4 dBu = 1.23 V_{rms}. This is the input.

$$A_v = -\frac{v_{load}}{e_{in}}$$

$$A_v = -\frac{14.1\ V_{rms}}{1.23\ V_{rms}} = -11.5$$

The gain of an inverting op amp amplifier is

$$A_v = -\frac{R_f}{R_i}$$

Solving this for R_f,

$$R_f = -A_v R_i$$

$$R_f = -(-11.5) \times 4.7\ k\Omega = 53.9\ k\Omega$$

You can build R_f with a 50 kΩ resistor in series with a 3.9 kΩ resistor.

The component values in the suspended supply second stage are fine. You can replace the high voltage bipolar transistors with the more common 2N3904 and 2N3906. They work well with $\pm 28\ V_{dc}$ supplies.

Resistors R5 and R6 in the push-pull amplifier must be calculated to set the gate voltages at $\pm 2\ V_{dc}$. The current through the divider string is

$$I = \frac{28\ V_{dc} - 2\ V_{dc}}{22\ k\Omega} = 1.18\ mA_{dc}$$

This current must drop no more than 2 V across R5 and R6.

$$R \leq \frac{2\ V_{dc}}{1.18\ mA_{dc}} = 1.69\ k\Omega$$

A smaller resistor drops a smaller voltage, assuring that the transistors are off. Pick

$$R5 = R6 = 1.5 \text{ k}\Omega$$

All of the components have been calculated. It's now time to look at the worst case power dissipation, and select the transistors' heat sinks. In Section 4.2 and Figure 4-11, you saw that the worst case power dissipation happens when

$$v_{\text{p load @ worstcase}} = 0.63 \times E_{\text{supply}}$$

$$v_{\text{p load @ worstcase}} = 0.63 \times 28 \text{ V} = 17.6 \text{ V}_{\text{p}}$$

$$i_{\text{p load @ worstcase}} = \frac{17.6 \text{ V}_{\text{p}}}{4 \Omega} = 4.4 \text{ A}_{\text{p}}$$

This current flows from either supply, through the transistor that is conducting and through the load. For the supplies and the transistors the current is a half-sine of 4.4 A$_{\text{p}}$. The currents from each transistor combine in the load to create a full sine wave for the load voltage and current waveform.

Each supply provides a steady voltage and a half-sine current. From Table 3-2, the power each delivers is

$$P_{\text{supply @ worstcase}} = E_{\text{supply}} \frac{I_{\text{supply p}}}{\pi}$$

$$P_{\text{supply @ worstcase}} = 28 \text{ V}_{\text{dc}} \frac{4.4 \text{ A}_{\text{p}}}{\pi} = 39 \text{ W}$$

The load carries a full sine wave.

$$P_{\text{load @ worstcase}} = \left(I_{\text{rms load @ worstcase}} \right)^2 \times R_{\text{load}}$$

$$P_{\text{load @ worstcase}} = \left(\frac{4.4 \text{ A}_{\text{p}}}{\sqrt{2}} \right)^2 \times 4 \Omega = 39 \text{ W}$$

Half of this power comes through each of the transistors. The transistors must dissipate the power that is provided by the supplies but not delivered to the load.

$$P_{Q@\,worstcase} = P_{supply@\,worstcase} - \frac{P_{load@\,worstcase}}{2}$$

$$P_{Q@\,worstcase} = 39\,W - \frac{39\,W}{2} = 19.5\,W$$

In Section 4.3 you saw the thermal characteristics of the transistor and its heat sink.

$$T_J = T_A + P\left(\Theta_{JC} + \Theta_{CS} + \Theta_{SA}\right)$$

Solve this for the thermal resistance of the heat sink.

$$\Theta_{SA\,max} = \frac{T_{J\,max} - T_A}{P} - \Theta_{JC} - \Theta_{CS}$$

For the IRF530, $\Theta_{JC} < 1.9°C/W$. Properly installing the heat sink *without* a mica wafer insulator sets $\Theta_{CS} = 0.2°C/W$.

$$\Theta_{SA\,max} = \frac{140°C - 40°C}{19.5\,W} - 1.9\frac{°C}{W} - 0.2\frac{°C}{W} = 3.0\frac{°C}{W}$$

So, you must provide a heat sink that fits a TO220 package (the IRF530 and IRF9530 package) with a thermal resistance of less than 3°C/W when passing 19.5 W. This is a very low thermal resistance. It may be difficult to find that heat sink. You may have to consider forced air cooling (see Figure 4-14.)

Practice: What effect would changing the power supply to ±56 V_{dc} have?

Answers: Use the 2N5555 and 2N5551 in the suspended supply second stage, IRF630 and IRF9630 in the push-pull stage. Set R5 = R6 = 820 Ω. $v_{load@\,worst\,case} = 35.3\,V_p$, $P_{load@\,worst\,case} = 156$ W, $P_{supply@\,worst\,case} = 156$ W, $P_{Q@\,worst\,case} = 78$ W. There is *no* heat sink able to keep the transistors from overheating.

Op Amp Selection

Up to this point the op amps have been ignored. At the low frequencies and relatively low gains of the examples, the 741 is an inexpensive, commonly available, through-hole part that works adequately for the first-stage amplifier. But it is this IC's characteristics that set the performance of the entire amplifier. For applications that require higher speed, lower dc offsets, surface mount, or very low cost, you may need to pick a different op amp.

The first-stage op amp's input offset voltage (V_{ios}) is multiplied by the overall gain of the amplifier, and shows up across the load as dc.

$$V_{out\ nonideal} = A_{amp} \times V_{ios}$$

Even though you have been told that *no* current flows into the input of an op amp, in reality there is a small current that *must* flow into the IC in order to bias its input transistors. Without this current, these transistors turn off, and the op amp's output goes into saturation. In the worst case this current flows through R_f, producing a voltage drop.

$$V_{out\ nonideal} = A_{amp} V_{ios} + I_{bias} R_f$$

An output dc voltage across the load of only two hundred millivolts *may* produce enough dc current through the load to disrupt its ac performance. This is particularly true of electromagnetic loads such as loudspeakers and transformers.

If the nonideal dc load current rises to this level, first lower R_f (and R_i to keep the gain correct). If this does not lower the output dc voltage and current enough, then you have to select an op amp with smaller input offset voltage and bias current. Be sure to use the manufacturer's worse case specifications for V_{ios} and I_{bias}, and use the values that account for variation with temperature.

Example 5-9

> For the circuit of Example 5-8, look up the specifications of the 741C op amp and calculate the worst case dc load current. Is this acceptable for a loudspeaker?

Solution

> The worst case specifications for the 741C across the entire temperature range are:

$$V_{ios} = 7.5\,mV_{dc}$$

$$I_{bias} = 800\,nA_{dc}$$

Combining these with the values from Example 5-8 gives

$$V_{out\,nonideal} = 11.5 \times 7.5\,mV_{dc} + 800\,nA_{dc} \times 53.9\,k\Omega = 129\,mV_{dc}$$

This voltage produces

$$I_{dc\,load} = \frac{129\,mV_{dc}}{4\Omega} = 32\,mA_{dc}$$

Although this does not seem like very much current, it may affect the magnetic characteristics of the loudspeaker. At this point you should refer to the manufacturer of the loudspeaker to verify that 32 mA$_{dc}$ does not adversely affect it.

Practice: If the loudspeaker can only tolerate 20 mA$_{dc}$, how much V_{ios} can the op amp have, if it has only I_{bias} = 300 pA$_{dc}$?

Answer: 7 mV$_{dc}$

There are two specifications that relate to the speed of the op amp. The gain bandwidth product is of concern if the input or output is below 1 V$_p$.

$$GBW = A_O \times f_H$$

where A_O = the amplifier's closed loop gain.

f_H = the amplifier's high frequency cutoff, the frequency at which the gain has dropped to 0.707 A_O.

For larger signals you must be concerned about the slew rate. This specification tells how rapidly the output of the op amp can change. *If the op amp's output is a pure sine wave, then the required slew rate is*

$$SR = 2\pi f_{max} V_{p\,opamp\,out}$$

Be careful of the powers of ten. Slew rate is specified in V/μs. But the calculation produces V/s.

A difficulty arises because the output of the op amp *steps* several volts as the sine wave passes through common, in order to turn one transistor off and turn the other on. Look back at Figure 5-20. Typically a 741C op amp's dc characteristics are low enough and its gain bandwidth is high enough to be used as the input stage in an audio power amplifier. But as the frequency increases from the 100 Hz in Figure 5-20 into the voice band, the 741C's slew rate is too slow for the jump needed to prevent crossover. As the frequency increases, you see a flat spot develop as the load waveform crosses common. An op amp with a faster slew rate is needed. The 741C has a slew rate of 0.5 V/μs. Op amps designed for audio applications often have slew rates of 10 V/μs, even though their gain bandwidth is no faster than the 741C's GBW.

The dc characteristics of the op amp in the suspended supply stage are not particularly important. They must not be so bad that they force that stage into saturation. But any dc added to the signal by the suspended supply only adds (and subtracts) slightly to the bias of the MOSFETs. Any shift in the output this causes is sent to the input op amp as *negative* feedback. That op amp then changes its output to compensate.

However, it is critical that the op amp used in the inner, suspended supply stage, be several times faster (GBW and SR) than the input, first stage op amp. Otherwise, the suspended supply op amp's output responds more slowly than its input, producing a phase lag. At some frequency this lag causes the negative feedback to the input op amp to actually be in-phase (positive), rather than negative, (out of phase) feedback. The amplifier oscillates. In the examples, a LF411 is used for the suspended supply stage. This allows the suspended supply stage to output 50 V_p signals at over 200 kHz. A less expensive, high speed op amp certainly would also work.

5.5 Parallel MOSFETs in a Linear Amp

In Example 5-8, delivering 50 W to a 4 Ω load requires a heat sink with a thermal resistance so low that forced air cooling is needed. More power to the load means a larger supply voltage, which forces the transistor to dissipate so much power that you may not be able to cool it enough to make the amplifier work. A single pair of MOSFETs just cannot deliver much more than about 30 W without inappropriately extreme measures to keep them cool.

To provide more power place several *n*-channel MOSFETs in parallel, instead of the single *n*-channel MOSFET. Of course the *p*-channel MOSFETs should be paralleled too. This certainly seems simple enough. Paralleling three transistors on each side now means that the 5 A_p that must be sent to the load is shared by the three transistors, each carrying only 1.7 A_p. Just solder the three drains together, the gates together, and the sources together. Bolt each transistor to a heat sink and away you go. Right?

When you power up this concoction, one of two things happens. If you are lucky, massive oscillations appear, completely overwhelming

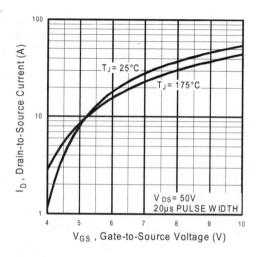

Figure 5-24 Thermal impact on MOSFET performance (*courtesy of International Rectifier*)

the signal that is supposed to be there. In the worse case, the transistors rapidly burn out, one at a time, though it probably happens so fast that you think they committed mass suicide.

Paralleling the transistors certainly *seemed* like a good idea. What did we miss? Look at Figure 5-24. It is a repeat of Figure 5-5, the gate-to-source voltage to drain current characteristic curve.

There are *two* curves on the graph, one at 25°C, the other at 175°C. As the transistor heats up, for the same gate-to-source voltage, the drain current goes *up*. At V_{GS} = 4.5 V_{dc}, 25°C, the transistor passes 3.8 A_{dc}. As it heats up, the current increases, until at 175°C (just before it burns up) it is passing 5.5 A_{dc}. That is a 45% increase in current, just because of a rise in temperature.

It is impractical to electrically and thermally match the transistors you wire in parallel well enough to force them to track each other as temperature rises. Instead, one of the trio will be more sensitive to temperature variations than the other two. As soon as the temperature begins to increase, the most sensitive transistor increases its current much more quickly than the others. Since the op amp is trying to drive a constant voltage (and current) to the load, when the sensitive transistor starts hogging the current, this current is stolen from the other transistors. The hog heats up more ($P = I^2R$), and the others cool. This drives more current through the hog and less to the others. The hog heats up more and steals more current. The other two transistors soon have no current while the most sensitive part is trying to carry all of the current you intended be evenly shared. It burns out. Now the same dance is performed by the remaining two transistors, until one of them burns up, leaving only one, which also quickly burns up.

The solution is a little negative feedback right at the transistors. Look at Figure 5-25.

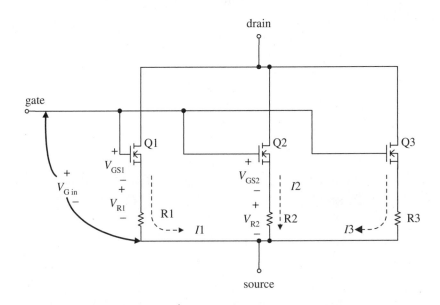

Figure 5-25 Paralleling three MOSFETs in a linear amplifier

The external circuit produces $V_{G\,in}$. In response, each transistor conducts ($I1$, $I2$, and $I3$). Each transistor's current flows through its source resistor, producing a voltage drop (V_{R1}, V_{R2}, and V_{R3}). But it is the transistor's own gate-to-source voltage (V_{GS1}, V_{GS2}, and V_{GS3}) that actually controls that transistor's current. Summing the loop for Q1 gives

$$V_{Gin} = V_{GS1} + I1 \times R1$$

Solve for V_{GS1}.

$$V_{GS1} = V_{Gin} - I1 \times R1$$

Remember, $V_{G\,in}$ is the fixed external voltage that is driving the amplifier, $I1$ is what we are trying to assure is not too large, and V_{GS1} is the voltage that directly sets $I1$ (from Figure 5-24).

Assume that Q1 begins to steal current, raising $I1$ and lowering $I2$ and $I3$. When $I1$ increases, the voltage dropped across R1 goes up. This leaves less of the fixed $V_{G\,in}$ for V_{GS1}. But when V_{GS1} goes down, $I1$ also decreases. The initial problem is that $I1$ had increased. This mismatch has been compensated for.

When Q1 steals current from the other transistors, $I2$ and $I3$ drop. This drop lowers V_{R2} and V_{R3}. With less voltage lost across these resistors, more of $V_{G\,in}$ is left between Q2 and Q3's gate and source. This increase in V_{GS2} and V_{GS3} sends their currents up, fighting the theft that Q1 is attempting. Properly sized, R1, R2, and R3 force poorly matched transistors to share the current.

A reasonable value for R1, R2, and R3 is

$$R1 = R2 = R3 > \frac{1}{g_{fs}}$$

where g_{fs} is the transistor's forward transconductance. This specification was introduced in the AC Operation part of Section 5.2. For the transistors used in the examples

$$g_{fs\ IR530N} > 6.4\,\frac{A}{V} \qquad g_{fs\ IRF9530N} > 3.2\,\frac{A}{V}$$

For all power MOSFETs R1, R2, and R3 fall in the 0.05 Ω to 1.0 Ω range. The larger these resistors, the more closely the transistors are forced to share the current. But remember that the load itself may only be 4 Ω to 8 Ω. So a 1 Ω resistor requires one fourth the power delivered to the load. Three of them dissipate almost as much power as the load. So considerations of efficiency, cost, and size mean you should set R1, R2, and R3 as small as possible.

The actual mismatch that these resistors are to correct is very poorly defined by typical manufacturers' specifications. A reasonable approach is to start with resistors twice the minimum suggested above. Build the amplifier and drive it at its worst case point, $V_{p\,load} = 0.63E_{supply}$ for a sine wave, then cycle the temperature over its full ambient range, and monitor the currents. Repeat this test for a variety of transistor samples. If the amplifier is stable, lower the resistors; if not, increase the resistors and repeat the test.

Example 5-10

Replace the single *n*-channel and single *p*-channel MOSFETs of Example 5-8 each with three transistors in parallel. Calculate:

1. The value and wattage of the required resistors.
2. The heat sink's thermal resistance.

Solution

1. Set the resistors at twice the recommended minimum.

$$R = 2 \times \frac{1}{g_{fs}} = \frac{2}{6.4 \frac{A}{V}} = 0.31\Omega$$

Pick $R = 0.33 \ \Omega$

At maximum load current, each transistor and each source resistor carries a half-sine current.

$$I_{max} = \frac{5\,A_p}{3} = 1.7\,A_p$$

This current, flowing through the source resistors, drops

$$V_{R\,max} = 1.7\,A_p \times 0.33\Omega = 0.56\,V_p$$

This half-wave rectified sine current, and half-wave rectified sine voltage require that the resistor dissipate

$$P_R = \frac{V_p I_p}{4}$$

$$P_R = \frac{0.56\,V_p \times 1.7\,A_p}{4} = 0.24\,W$$

This assumes that the current is evenly shared. To account for variation, use ½ W, 0.33 Ω resistors.

2. In Example 5-8, it was calculated that a single transistor would have to dissipate up to 19.5 W, in the worst case. With three transistors, ideally each would dissipate one third of this power. To account for some mismatch, assume that one of the transistors takes half of the power.

$$P_{Q\,@\,worstcase} = \frac{19.5\,W}{2} = 9.8\,W$$

Each of the transistors now needs a heat sink with a thermal resistance no larger than

$$\Theta_{SA\,max} = \frac{T_{J\,max} - T_A}{P_Q} - \Theta_{JC} - \Theta_{CS}$$

$$\Theta_{SA\,max} = \frac{140°C - 40°C}{9.8\,W} - 1.9\frac{°C}{W} - 0.2\frac{°C}{W} = 8.1\frac{°C}{W}$$

A 8.3°C/W TO220 heat sink is much more practical than the 3°C/W heat sink originally calculated in Example 5-8.

Practice: With a 56 V supply and six paralleled transistors, calculate the maximum load power and the transistor heat sinks.

Answers: $v_{p\,load} \approx 46\,V_p$, $P_{load\,max} = 264\,W$, $\Theta_{SA} = 5.6°C/W$

5.6 Amplifier Protection

In the preceding sections you have seen how to build a linear amplifier capable of delivering hundreds of watts to the load. The traditional way of connecting a load is to turn the amplifier on, set the volume up, then grab a screwdriver and some wire. The owner fiddles with the metal screwdriver, the output terminals, and the loudspeaker until sound comes blasting out. Of course, during this operation the screwdriver and connecting wires short out the amplifier repeatedly. Once the amplifier has survived this operation, the wires are run under a rug, and books. There the wires are subjected to repeated rubbing, chewing from pests, and overheating. An eventual short circuit somewhere along these wires is not unusual.

But under any of these conditions the amplifier must *not* fail, shock the user, spark, overheat, or cause a fire. That, however, is precisely what would happen with the amplifier you have seen so far. So protection must be added. It should take two forms; current limiting to protect the amplifier and the user from excessive temporary shorts, and thermal shutdown that takes over if the fault continues. Both forms of protection are built into the OPA548 from Chapter 4. Look back over that section.

Current Limiting

Current limiting is illustrated in Figure 5-26. This is a copy of Figure 4-15. When the current is low, the output current is dictated by the output voltage and the load resistance, as set by Ohm's law. Once I_{LIM} is reached, the output current is fixed. Since Ohm's law still applies, once current limiting is entered, further reductions of the load resistance causes the load *voltage* to drop proportionally.

Select

$$I_{LIM} \approx 120\% \times I_{p \text{ max load}}$$

This assures enough current to the load at maximum power, but limits the current under a fault.

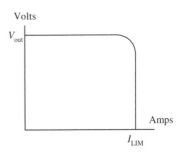

Figure 5-26 Current limiting characteristic curve.

Example 5-11

Select an appropriate current limit level for an amplifier that delivers up to 200 W into an 8 Ω load.

Solution

Assuming a sinusoidal wave form,

$$P_{load} = \left(I_{rms \text{ load}}\right)^2 \times R_{load}$$

Solve for the load current at the maximum load power.

$$I_{rms \text{ load}} = \sqrt{\frac{P_{load \text{ max}}}{R_{load}}}$$

$$I_{load} = \sqrt{\frac{200\,\text{W}}{8\,\Omega}} = 5.0\,\text{A}_{rms} = 7.1\,\text{A}_{p}$$

Set I_{LIM} to be

$$I_{LIM} = 1.2 \times 7.1\,\text{A}_{p} = 8.5\text{A}$$

Practice: Select an appropriate current limit for $P_{load \text{ max}} = 50$ W, 4 Ω. (Example 5-8)

Answer: Set $I_{LIM} = 6.0$ A

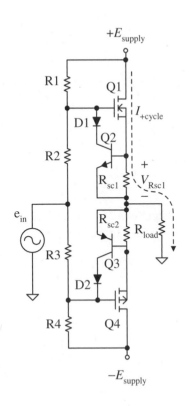

Figure 5-27 Current limiting added to a push-pull amplifier

Short circuit resistor value

Current limiting can be installed in the push-pull amplifier with the addition of three components on each side. These are shown in Figure 5-27. On the positive side, D1, Q2, and R_{sc1} have been added between the gate of Q1 and the load. A similar arrangement is installed on the negative side around the *p*-channel MOSFET, Q4. Carefully notice that Q2 is an *npn* transistor while Q3 is *pnp*. Also, the direction of D1 and D2 must be observed. Resistors R_{sc1} and R_{sc2} control the current limit level.

At low current on the positive cycle, the voltage developed by the load current flowing through R_{sc1} is small. Transistor Q2 and diode D1 are off. They have no effect. The amplifier operates as it did without current limiting.

When the current becomes large enough to cause V_{Rsc1} to be greater than about 0.6 V, Q2 turns on. During that part of the cycle, input current flows from the generator, through R2, D1, and Q2. This drops more voltage across R2, lowering the signal at the gate. Transistor Q2 is beginning to short out the MOSFET's gate signal.

When the gate voltage drops, the MOSFET's drain (and source) current drops. This lowers the voltage across R_{sc1}, causing Q2 and D1 to conduct less. With less of the input signal current flowing through Q2, less input signal voltage is dropped across R2, and the gate voltage goes up.

But an increase in gate voltage sends the current to the load back up. This increases the voltage across R_{sc1}, turning Q2 and D1 on harder. This draws a little more current from the generator, dropping more voltage across R2, lowering the gate signal, dropping the load current.

The net result of this iterative process is that the load current is controlled at a constant level, independent of the generator voltage, the load resistance, or any op amps between the generator and R2.

Pick

$$R_{sc} = \frac{0.6\,\text{V}}{I_{LIM}}$$

The voltages around the transistors and the currents through them are small enough that any commonly available BJT should work. Typically the 2N3904 (*npn*) and 2N3906 (*pnp*) may be used.

The diodes' purpose may not be obvious. When the output voltage is driven to the negative extreme, a large negative voltage is placed on the base of the *npn* transistor, Q2. The voltage from the generator (or suspended supply second stage) is even more negative, to overcome the losses in the voltage divider and the MOSFET. Depending on the resistors in the divider and the size of the dc supplies, it is possible that the gate of Q1 may be driven more negative than the output. This has no effect on Q1. But, without D1, driving Q2's collector more negative than its base would forward bias Q2's collector-base junction. On the negative half-cycle the intention is that Q1 and Q2 be off. So, D1 is placed in the circuit. When the gate of Q1 is driven more negative than the output, D1 is reverse biased, keeping Q2 off. This diode should have a reverse voltage breakdown rating greater than the load peak voltage, but turn on in a fraction of the highest frequency's half cycle.

Power dissipation calculations for Q1, Q4, and R_{sc1} and R_{sc2} during current limiting are a little different from those you have done so far. There are two major points. When Q1 and Q4 go into current limiting, the load is no longer dissipating any power. It has been shorted. *All* of the power provided by the supplies must be dissipated by the MOSFETs. Secondly, the current wave shape is no longer a half-sine. The current is being forced to a *constant* value. So the current is at a steady value of I_{LIM} for half of the time.

Example 5-12

For Example 5-8, add current limiting with $I_{LIM} = 6.0$ A. Calculate

1. R_{sc}
2. P_{Rsc}
3. P_Q with $E_{supply} = \pm28$ V$_{dc}$
4. T_J with $T_A = 40°C$, $\Theta_{SA} = 3°C/W$

Solution

1.
$$R_{sc} = \frac{0.6V}{6A} = 0.1\Omega$$

Be careful with resistances this small. You must also take into account the trace resistance in series with the resistor.

2. The current through the resistor is a 50% duty cycle square wave with a peak amplitude of 6 A$_p$ and 0.6 V$_p$. Look back at Table 3-2.

$$P_{Rsc} = DV_p I_p$$

$$P_{Rsc} = 0.5 \times 0.6V_p \times 6.0A_p = 1.8W$$

This is considerably more than the ¼ W that you might have used out of habit. Pick a 0.1 Ω, 2 W resistor.

3. The same calculation applies for the MOSFETs, except they each have the entire supply voltage across them.

$$P_{MOSFET} = DV_p I_p$$

$$P_{MOSFET} = 0.5 \times 28\,V_p \times 6\,A_p = 84\,W$$

4. So, what happens to the MOSFETs when current limiting takes over to protect the load?

$$T_J = T_A + P\left(\Theta_{JC} + \Theta_{CS} + \Theta_{SA}\right)$$

$$T_J = 40°C + 84\,W\left(1.9\frac{°C}{W} + 0.2\frac{°C}{W} + 3.0\frac{°C}{W}\right) = 468°C$$

In response to a shorted load, the circuit limits the current, saving the wiring and the load, but causes the power transistors to burn up!

Practice: Work steps 4 and 3 backwards to determine the largest short circuit current that would not cause the transistors to burn up. At that I_{LIM}, what is the maximum sinusoidal load power?

Answers: $P_{Q\,@140°C} = 20$ W, $I_{LIM\,@\,20\,W} = 1.5$ A$_p$, $P_{sine\,@\,1.5Ap} = 4.5$ W

Thermal Shutdown

The MOSFETs fail because they become too hot. So the most direct way to protect them is to measure their temperature, and turn the power supply off if the transistors get too hot. In Chapter 4 you saw that the OPA548 power op amp has thermal shutdown built into the IC.

You cannot directly measure the junction temperature of discrete transistors. But the temperature of the heat sink, *immediately adjacent* to the transistor, can be sensed. This gives a good indication of the transistor's junction temperature. In the last two chapters you have often seen that the junction temperature is related to the *ambient* temperature.

$$T_J = T_A + P\left(\Theta_{JC} + \Theta_{CS} + \Theta_{SA}\right)$$

In terms of the heat sink's temperature, T_S, this becomes

$$T_J = T_S + P\left(\Theta_{JC} + \Theta_{CS}\right)$$

Example 5-13

Calculate the maximum safe temperature of the heat sink for the amplifier in Example 5-8, holding the junction temperature (T_J) at 130°C when the transistors are dissipating their worst case power of 19.5 W.

Solution

$$T_J = T_S + P(\Theta_{JC} + \Theta_{CS})$$

or

$$T_S = T_J - P(\Theta_{JC} + \Theta_{CS})$$

$$T_S = 130°C - 19.5W\left(1.9\frac{°C}{W} + 0.2\frac{°C}{W}\right) = 89°C$$

Practice: For an ambient temperature of 40°C and a heat sink thermal resistance of 3°C/W, calculate the heat sink temperature and the junction temperature if the transistors are forced to dissipate 21 W.

Answers: $T_J = 147°C$ (on the verge of destruction), $T_S = 103°C$

When dissipating the rated worse case power, the transistors' heat sinks should be no more than 93°C. As soon as the transistors are required to dissipate a little more power, as the result of a fault, the junction temperature rises to the point of semiconductor meltdown, and the heat sink temperature goes over 103°C.

The simplest thermal protection is an IC temperature switch, bonded directly to the heat sink, and rated at 95°C. When the sink reaches that temperature, the IC changes logic levels on its output. You can then use this logic level change to turn the power supply off.

The MAX6501 is such a family of temperature switches. Look at Figure 5-28. The sense temperature is fixed. You just buy the temperature rating that you need. Standard values are 45°C, 55°C, 65°C, 75°C, 85°C, 95°C, 105°C, and 115°C. Below its rated temperature, the IC outputs an open. Connect a 3.3 kΩ pull-up resistor between the switch's output and +5 V to produce a legal TTL or CMOS logic high at the IC output pin. When the temperature exceeds its rating, the output pin is shorted to common. This open drain arrangement allows you to tie the output pin of several switches to the same pull-up resistor. If any of the switches sense an over temperature, the line is pulled low to warn you to remove power.

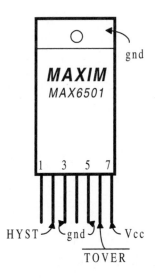

Figure 5-28 MAX6501 TO220 pin diagram

There are two package styles. The surface mount package is designed to monitor the ambient temperature of a printed circuit board. The TO220 package, shown in Figure 5-28, can be connected directly to the same heat sink that contains the MOSFETs. In fact, you could even connect it on the back side of the sink, directly behind the transistor, using the same mounting hardware. However, be sure to use a mica wafer and nylon hardware since the case of the MOSFET is connected to the supply voltage, and the case of the MAX6501 is tied to common.

Always **provide the proper current limiting and thermal shutdown.**

The MAX6501 also has a hysteresis pin. Connecting this pin to +V provides a 10°C hysteresis. That means that once the IC has output a low in response to a 95°C temperature, the temperature must fall to 85°C before the output returns to an open (that is pulled to +5 V by the pull-up resistor). This allows the system to cool more than just a little before changing the signal that may automatically turn the power supplies back on.

5.7 Driving a Reactive Load

All of your calculations so far, in every section of this book, have assumed that the load is purely resistive. In reality, *every* load has some reactive element. It may be as simple as the inductance contained in the wires that tie the resistor to the amplifier. Or it may be as complicated as a loudspeaker with a crossover network. Loudspeakers alone change their characteristics as frequency varies. At some frequencies, the loudspeaker looks like an 8 Ω resistor; at others it may be purely inductive; and at others purely capacitive; while in between the loudspeaker is a complex reactance. Add the crossover network and the load is definitely complex. Shifting the phase relationship of voltage to current drastically alters the power delivered to the load and the power that the MOSFETs must dissipate. Of course, that alters all of the heat sink and protection calculations.

From your ac circuits course you learned that a complex load shifts the phase relationship of the current through the load with respect to the voltage across it:

$$P_{\text{complex load}} = v_{\text{rms}} i_{\text{rms}} \cos\theta$$

where θ = angle of the current through the load with respect to the voltage across the load.

But for a resistor, or the *resistive part* of the load, current and voltage are in phase. So cos θ = 1. That's simple enough.

$$P_{\text{resistive load}} = v_{\text{rms}} i_{\text{rms}} \cos\theta = v_{\text{rms}} i_{\text{rms}}$$

For the capacitive or inductive *part* of the load, the current leads or lags the voltage across the capacitor or inductor by 90°. The cos ±90° = 0. So the reactive part of the load dissipates no power. Only the resistive part dissipates power.

With those facts, power calculations around a push-pull power amplifier driving a reactive load are done just as you have done it for a resistive load. You begin by figuring out how much voltage is across the resistive part of the load. After that, you use the same steps and equations as you have done for the OPA548 power op amp or for the discrete MOSFET push-pull amplifier.

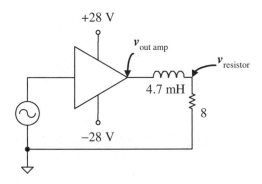

+28 V

$v_{\text{out amp}}$

v_{resistor}

4.7 mH

8

−28 V

Figure 5-29 Amplifier driving a woofer crossover and loudspeaker

Example 5-14

Calculate the three powers for the circuit in Figure 5-29, given that the output voltage is 15 V_{rms} at 300 Hz.

Solution

First, find the reactance of the inductor.

$$X_L = 2\pi f L$$

$$X_L = 2\pi \times 300\,\text{Hz} \times 4.7\,\text{mH} = 8.9\,\Omega$$

Phasor calculations are needed for the next few steps.

$$\overline{Z_{\text{load}}} = 8\,\Omega + j8.9\,\Omega$$

$$\overline{V_{\text{amp out}}} = 15\,\text{V}_{\text{rms}}\angle 0°$$

Calculate the voltage across the resistor.

$$\overline{V_R} = \left(15\,\text{V}_{\text{rms}}\angle 0°\right)\frac{8\,\Omega\angle 0°}{8\,\Omega + j8.9\,\Omega} = 10.0\,\text{V}_{\text{rms}}\angle -48°$$

The resistive part of the load dissipates

$$P_R = \frac{\left(V_R\right)^2}{R}$$

$$P_R = \frac{\left(10\,\text{V}_{\text{rms}}\right)^2}{8\,\Omega} = 12.5\,\text{W}$$

Since the inductive part of the load dissipates no power, the entire load dissipates

$$P_{\text{load}} = P_R = 12.5\,\text{W}$$

The current through the load is the current through the resistance, which is

$$I = \frac{10.0\,\text{V}_{\text{rms}}}{8\,\Omega} = 1.25\,\text{A}_{\text{rms}} = 1.77\,\text{A}_{\text{p}}$$

This current flows from the 28 V_{dc} supply, a positive half-cycle from the positive supply and a negative half-cycle into the negative supply. *Each* of these dc supplies provides

$$P_{\text{supply}} = V_{\text{dc}}\frac{I_{\text{p}}}{\pi}$$

$$P_{\text{supply}} = 28\,\text{V}_{\text{dc}}\frac{1.77\,\text{A}_{\text{p}}}{\pi} = 15.8\,\text{W}$$

Each transistor must dissipate the power provided by its supply that is not passed on to the load.

$$P_Q = P_{\text{supply}} - \frac{P_{\text{load}}}{2}$$

$$P_Q = 15.8\,\text{W} - \frac{12.5\,\text{W}}{2} = 9.6\,\text{W}$$

So, only the first step is different from the power calculations that you have previously done. Use phasor math to calculate the voltage across the resistive part of the load.

Practice: Calculate P_{load}, P_{supply}, and P_Q for a push-pull amplifier powered from ± 56 V_{dc}, driving a 10 μF capacitor in series with a 8 Ω resistance. The output from the amplifier is 35 V_{rms} at 3 kHz.

Answers: $P_{\text{load}} = 106.3$ W, $P_{\text{supply}} = 92.7$ W, $P_Q = 39.5$ W

Summary

Enhancement mode MOSFETs are normally *off*. There are two types, *n*-channel and *p*-channel. The *n*-channel devices are turned on with a positive gate-to-source voltage. *P*-channel MOSFETs bias on with a negative gate-to-source voltage. Once this gate-to-source voltage is above the threshold, drain current increases as the square of gate-to-source voltage. So very little change in input voltage results in a large output current.

The common drain class A amplifier takes advantage of this relatively large transconductance ($\Delta I_D / \Delta V_{GS}$). The transistor is biased well on so that the input voltage can increase and decrease current flow, but cannot turn the transistor off. This produces a low distortion output voltage slightly smaller than the input voltage, but with enough current to drive heavy loads. But the bias current flows through the load requiring it to dissipate a large dc power even when there is no signal. The transistor also must waste a lot of power just idling.

There are two transistors in the class B push-pull amplifier. Each is biased off, but on the edge of conduction. The positive half of the input turns the *n*-channel MOSFET on. The negative half is processed by the *p*-channel. To accommodate the 2 V variation in threshold voltage from part to part, biasing must be set at the lowest specified level. So there must be an increase of several volts at the input before the output begins to conduct. The result is crossover distortion. The signal voltage also is attenuated by the bias network and by the MOSFET itself.

Adding an op amp *around* the push-pull stage solves both of these problems. The negative feedback for this input op amp stage is taken directly from the load. The op amp's output steps at the crossover and is larger at the peak than the load voltage. This provides a clean signal at the load A suspended supply stage is needed for voltages above 15 V.

Power calculations are done similarly to the power op amp. But each transistor provides only half of the load power. For a sinusoid,

$$P_{\text{load}} = \frac{\left(V_{\text{rms load}}\right)^2}{R_{\text{load}}} = \left(I_{\text{rms load}}\right)^2 R_{\text{load}}$$

$$P_{\text{each supply}} = E_{\text{dc supply}} \frac{I_{\text{p supply}}}{\pi}$$

$$P_{Q} = P_{\text{supply}} - \frac{P_{\text{load}}}{2}$$

These calculations should be done at maximum load power to be sure that the supply and the load are adequate, and at the worst case for the transistor ($V_{\text{p load}} = 0.63\ E_{\text{supply}}$ for a sine wave). Heat sinking calculations are performed at transistor worst case.

MOSFETs can be paralleled to provide load power over 50 W. But to assure that each transistor handles its fair share of current, place a small resistor between the transistor's source and their common connection to the load. These resistors should be a little larger than $1/g_{\text{fs}}$.

Current limiting is implemented by sensing the current leaving the amplifier with a small resistor. When that current develops enough voltage to turn on a bipolar transistor, the gate voltage is reduced, holding the current at a constant level. Under a prolonged short circuit, the power transistors overheat. A temperature switch must be added to the heat sink to turn off the power supply if the sink gets too hot.

Reactive loads force the load current out of phase with the load voltage. The power calculations still work. But you must first use phasor algebra to determine the voltage across the resistive part of the load.

If the objective is to provide more or less power to a load, and you can tolerate distortion levels of a few percent, then the transistors can be operated as switches. Turned on the switch provides full power to the load, and dissipates very little itself. The circuit may be 98% efficient.

During the next chapter you will look at the ideal and nonideal characteristics of MOSFETs used as switches. Paralleling MOSFET switches is easy. The transistors can be used to apply and remove power (high side) or to connect to circuit common (low side). Special gate drive circuits are needed to realize the advantages of high speed power switching

Problems

Enhancement Mode MOSFETs

5-1   Locate data sheets for the IRF530, IRF630, IRF9530, and IRF9630. Construct a table with a column for each of these transistors and a row for each of the following specifications:

V_{DSS}, $I_{D\,max}$, $V_{GS\,max}$, $T_{J\,max}$, $\Theta_{JC\,max}$, V_{th}, g_{fs}

5-2   **a.** For the IRF530 data determine I_D for V_{GS} = 4.5 V and T_J = 25°C.
   b. For problem 5-2a, what threshold voltage is assumed? How do you know?
   c. If V_{th} = 2.5 V, find I_D for V_{GS} = 4.5 V and T_J = 25°C.
   d. Compare your answers to problem 5-2a and problem 5-2c. Discuss the effect of this variation.

5-3   **a.** For the IRF9530 data determine I_D for V_{GS} = −4.5 V and T_J = 25°C.
   b. Compare your answers to problem 5-2a and problem 5-3a. Do the two transistors appear to be complements?

Class A Operation

5-4   Design a class A power amplifier using an IRF530. Set E_{supply} = 18 V_{dc}, R_{load} = 2 Ω, $e_{in\,max}$ = 5 V_p

5-5   For your design of Problem 5-4,
   a. assuming no input signal, calculate P_{load}, P_{supply}, and P_Q.
   b. with e_{in} = 5 V_p, calculate $v_{G\,rms}$, $v_{rms\,load}$, and $P_{load\,from\,ac\,signal}$.
   c. discuss the efficiency of this amplifier.

Class B Push-Pull Amplifier

5-6   Given that R1 = R4 = 33 kΩ, with ±28 V_{dc} power supplies, calculate the value of R2 and R3 to properly bias a class B amplifier using an IFR530 and IRF9530.

5-7   For the circuit designed in Problem 5-6, with e_{in}= 0 V_{rms}, calculate:

V_{G1}, I_{Q1}, V_{G2}, I_{Q2}, V_{load}, P_{supply}, P_{Q1}, and P_{Q2}

5-8   For the circuit designed in Problem 5-6, with R_{load} = 4 Ω, and e_{in} = 10 V_{rms}, calculate:

$v_{G\,rms}$, $v_{rms\,load}$, P_{load}, P_{supply}, and P_Q

Class B Amplifier with Op Amp Driver

5-9   Design a class B amplifier with a noninverting gain of 48, capable of delivering 8 W into an 8 Ω resistive load. Determine the following:

> supply voltage (±18 V, ±28 V, or ±56 V)
> bias components
> R_f and R_i
> $P_{Q @ worse case}$ (assume a sine wave input)
> heat sink thermal resistance ($T_A = 50°C$)

5-10   Design a class B amplifier with an inverting gain of −100, capable of delivering 60 W into an 8 Ω resistive load. Is a suspended supply second stage necessary? Explain.

5-11   Calculate the gain bandwidth and the slew rate needed for the op amp in Problem 5-9 assuming that the amplifier is to be used across the entire audio range.

5-12   Calculate the gain bandwidth and the slew rate needed for the op amp in Problem 5-10 assuming that the amplifier is to be used up to 3 kHz (the voice band).

Parallel MOSFETs in Linear Applications

5-13   A class B three-stage amplifier has a ±72 V_{dc} supply, an 8 Ω load and a sinusoidal input. If each transistor has a 10°C/W heat sink, and the ambient temperature is 60°C, determine the:

a.   number of *n*-channel and *p*-channel transistors.
b.   power available to the load.
c.   resistance and wattage of the source resistors.

5-14   Four IRF530 transistors are paralleled as the push element in a dc amplifier that outputs 0 V_{dc} to +12 V_{dc} at 5 A_{dc} from an 18 V_{dc} supply. Assuming an ambient temperature of 40°C, calculate:

a.   the thermal resistance of the transistors' heat sinks.
b.   the resistance and wattage of the source resistors.

Amplifier Protection

5-15   **a.**   Add the components necessary to the amplifier in Problem 5-14 to limit the load current to 5 A_{dc}.
  b.   Calculate the wattage of the current limit resistor.

5-16 **a.** If the load for Problem 5-15 shorts, calculate the power that each transistor must dissipate.

    b. Using the heat sink selected in Problem 5-14, calculate the junction temperature of each transistor if the load shorts.

5-17 An IRF9530 is dissipating 10 W, with a case temperature of 105°C. Calculate the junction temperature.

5-18 An IRF9530 has a case temperature of 85°C. Determine:

    a. a *safe* junction temperature.

    b. the power the transistor can dissipate at these temperatures.

Reactive Loads

5-19 A class B three-stage amplifier has a ±72 V_{dc} supply, and a complex load made from a 2.2 mH inductor in series with a 12 Ω resistance. For a 1 kHz sine wave, maximum output amplitude, calculate P_{load}, $P_{each\ supply}$, and $P_{each\ transistor}$.

5-20 A class B three-stage amplifier has a ±56 V_{dc} supply, and a complex load made from a 68 μF capacitor in series with a 4 Ω resistance. For a 500 Hz sine wave, maximum output amplitude, calculate P_{load}, $P_{each\ supply}$, and $P_{each\ transistor}$.

Class B Amplifier Lab Exercise

A. *N*-Channel Enhancement Mode MOSFET Characteristics

    1. Build the circuit in Figure 5-30.

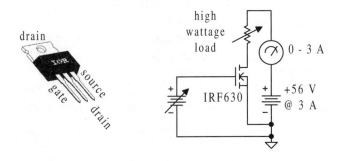

Figure 5-30 V_{GS} versus I_D test circuit

2. Assure that :
    a. the TO220 heat sink is properly installed.
    b. the load rheostat is set to at least 110 Ω.
    c. the +56 V supply is turned *off*.
    d. the transistor's source returns to the +56 V_{dc} supply's single point common in its own lead.
    e. the low voltage adjustable supply's common returns to the +56 V_{dc} supply's single point common in its own lead.
    f. the low voltage adjustable supply is set to 0 V_{dc}.

3. While watching the ammeter in the +56 V_{dc} line, turn on the +56 V_{dc} supply. If its current exceeds 100 mA_{dc}, turn the supply *off* and determine the error.

4. Measure and record the drain current.

5. Lower the load rheostat to the value indicated below, then increase V_{GS} until the drain current equals the indicated current. Record V_{GS}. Repeat the procedure for the five points.

110 Ω, 0.1 A_{dc}	70 Ω, 0.5 A_{dc}	40 Ω, 1.0 A_{dc}
25 Ω, 1.5 A_{dc}	20 Ω, 2.0 A_{dc}	

6. Complete a plot of V_{GS} (*x* axis) versus I_D (*y* axis).

B. Class B Operation
1. Build the circuit in Figure 5-31. Neatness counts. Carefully follow the steps from Chapter 2.

2. Assure that
    a. the ±56 V_{dc} power supply is turned *off*.
    b. R_{load} is set to its maximum value.
    c. the load's return is run *directly* back to the power supply's common (single common point), *not* back to the amplifier's or generator's common.
    d. the amplifier's and the generator's common returns to the power supply's common in its own, separate wire.

3. Set the input signal generator to provide an input signal of:

    $$0\ V_{dc}, \quad 0\ V_{rms}, \quad 100\ Hz$$

4. Connect the generator to the circuit. Do *not* leave the node between R2 and R3 open.

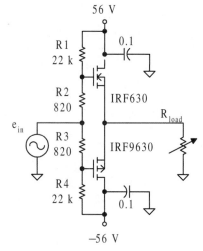

Figure 5-31 Class B push-pull amplifier

5. Monitor the load voltage with the oscilloscope and digital multimeter. Measure and record the load dc and rms voltages and the current from the +56 V_{dc} supply. If any of these values are more than a few tenths of a volt or amp, stop and determine the problem.

6. Set the input signal generator to provide an input signal of

$$0 \ V_{dc}, \quad 20 \ V_{pp}, \quad 100 \ Hz$$

7. Monitor the load voltage with the oscilloscope and digital multimeter. Measure and record the load dc and rms voltages and the current from the +56 V_{dc} supply.

8. Compare the results with theory. If there are any notable errors, stop and determine the mistake before continuing.

9. Lower the load resistance to 12 Ω.

10. Monitor the load voltage with the oscilloscope and digital multimeter. Measure and record the load dc and rms voltages and the current from the +56 V_{dc} supply.

11. Compare the results with theory. If there are any notable errors, stop and determine the mistake before continuing.

12. When the results are numerically correct, record and explain the wave shape.

13. Set the signal generator to 0 V_{dc}, 0 V_{rms}. Turn the ±56 V_{dc} supply off.

C. Push-Pull Amplifier with Op Amp Driver
1. Build the circuit in Figure 5-32. Neatness counts. Carefully follow the steps from Chapter 2.

2. Set R_{load} to its maximum resistance.

3. Verify that the signal generator is set to 0 V_{dc} and 0 V_{rms}.

4. Turn on the ±18 V_{dc} and ±56 V_{dc} power supplies.

5. Verify that the load voltage (dc and ac) and the current from the +56 V_{dc} supply are all negligible. Record their actual values.

6. Set the input signal generator to 0 V_{dc}, 100 mV_{rms}, 100 Hz.

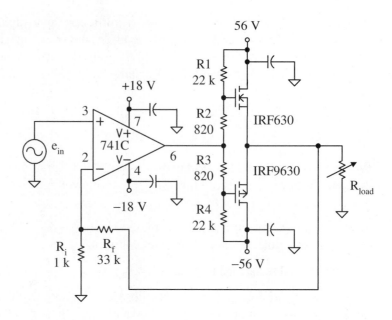

Figure 5-32 Push-pull power amplifier with an op amp driver

7. Monitor the load voltage with the oscilloscope and digital mul-
    timeter. Measure and record the load dc and rms voltages and
    the current from the +56 V_{dc} supply.

8. Compare the results with theory. If there are any notable errors,
    stop and determine the mistake before continuing.

9. Gradually lower the load resistance while carefully monitoring
    the supply current. Be sure that the supply current does not ex-
    ceed 0.3 A_{dc}. Lower the load resistance to 12 Ω.

10. Measure and record the load dc and rms voltages and the current
    from the +56 V_{dc} supply.

11. Compare the results with theory. If there are any notable errors,
    stop and determine the mistake before continuing.

12. Record the waveform at the load and at the output of the op amp.
    Explain each.

13. Increase the input amplitude until the load voltage clips. Record this peak load voltage. Explain the reason for the limiting.

14. Set the signal generator to 0 V_{dc}, 0 V_{rms}. Turn the ±56 V_{dc} supply off.

D. Three-Stage Push-Pull Amplifier
    1. Build the circuit in Figure 5-33. Neatness counts. Carefully follow the steps from Chapter 2.

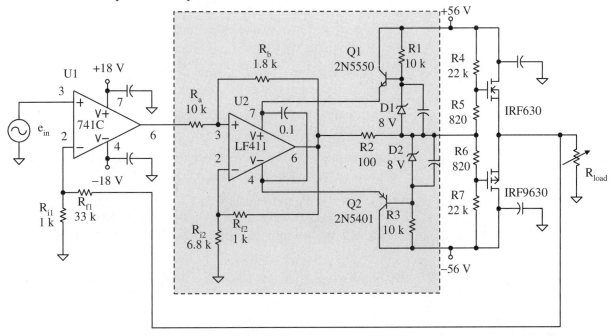

Figure 5-33 Three-stage push-pull amplifier

2. Set R_{load} to its maximum resistance.

3. Verify that the signal generator is set to 0 V_{dc} and 0 V_{rms}.

4. Turn on the ±18 V_{dc} and ±56 V_{dc} power supplies.

5. Verify that the load voltage dc and ac and the current from the +56 V_{dc} supply are all negligible. Record their actual values.

6. Set the input signal generator to 0 V_{dc}, 100 mV_{rms}, 100 Hz.

7. Monitor the load voltage with the oscilloscope and digital multimeter. Measure and record the load dc and rms voltages and the current from the +56 V_{dc} supply.

8. Compare the results with theory. If there are any notable errors, stop and determine the mistake before continuing.

9. Gradually lower the load resistance while carefully monitoring the supply current. Be sure that the supply current does not exceed 0.5 A_{dc}. Lower the load resistance to 24 Ω.

10. Measure and record the load dc and rms voltages and the current from the +56 V_{dc} supply.

11. Compare the results with theory. If there are any notable errors, stop and determine the mistake before continuing.

12. Increase the input amplitude until the load voltage clips. Record this peak load voltage.

13. Lower the input signal amplitude until the output is as large as possible without clipping either peak.

14. Increase the input frequency until the output amplitude drops or until the output wave shape begins to distort. Explain the cause of the drop in amplitude or the distortion caused by an increase in frequency.

15. Set the signal generator to 0 V_{dc}, 0 V_{rms}. Turn the ±56 V_{dc} supply off.

6

Power Switches

Introduction

Linear amplifiers can deliver hundreds of watts to the load, at low distortion. But often more than *half* of the power provided by the supply must be dissipated by the active device. This means that the linear amplifier spends much of its existence converting dc voltage into heat. It only happens to drive the load as a by-product.

If you can tolerate errors or distortion of 0.5%, then the *linear* amplifier can be replaced with a switch. When the switch is *off*, it dissipates *no* power. When it is *on*, its voltage drop is very small, so it wastes very little power. Delivering power to a load through a switch means that 90% or more of the power actually makes it to the load.

To realize this efficiency and this accuracy, the switch must be operated *very* quickly and very precisely. First the switching characteristics of diodes and transistors must be carefully examined. Do you want to switch circuit common, allowing the load to float (low side switch), or do you want to switch the power *on* and *off* (high side)? Driving the main power transistors at high speed and current requires a special switch driver interface. Properly combining all of these circuits allows you to control the level and polarity of the power delivered to a load, driving current in either direction through a load from four switches and a single dc supply.

Objectives

By the end of this chapter, you will be able to:

- Describe the performance of a *pn* diode, Schottky diode, bipolar junction transistor, and MOSFET as a switch, defining key parameters.

- Properly build a power switch circuit to account for the nonideal high frequency parasitics.

- Design and analyze a low side and a high side switch using paralleled MOSFETs. Determine all voltages, currents, power, temperatures, and heat sink requirements.

- Define switch driver requirements and select an appropriate circuit.

- Design an H bridge circuit from discrete transistors and with an IC.

6.1 Switching Characteristics

Diodes

If you are willing to wait, a *pn* silicon diode, such as the 1N4004 rectifier or the 1N914 small signal diode, can be turned on or off by simply applying or removing the 0.6 V bias needed to overcome its barrier potential. But if the load voltage must change quickly, it is possible that the diode simply cannot keep up.

Figure 6-1 is the simulation of a 1N4002 power rectifier passing the positive part of a rectangular input that swings from +10 V_p to −10 V_p. The rectifier is designed for line driven power supplies. It usually only sees 60 Hz and 120 Hz sine waves. At 100 Hz, (not shown) the output is the proper +10 V_p signal, with the negative half blocked.

But in Figure 6-1b and c the input's frequency has been raised to 100 kHz. There are two problems illustrated by these waveforms. When the diode is turned on, only 9 V is passed to the load. Ten percent of the

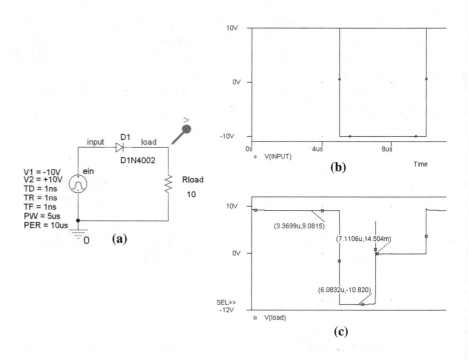

Figure 6-1 Rectification with a 1N4004

input is being dropped by the diode. Depending on the diode and the current level, that loss may be as much as 2 V. In a line powered application, 2 V out of 160 V_p is insignificant, but in a logic level circuit, that 2 V drop is 40% of the 5 V supply. The load voltage may be seriously reduced by the diode. Also, the voltage dropped across the diode means that the diode must dissipate quite a bit of power. You can no longer assume that the diode is an ideal switch. It attenuates the signal and dissipates power that you must worry about.

$V_{on} \approx 1.5$ V @ I = several amps

At 100 kHz, shown in Figure 6-1c, there is an even bigger problem. When the input signal swings negative, the diode *should* turn off, blocking the current from the load. That is what happened at 100 Hz (Figure 6-1b). But, at 100 kHz, the negative step passes to the load for awhile. Load current reverses direction but continues to flow. The diode switch is certainly *not* passing current in only one direction.

When the diode is forward biased, massive numbers of electrons and holes are driven into the depletion region around the *pn* junction, to create several amps of load current. The depletion region disappears. Forward conduction occurs. When the input signal jumps negative, the electrons and holes *begin* to reverse direction in response to the reversed voltage. *Eventually* the current carriers are pulled back far enough for the junction to be depleted of charge. Only then is the diode off. The larger the forward current, the more electrons and holes there are, and the longer it takes to remove them from the junction.

This time is the **reverse recovery time** (t_{rr}). For the simulation of the 1N4002 in Figure 6-1 it is 2.1 microseconds. In practice it may be much longer. For 60 Hz circuits this time is such a small part of the signal period, that little or no effect is observed. But at 100 kHz, a significant part of the input signal is passed on to the load, before the diode can finally go off. During that time, the rectifier is *not* rectifying. Select a diode whose t_{rr} is much smaller than the signal's off pulse width.

Select a diode whose
t_{rr} << $t_{signal\ off}$

The **Schottky barrier diode** is built *without* a traditional *pn* junction. Rectification is accomplished by the way the metal contact is attached to the *n*-type semiconductor material. There is no *p*-type material. By eliminating the *pn* junction, both of the problems shown in Figure 6-1 are significantly reduced. The 0.6 V barrier voltage drop is gone. So the voltage across the diode is much smaller, passing more of the signal to the load, and requiring the diode to dissipate less power. Without having to fill holes to turn the diode off, the Schottky diode's reverse recovery time is much lower. Switch high speed power with a Schottky diode rather than with the traditional *pn* power diode.

The Schottky diode has much
lower V_{on} and t_{rr}.

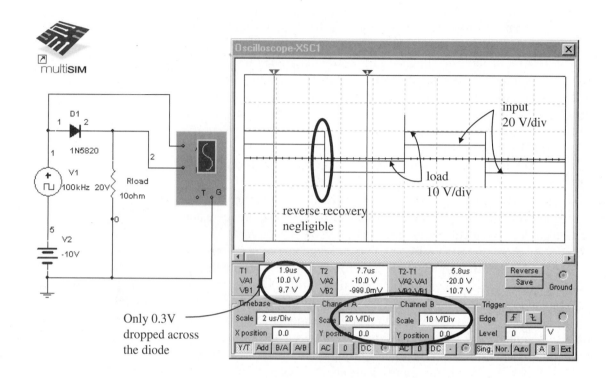

Figure 6-2 Simulation of a Schottky diode response

Figure 6-2 shows the response of a Schottky diode at 100 kHz. The vertical sensitivity for the two oscilloscope channels is different to more easily display the two waveforms. At cursor T1, the input has an amplitude of 10 V, and the load is at 9.7 V. So the Schottky diode only drops 0.3 V. That is a *big* improvement over 1.5 V for the 1N4002. Also, the reverse recovery time is insignificant. When the input swings negative, there is no current through the load.

Schottky diodes' PIV *may* be small. Check the specs and your circuit carefully.

There is one precaution you must observe when using Schottky diodes. The **reverse voltage breakdown** is usually less than 60 V. So you must carefully examine the diode's **PIV** rating and the performance of your circuit. If the load contains inductance, filter capacitance, batteries, or other energy storage elements, the circuit may place a reverse voltage across the diode that is much larger than the input signal's negative peak.

Transistors

An *npn* bipolar junction transistor switch is shown in Figure 6-3.

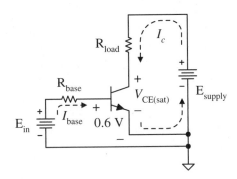

Figure 6-3 Bipolar junction transistor switch

When the signal generator goes positive, the transistor's base-emitter junction is forward biased. Base current is limited by the base resistor.

$$I_{base} = \frac{E_{in} - 0.6V}{R_{base}}$$

This base current then establishes collector current. For *linear* applications,

$$I_{collector} = \beta I_{base}$$

For small signal transistors at low current levels, β may be several hundred. So you do not have to drive much current into the base to provide a usable collector current. However, for power transistors with collector currents of an ampere or more, β may be as low as 20.

To use the transistor as a switch, it must be turned *on*, passing all of the current it possibly can. The collector current is then limited by the collector supply voltage and the load.

$$I_{collector} = \frac{E_{supply} - V_{CE(sat)}}{R_{load}}$$

where $V_{CE\,(sat)}$ is the transistor's saturation voltage. Its precise value depends on the transistor, the amount of collector current, and the junction temperature.

$$0.2V \leq V_{CE(sat)} \leq 1.0V$$

The higher the collector current, the larger the voltage across the transistor switch.

To be sure that the transistor is turned *on* as hard as possible, you must drive more base current into it than is needed to set the collector current in a linear application.

$$I_{base} > \frac{I_{collector}}{\beta}$$

Example 6-1

For the bipolar transistor switch in Figure 6-3, with

$$E_{supply} = 28 \text{ V}_{dc}, \quad E_{in} = 5 \text{ V}_{dc}, \quad R_{load} = 15 \text{ } \Omega$$
$$V_{CE\,(sat)} = 0.8 \text{ V}, \quad \beta = 20$$

calculate the following:

$$I_C, \quad P_{load}, \quad P_Q, \quad I_{base}, \quad R_{base}$$

Solution

With the transistor turned hard *on*,

$$I_{collector} = \frac{E_{supply} - V_{CE(sat)}}{R_{load}}$$

$$I_{collector} = \frac{28 \text{ V}_{dc} - 0.8 \text{ V}_{dc}}{15 \Omega} = 1.81 \text{ A}_{dc}$$

Since this is dc, while the switch is *on*

$$P_{load} = \left(I_{load}\right)^2 R_{load}$$

$$P_{load} = \left(1.81 \text{ A}_{dc}\right)^2 \times 15 \Omega = 49.1 \text{ W}$$

The transistor switch dissipates

$$P_Q = I_{collector} V_{CE}$$

$$P_Q = 1.81 \text{ A}_{dc} \times 0.8 \text{ V}_{dc} = 1.45 \text{ W}$$

The load receives almost 50 W while the transistor wastes only 1.5 W. This is *much* better than delivering power with the push-pull *linear* amplifier.

But to be sure the switch is really on,

$$I_{base} > \frac{I_{collector}}{\beta}$$

$$I_{base} > \frac{1.81 A_{dc}}{20} = 90.5 \, mA_{dc}$$

This current must be provided by E_{in}. It is limited by R_{base}.

$$I_{base} = \frac{E_{in} - 0.6V}{R_{base}}$$

Solve for R_{base}.

$$R_{base} = \frac{E_{in} - 0.6V}{I_{base}}$$

$$R_{base} < \frac{5 V_{dc} - 0.6 V}{90.5 \, mA_{dc}} = 48.6 \Omega$$

To be sure there is enough base current, pick $R_{base} = 33 \, \Omega$, 1W

Practice: Repeat the calculations for Example 6-1 for $E_{supply} = 12 \, V_{dc}$, $R_{load} = 6 \, \Omega$, $\beta = 300$.

Answer: $I_C = 1.87 \, A_{dc}$, $P_{load} = 20.9$ W, $P_Q = 1.5$ W, $I_{base} > 6.2 \, mA_{dc}$, pick $R_{base} = 470 \, \Omega$, ¼ W

The 90 mA_{dc} base current in Example 6-1 is so large because the power transistor's β is so low. In the practice problem, the β is raised to 300, and the base drive requirement drops. You *can* have both high collector current and high β by using a small signal bipolar junction transistor to drive the power transistor. This is called a **Darlington pair** configuration. In fact, you can even buy the combination in a single power package, with one collector, one base, and one emitter connection. This is illustrated in Figure 6-4.

This would seem to solve the problems of using a bipolar junction transistor as a switch. If all of your applications are dc, then you are indeed finished. But the strength of switching transistors comes from us-

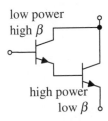

Figure 6-4 Darlington pair

ing them to deliver a rectangular waveform to the load. Varying the duty cycle of that signal changes the average value, the rms, and the power delivered to the load. Look back at Table 3-1 (page 137). So the dynamics of the transistor switch (how fast it is) are also important.

Example 6-2

Use simulation to investigate the speed of a Darlington pair in response to a 10 kHz, 50% duty cycle, 5 V_p input. Set

$$E_{supply} = 12 \ V_{dc}, \ R_{load} = 6 \ \Omega, \ and \ R_{base} = 470 \ \Omega$$

Solution

The schematic and probe output are shown in Figure 6-5 for the OrCAD simulation.

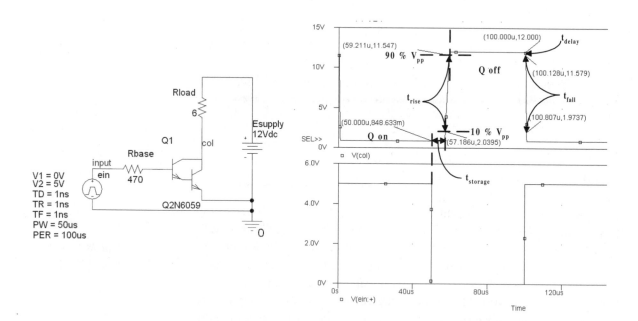

Figure 6-5 Darlington pair switching speed simulation

There are several key points. First, V_{col} is taken *across* the transistor switch. A high input turns the switch *on* and outputs a low voltage, 0.849 V in this simulation. When the input goes low, the transistor turns *off*, and the voltage at its collector with respect to common is the supply voltage, 12 V.

When the input falls to 0 V, at 50 μs, the transistor should immediately turn off and the collector voltage should head toward 12 V. But in the simulation, the transistor continues to stay *on* for a while. This is the **storage time**. It is measured from the middle of the input edge until the output voltage rises 10%.

Then it takes time for the collector voltage to go from 1.2 V to 10.8 V. This is the **rise time**. It is measured from the 10% point to the 90% point. The rise time is largely determined by the load resistance and any parasitic capacitance between the collector and common.

When the input jumps up, to turn the transistor *on*, there is a wait while electrons and holes move across the junctions and begin to combine. This is the **delay time**. It is measured from the middle of the rising edge of the input to the 90% point on the output.

Finally, those parasitic capacitances must be discharged through the transistor. This sets the **fall time**, measured from the 90% to the 10% point.

Practice: For Figure 6-5, calculate $t_{storage}$, t_{rise}, t_{delay}, and t_{fall}.

Answers: $t_{storage} = 7.19$ μs, $t_{rise} = 2.03$ μs, $t_{delay} = 128$ ns, $t_{fall} = 679$ ns

---

If you build this circuit, you would hear a high pitched whine, even though there is no loudspeaker. The switching frequency is 10 kHz. At 2 A_p, enough interference is created to be very annoying. The solution is to run the entire circuit much faster, perhaps at 100 kHz. But at that frequency, the storage time (7.19 μs) is longer than the pulse width (5 μs). The transistor never goes all the way off. The Darlington pair's bipolar transistors cannot switch above the audio range.

The enhancement mode MOSFET can also be used as a switch. The larger the gate-to-source voltage, the larger the channel between the source and the drain, and therefore the lower the channel resistance.

IRF530N

International
IΩR Rectifier

Electrical Characteristics @ T$_J$ = 25°C (unless otherwise specified)

	Parameter	Min.	Typ.	Max.	Units	Conditions
V$_{(BR)DSS}$	Drain-to-Source Breakdown Voltage	100	—	—	V	V$_{GS}$ = 0V, I$_D$ = 250µA
ΔV$_{(BR)DSS}$/ΔT$_J$	Breakdown Voltage Temp. Coefficient		0.13		V/°C	Reference to 25°C, I$_D$ = 1mA
R$_{DS(on)}$	Static Drain-to-Source On-Resistance	—	—	0.11	Ω	V$_{GS}$ = 10V, I$_D$ = 9.0A ④
V$_{GS(th)}$	Gate Threshold Voltage	2.0	—	4.0	V	V$_{DS}$ = V$_{GS}$, I$_D$ = 250µA
g$_{fs}$	Forward Transconductance	6.4	—	—	S	V$_{DS}$ = 50V, I$_D$ = 9.0A

t$_{d(on)}$	Turn-On Delay Time	—	6.4	—		V$_{DD}$ = 50V
t$_r$	Rise Time	—	27	—	ns	I$_D$ = 9.0A
t$_{d(off)}$	Turn-Off Delay Time	—	37	—		R$_G$ = 12Ω
t$_f$	Fall Time	—	25	—		R$_D$ = 5.5Ω, See Fig. 10 ④

Figure 6-6 IRF530N electrical characteristics (*courtesy of International Rectifier*)

Figure 6-6 is part of the data sheet for the IRF530N. It indicates that at V_{GS} = 10 V, while carrying 9 A at 25°C, the drain-to-source resistance is no more than 0.11 Ω. That's a pretty good closed switch. A little further down the same data table, the four response times are listed. They are *all* less than 40 ns.

The enhancement mode MOSFET requires no gate current to turn it on. But the manufacturer suggests that to get the transistor hard *on*

MOSFET switch turn *on* requirements

$$10\,\text{V} \le V_{GS} < 20\,\text{V}$$

multi**SIM**

Example 6-3

Use simulation to investigate the speed of an *n*-channel enhancement mode MOSFET (IRF530N) in response to an input of 500 kHz, 50% duty cycle, 12 V$_p$. Set

$$E_{supply} = 12\ V_{dc}, \text{ and } R_{load} = 6\ \Omega$$

This circuit is comparable to the bipolar junction Darlington pair of Example 6-2, but is running fifty times faster.

Solution

The Multisim schematic and simulation are shown in Figure 6-7.

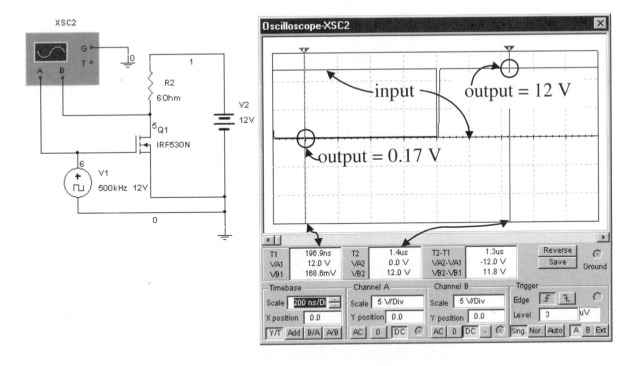

Figure 6-7 IRF530N switching speed simulation

Several observations can be made in comparing the bipolar junction Darlington pair with the MOSFET. When the bipolar transistors are turned on, there is still 0.85 V across them, resulting in 1.5 W of wasted power. The MOSFET has only 0.17 V across it. If it were on 100% of the time, this would only require the IRF530N to dissipate 0.33 W, four times better.

Even though the MOSFET is running 50 times faster than the bipolar transistors, there is very little storage time (turn off delay). The rise and fall times appear to be negligible.

The MOSFET switch is much faster and drops much less voltage than the bipolar junction transistor switch.

Practice: Run the simulation shown in Figure 6-7. Adjust the oscilloscope controls and the cursors to measure $t_{storage}$, t_{rise}, t_{delay}, and t_{fall}.

Answers: $t_{storage}$ = 9.2 ns, t_{rise} =8.0 ns, t_{delay} = 0.9 ns, t_{fall} = 2.6 ns

6.2 Parallel MOSFET Switches

For many battery operated applications (cell phones, laptop computers, pagers, radio controlled models) even the 330 mW that the IRF530N dissipates is far too wasteful. For these, the switch needs less resistance. One solution is to buy a better MOSFET, *if* one is available. Often a much easier and less expensive solution is to wire several transistors in parallel, drain-to-drain-to-drain- ... gate-to-gate-to-gate- ... source-to-source-to-source- ... . Some radio controlled model airplanes and cars have as many as eight transistors wired in parallel. This drops the *on* resistance from 0.11 Ω to 0.014 Ω, and the power lost in the switch from 330 mW (in Example 6-3) to 56 mW. In addition to lowering the switch resistance, so there is less of the precious battery power lost in the switch, each transistor stays cooler. Often it costs less to add transistors in parallel than it does to buy a heat sink for one transistor. And the paralleled transistors probably take up less space than the single transistor with its monster heat sink.

When paralleling MOSFETs in Section 5.5 you saw that the MOSFET has a *positive* temperature coefficient. As the junction warms up, the transistor passes more current, which causes it to get hotter, and pass more current, which makes it hotter still, until the transistor melts. But that only happens at relatively low levels of V_{GS}, like those used in class B amplifiers. Look at Figure 6-8.

This is the same V_{GS} versus I_D graph that you have seen several times before. At low values of V_{GS}, the 175°C curve is on top,. As the junction heats up, performance moves upward, producing more current and eventual thermal runaway. But at just under $V_{GS} = 5.3$ V, the two temperature curves cross.

At high levels of V_{GS}, the 175°C line is *below* the 25°C curve. That means that as the transistor heats up, the current *drops*. This smaller amount of current then allows the transistor to cool. When driven *on* as a switch, the MOSFET's V_{GS} is typically greater than 10 V. This puts the operation at the extreme right of the graph. Paralleled MOSFET switches are inherently temperature stable, exhibiting a *negative* temperature coefficient that encourages them to equally share the current. No compensating source resistors are needed.

Parallel MOSFETs lower:
the voltage lost across switch
the need for a heat sink

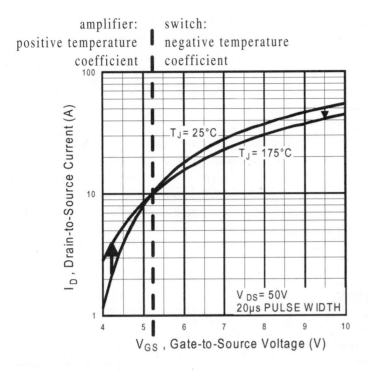

Figure 6-8 MOSFETs have both positive and negative temperature coefficients (*courtesy of International Rectifier*)

6.3 Low Side Switches

The switches in Section 6.1 are defined as low side switches. They are placed between circuit common and the load. The other end of the load is pulled up to the supply.

If the load can be tied to the supply, and common connected and removed from its lower end, this configuration is convenient. The signal that drives the switch is referenced to circuit common. Also connecting the switch's source or emitter to common means that the input (controlling) signal may directly drive the switch.

Bipolar Junction Transistor

The bipolar transistors from Section 6.1 make poor high speed, high current switches. However, there are many applications that require load current of 100 mA$_p$ or less. For these circuits, the small signal bipolar junction transistor works well. In fact, often it is preferable to a MOSFET. It can be turned hard *on* from a supply voltage of 3.3 V$_{dc}$ to 5 V$_{dc}$. These are the two most common digital logic supplies. Although there are a few MOSFETs that turn hard *on* at these low voltages, most require 10 V. For 3.3 V$_{dc}$ and 5 V$_{dc}$ systems, that means an additional, higher power supply voltage.

Small signal bipolar junction transistor switches are commonly available in the small plastic TO-92 package. Most MOSFETs are housed in the larger TO-220 power package. So, if you are building the circuit using through-hole parts, rather than surface-mount, the small signal bipolar junction transistor switch allows a much smaller and lighter board.

Key switching parameters for the 2N3904 are shown in Figure 6-9.

SWITCHING CHARACTERISTICS (except MMPQ3904)

t$_d$	Delay Time	V$_{CC}$ = 3.0 V, V$_{BE}$ = 0.5 V,		35	ns
t$_r$	Rise Time	I$_C$ = 10 mA, I$_{B1}$ = 1.0 mA		35	ns
t$_s$	Storage Time	V$_{CC}$ = 3.0 V, I$_C$ = 10mA		200	ns
t$_f$	Fall Time	I$_{B1}$ = I$_{B2}$ = 1.0 mA		50	ns

*Pulse Test: Pulse Width ≤ 300 μs, Duty Cycle ≤ 2.0%

Figure 6-9 2N3904 switch specifications (*courtesy of Fairchild*)

For the 2N3904 switch
I$_{C \, max}$ = 100 mA$_p$
V$_{CE \, on}$ = 0.3 V$_p$
V$_{BE \, on}$ = 0.9 V$_p$
β$_{@ \, Ic = 100 \, mA}$ = 100

When driven hard *on*, with I_C = 50 mA$_p$, the voltage across the transistor is V_{CE} = 0.3 V$_P$ and requires V_{BE} = 0.9 V$_p$. To turn the transistor *on* this hard, there must be

$$I_B > \frac{I_C}{\beta}$$

$$I_B > \frac{50 \, mA_p}{100} = 500 \mu A_p$$

Example 6-4

1. Design a low side switch using a 2N3904, a supply voltage of 28 V_{dc} with a rectangular pulse input of 0 V to 2.4 V_p, 100 kHz. Use the smallest practical load resistance.

multi**SIM**

2. Calculate P_{Rload}, and P_Q.

3. Verify performance with a simulation.

Solution

1. Although the manufacturer indicates that the 2N3904 can handle 100 mA_p, the specifications are tested at 50 mA_p. So it seems wise to set

$$I_C = 50 \text{ mA}_p$$

This current is limited by the collector load resistor.

$$R_{load} = \frac{E_{supply} - V_{CE}}{i_C}$$

$$R_{load} = \frac{28 \text{ V}_{dc} - 0.3 \text{ V}_p}{50 \text{ mA}_p} = 554\Omega$$

Pick $R_{load} = 560 \ \Omega$

A base current of at least 500 μA_p is needed to produce the 50 mA_p collector current. This base current comes from the input rectangular wave generator, when it is outputting 2.4 V_p. The base resistor sets this base current.

$$R_{base} < \frac{e_{inp} - V_{BE}}{i_B}$$

$$R_{base} < \frac{2.4 \text{ V}_p - 0.9 \text{ V}_p}{500 \mu A_p} = 3k\Omega$$

Pick $R_{base} = 2.2 \ k\Omega$

2. The load resistor has a rectangular 50% duty cycle voltage across it. Its value is

$$v_{p \ load} = 28 \text{ V}_{dc} - 0.3 \text{ V}_p = 27.7 \text{ V}_p$$

This produces a load current of

$$i_{\text{p load}} = \frac{27.7\,V_p}{560\Omega} = 49.5\,mA_p$$

The voltage and the current are both rectangular 50% pulses. From Table 3-2,

$$P_{\text{load}} = D v_p i_p$$

$$P_{\text{load}} = 0.5 \times 27.7\,V_p \times 49.5\,mA_p = 686\,mW$$

During the time that the transistor is *on*, there is 0.3 V_p across it and 49.5 mA_p through it. When the transistor is *off*, there is no current through it, so for that time it dissipates no power.

$$P_Q = 0.5 \times 0.3\,V_p \times 49.5\,mA_p = 7.4\,mW$$

Even though well over half a watt is being delivered to the load, the transistor dissipates virtually no power. This is because the transistor only carries current when there is very little voltage across it.

3. The Multisim simulation is shown in Figure 6-10. There is good agreement with the calculations. Remember that the wattmeters have two sets of connections. The voltage terminals must be connected *across* the element. The current terminals must be in *series* with the element. The only surprise is the voltage across the transistor when it is *on*. The oscilloscope, cursor 2, indicates 201 mV not 300 mV. But the 300 mV specification is a manufacturer's worst case maximum.

    The circuit designed for Example 6-4 produces as much load current as is prudent to expect from a 2N3904. But that 50 mA_p requires 500 μA_p of current be driven into the base from the input generator. A logic gate or microprocessor output may be unable to provide that much current.

Practice: Rework Example 6-4, selecting R_B then R_{load} assuming that the input signal is 2.4 V_p at 100 μA_p.

Answer: $R_B < 15$ kΩ (i.e., 13 kΩ), $I_C \leq 10$ mA. $R_C > 2.8$ kΩ, (or 3.3 kΩ)

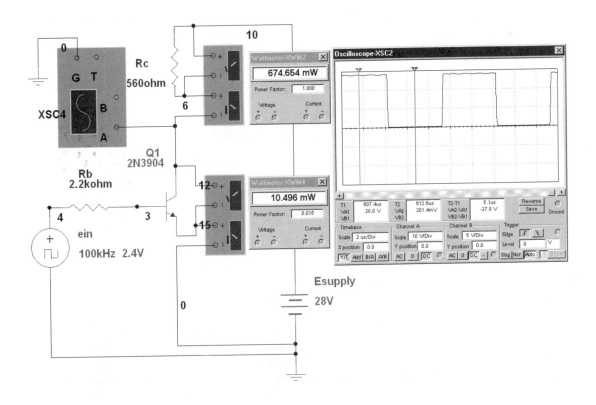

Figure 6-10 Simulation of a bipolar junction transistor switch for Example 6-4

One of the major concerns with a low side transistor switch is the effect of capacitive loading. Only rarely is a capacitor actually wired between the collector and common. However, the wires, circuit board traces, and inputs to following stages all present *some* capacitance

Look at Figure 6-11. When the switch turns *on*, any capacitance between the collector and common discharges through the switch. Since the switch's *on* resistance is very low, discharge occurs quickly and the fall time is not seriously affected. But, when the switch is turned *off*, the capacitance must charge to the supply voltage, through the load. The larger R_{load}, the longer it takes for that voltage to climb. During this time, current continues to flow through the load (into the capacitance) even though the switch is *off*. Rise time is measured from the 10% to the 90% point.

$$t_{rise\,RC} = 2.2R_{load}C_{parasitic}$$

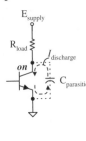

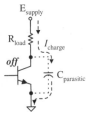

Figure 6-11 Capacitance discharge and charge

Example 6-5

Determine the rise time for C = 640 pF, R_{load} = 560 Ω (Example 6-4) and for C = 640 pF, R_{load} = 3.3 kΩ (Practice 6-4).

Solution

For the circuit from Example 6-4, when the transistor turns *off*, the 640 pF parasitic capacitance must charge through the 560 Ω.

$$t_{rise\ RC} = 2.2 R_{load} C_{parasitic}$$

$$t_{rise} = 2.2 \times 560\,\Omega \times 640\,pF = 788\,ns$$

For a pulse width of 5 μs (50% of 100 kHz), this is 16%, not good, but possibly tolerable.

For the bipolar switch, driven from a logic circuit, in Practice 6-4,

$$t_{rise} = 2.2 \times 3.3\,k\Omega \times 640\,pF = 4.7\,\mu s$$

This is the *entire* pulse width. By the time that the capacitance allows the voltage to reach 90% of E_{supply}, the pulse is over and it is time to discharge.

Practice: For the switch driven from the logic signal, determine the *frequency* that allows a duty cycle of 20% and the rise time to be only 10% of the pulse *on* time.

Answer: t_{rise} = 4.7 μs, t_{high} = 47 μs, T = 235 μs, f = 4.3 kHz (right in the middle of the audio range!)

Open Collector Inverter

To charge this capacitance more quickly, you must provide more current by lowering R_{load}. But the 2N3904 cannot provide more current from the 100 μA_p available as base current from a logic signal. Open collector logic gates can help. Their input is a single logic load, properly driven by standard logic signals. Their output, though, is a bipolar transistor with the emitter connected to common and the collector brought directly to the output pin. To use these gates you must connect a pull-up resistor between the gate's output and the supply voltage. Look at Figure 6-12.

The 7406 is a hex inverting buffer with high voltage open-collector outputs. When the input goes to a logic high, the output transistor is driven *on*. It can sink up to 40 mA$_p$. That is four times the current that you could count on from a 2N3904. When the input is a logic low, the output transistor is open. The output voltage then rises to +E$_{pull-up}$. For the 7406 that may be as large as 30 V$_{dc}$.

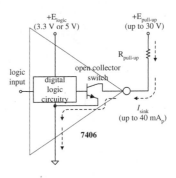

Figure 6-12 7406 open collector logic switch

Example 6-6

Using a 7406:

1. Select R$_{pull-up}$ to give the fastest possible output rise time and a 28 V$_p$ high level.

2. Calculate that rise time with 640 pF of parasitic capacitance.

4. Verify your calculations with a simulation.

Solution

1. For an output with a 28 V$_p$ high level, pick E$_{pull-up}$ = 28 V$_{dc}$

$$R_{pull-up} = \frac{E_{pull-up} - V_{OL}}{I_{OL}}$$

$$R_{pull-up} = \frac{28\,V_{dc} - 0.7\,V_p}{40\,mA_p} = 683\,\Omega$$

Pick R$_{pull-up}$ = 680 Ω

2. The 640 pF charges through this 680 Ω.

$$t_{rise} = 2.2 \times 680\,\Omega \times 640\,pF = 957\,ns$$

3. The results of the OrCAD simulation are shown in Figure 6-13. It indicates a rise time of

$$t_{rise} = 6.0188\,\mu s - 5.066\,\mu s$$

7406

e$_{in}$ = **logic high**
Q is *on* (**short to common**)

e$_{in}$ = **logic low**
Q is *off* (**open**)

E$_{pull-up}$ ≤ 30 V$_{dc}$
i$_{sink}$ = I$_{OL}$ ≤ 40 mA$_p$
v$_{OL}$ ≤ 0.7 V at 40 mA$_p$

Figure 6-13 OrCAD simulation for Example 6-6

MOSFET Low Side Switch

The open collector gate, such as the 7406, can barely sink 40 mA$_p$. But it *is* TTL compatible. The 2N3904 bipolar junction transistor can do only a little better. To switch several amps of current or more with switching speeds less than 100 ns, you need *n*-channel enhancement mode MOSFETs.

To turn the IRF530 hard *on* you must set 10 V$_p$ ≤ V_{GS} ≤ 20 V$_p$. This drives the channel wide open, lowering its resistance to less than 0.11 Ω, at 25°C. As the channel heats up, this resistance increases. The **normalization factor** is given in Figure 6-14. To determine the effect of temperature, multiply the appropriate factor from the vertical axis by the $R_{on\ @\ 25°C} = 0.11$ Ω. The resistance at 150°C is

$$R_{on\ @\ 150°C} = 2.25 \times 0.11\ \Omega = 0.25\ \Omega$$

For the IRF530 switch
$i_{○\ max} = 12\ A_p$
$R_{on} = 0.11\ \Omega$ @ 25°C
      $= 0.25\ \Omega$ @ 150°C
$10\ V \leq V_{GS} \leq 20\ V$
$\Theta_{JA\ bare\ case} = 62°C/W$

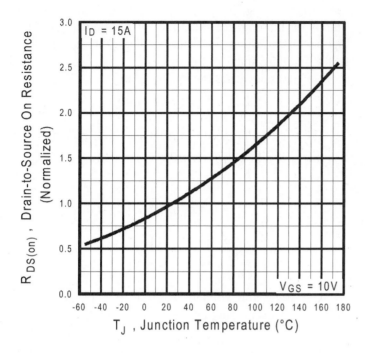

Figure 6-14 IRF530 channel resistance normalization factor versus temperature (*courtesy of International Rectifier*)

Example 6-7

For the circuit in Figure 6-15, calculate the following:

$R_{on\ worst\ case}$, $i_{p\ load}$, P_{load}, P_Q, T_J, with $T_A = 40°C$ and *no* heat sink

Solution

The worst case junction temperature for the IRF530 is 175°C. According to Figure 6-14, at 175°C, the channel resistance has gone up by a factor of 2.6 over its resistance at 25°C.

$$R_{on\ worstcase} = 2.6 \times 0.11\,\Omega = 0.29\,\Omega$$

The current is

$$i_p = \frac{28\,\mathrm{V_{dc}}}{6\,\Omega + 0.29\,\Omega} = 4.45\,\mathrm{A_p}$$

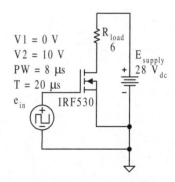

Figure 6-15 MOSFET low side switch for Example 6-7

This is far more current than the 7406 logic gate or the 2N3904 bipolar junction transistor can handle. However, it is less than half of the maximum for the IRF530.

This current causes the load to drop

$$v_{\text{load}} = 4.45\,\text{A}_p \times 6\,\Omega = 26.7\,\text{V}_p$$

The load voltage and current are both rectangular pulses, with a 40% duty cycle. From Table 3.2,

$$P_{\text{load}} = Dv_p i_p$$

$$P_{\text{load}} = 0.4 \times 26.7\,\text{V}_p \times 4.45\,\text{A}_p = 47.5\,\text{W}$$

The current causes the MOSFET to drop

$$v_Q = 4.45\,\text{A}_p \times 0.29\,\Omega = 1.29\,\text{V}_p$$

The transistor voltage and current are both rectangular pulses, with a 40% duty cycle. From Table 3.2,

$$P_Q = 0.4 \times 1.29\,\text{V}_p \times 4.45\,\text{A}_p = 2.3\,\text{W}$$

Without a heat sink $T_J = T_A + P\Theta_{\text{JA}}$

$$T_J = 40°\text{C} + 2.3\,\text{W} \times 62\frac{°\text{C}}{\text{W}} = 183°\text{C} > T_{\text{max}}$$

This is too hot. You must either add a heat sink or parallel several MOSFETs.

Practice: Repeat the calculations for Example 6-7 if the duty cycle is increased to 75%.

Answer: $i_{p=} = 4.45$ A$_p$, $v_{\text{load}} = 26.7$ V$_p$, $P_{\text{load}} = 89$ W, $P_Q = 4.3$ W

The schematic in Figure 6-15 is rather simplistic. Several improvements have been made in Figure 6-16. First, often the input pulse comes from a logic circuit or microprocessor. Its amplitude is only 2.4 V$_p$. Most MOSFETs need 10 V$_p$ to turn hard *on*. The 7406 has been added to convert the logic level at its input to 10 V$_p$ to 20 V$_p$. Be sure to include the 0.1 µF decoupling capacitor *close* to the 7406's power pin.

The IRF530 needs a gate voltage between 10 V$_p$ and 20 V$_p$. But the only pull-up supply is 28 V$_{\text{dc}}$. Resistors R1 and R2 divide the 28 V$_{\text{dc}}$ in half when the 7406's output goes *open*. Select R1 = R2 = 680 Ω. This gives the fastest rise time from the 7406. Also decouple this circuit with a 0.1 μF capacitor, C2, as close to R1 as practical.

From the junction of R1 and R2 the signal travels to the gate. That looks like an open, right? Well, actually, there is about 640 pF of parasitic capacitance between the gate and source. Also, the leads and traces contain inductance. The result is a series resonant circuit. When the 7406 hits it with a pulse, it rings, and rings, and rings, completely confusing the MOSFET. Resistor R3 dampens that parasitic tank. Set R3 in the 10 Ω to 100 Ω range.

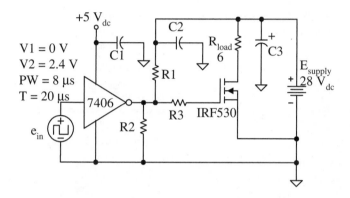

Figure 6-16 Additions to the simple low side MOSFET switch

Capacitor C3 should be connected as close to the power end of the load as practical. The rule-of-thumb is about 100 μF for each amp of current. That calls for a 470 μF electrolytic capacitor. Since electrolytic capacitors are very slow, place a 0.1 μF film capacitor immediately across its terminals.

Finally, look at the common return connections. The load current from the MOSFET runs *directly* back to the main power supply's common in its own lead. The low level currents from e_{in}, the 7406, the logic +5 V supply, and the biasing network may be bussed together and run back to the same point. Since the purpose of the capacitors is to remove noise, it is wise to provide these with their own separate return to the circuit common at the + 28 V_{dc} supply.

In your body, muscles convert signals into motion. In the electrical world this is most often done with electromagnetics. Relays close switch contacts. Solenoids open and close hydraulic and pneumatic valves. Loudspeakers make sound by moving air. Motors result in a turning shaft. Transformers step voltage up and down.

All of these loads have significant inductance. The relationship between voltage and current for an inductor is

$$v_L = L \frac{di}{dt}$$

During the time current flows through the inductance, magnetic field lines build up. The inductor is acting as a load, storing energy. But when the switch turns *off*, the magnetic field collapses. As the field falls, it cuts the wires of the inductance, *generating* a voltage. This induced voltage is opposite in polarity to that across the inductor when the field was growing. That is, the voltage across the load reverses, and is created by the load itself. This voltage induced as the fields collapse is called the **flyback** voltage and may be well over 100 V.

Example 6-8

Calculate the flyback voltage created by a load with a 10 μH inductance carrying 4.5 A, if the switch driving the load turns *off*, sending the current to 0 A in 300 ns.

Solution

$$v_L = 10\,\mu H \frac{4.5\,A_p}{300\,ns} = 150\,V_p$$

These levels of flyback voltage can destroy the transistor switch, as well as other electronics tied to the inductive load. They may also be lethal. A simulation of the improved MOSFET switch in Figure 6-16 is shown in Figure 6-17. A 10 μH inductance has been added to the load.

Since the maximum drain-to-source voltage is 100 V_p for a IRF530, the first time you try to test your circuit, before you can even capture a waveform, the MOSFET fails, unless some protection is provided. Many MOSFETs have a zener diode inherent to the transistor, between drain and source. With this internal zener, the MOSFET survives, and the peaks are trimmed to 100 V_p. But these are still almost four times larger than the supply voltage.

A better solution is to connect a Schottky diode (called a flyback diode) directly across the terminals of the inductive load. Look at Figure 6-18.

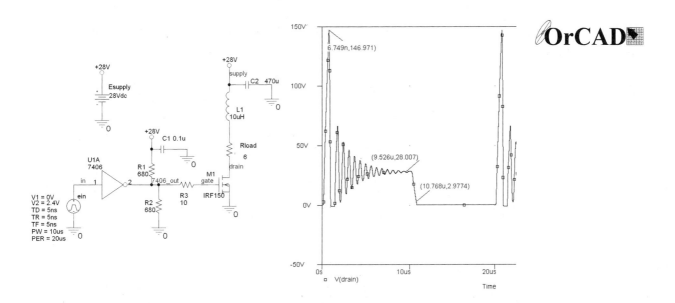

Figure 6-17 Flyback caused by inductive kick when the switch turns *off*

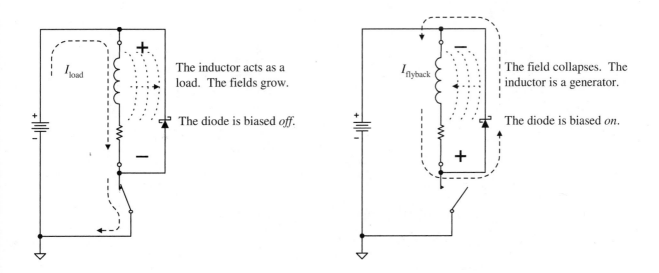

Figure 6-18 Inductive load and flyback diode

During the time the switch is *on* the inductance is a load, dropping voltage and storing energy. The Schottky diode is reverse biased. When the switch goes *off*, the magnetic field falls, cutting the inductance, generating an opposite polarity voltage. That polarity turns the diode *on,* giving the induced current a complete path to the other side of the generator (inductor). A Schottky diode is needed to handle the high currents in the very short time that the flyback spike occurs. The result of including the flyback diode is shown in Figure 6-19.

Do not assume that just because a load is not defined as an inductor, there is no inductance associated with it. Every wire has inductance, even the leads that connect the load to the switch and the supply. So even if you are driving a pure resistor, there *is* inductance there too.

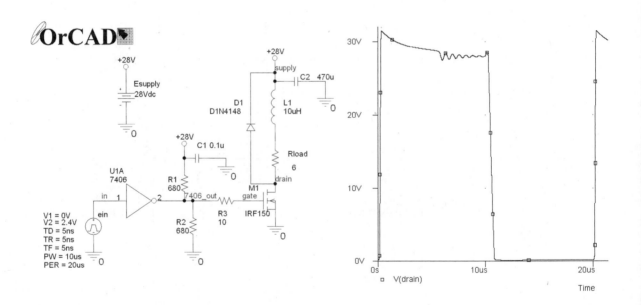

Figure 6-19 Results of adding a flyback diode

6.4 High Side Switches

All of the switches you have seen so far are connected to circuit common, and the loads are connected to the supply. This is certainly convenient for controlling the switch. But many loads *must* be tied to common (perhaps even earth). The switch then applies and removes the voltage from the supply. That is, the switch sits on the **high side** of the load.

N-Channel High Side Switch

So, let's just swap the switch and the load. Unfortunately, just that one simple change assures that the circuit will fail. Initially there is no current flowing through the load, so it drops no voltage. The top of the load and the source of the MOSFET are at 0 V. When the input drives the gate to 10 V_p, this sets $V_{GS} = 10$ V_p. The MOSFET starts to go *on*. Turning the switch *on* sends current through the load, producing a voltage drop across the load. The voltage at the top of the load climbs positive. Raising this voltage increases the source voltage (where the load is tied). But if the voltage on the source goes up, then the *difference* between the gate voltage and the source voltage decreases. V_{GS} drops, and the transistor turns *off*.

The solution is to provide a separate, higher gate pull-up supply. This is shown in Figure 6-20. The load supply voltage has been dropped to 12 V_{dc}, and the voltage divider has been removed from the MOSFET's gate. When the input is high, the 7406 outputs a low, turning the switch *off*. When the input goes low, the 7406's output transistor goes open. The gate is then pulled up toward +28 V_{dc}. As the gate capacitance charges through R1 and R2, the MOSFET turns *on*, and its source voltage rises too. By the time that the gate is at +28 V_p, the MOSFET is hard *on*, and there is +12 V_p at the source, and across the load. This still leaves

$$v_{GS} = 28\,V_p - 12\,V_p = 16\,V_p$$

This is more than enough to keep the *n*-channel enhancement MOSFET switch turned hard *on*, but does not violate the 20 V maximum.

Three separate supplies are needed. The voltage to the load has been reduced to +12 V to allow head room for $E_{\text{gate pull-up}}$. Also notice that the Schottky diode is still connected across the load.

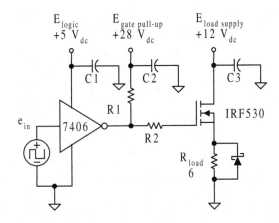

An *n*-channel MOSFET high side switch requires a third supply.

E_{gate pull-up} > E_{load supply} + 10 V

$E_{gate\ pull-up} > E_{load\ supply} + 10\ V$

Figure 6-20 *N*-channel high side switch with a boosted gate supply

P-Channel High Side Switch

P-channel MOSFETs are slower than their *n*-channel complements, and more expensive. However, they can be used to build a high side switch with only one power supply. A simple version is shown in Figure 6-21.

Remember, to turn a *p*-channel MOSFET *on*, the gate voltage must be 10 V more negative than the source. Also, channel current flows into the source, and out of the drain. This is the same direction that the semi-conductor current carrier holes travel.

In Figure 6-21, the MOSFET has been inverted. Its source is connected to the supply, and its drain is down, tied to the load. When the voltage at e_{in} = +E_{supply}, the voltage at the gate is the same as the voltage at the source.

$$v_{GS} = v_G - v_S$$

$$v_{GS} = e_{in} - E_{supply} = 0$$

The MOSFET is *off*.

On the other part of the input cycle, e_{in} = 0 V. Watch the signs carefully.

$$v_{GS} = e_{in} - E_{supply}$$

$$v_{GS} = 0\,V - E_{supply} = -E_{supply}$$

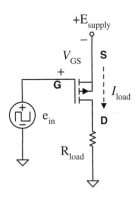

Figure 6-21 Simple *p*-channel high side switch

Pulling the gate to common while the source is held at $+E_{supply}$ makes the gate negative with respect to the source. This negative potential difference repels electrons in the channel, and attracts holes, completing the channel and turning the *p*-channel MOSFET hard *on*.

Both the *n*-channel low side switch and this *p*-channel high side switch cause the signal at the drain to be inverted from the input. But the *n*-channel switch goes *on* when the input is high. The *p*-channel switch is *off* when the input is driven up to the supply voltage, and goes *on* when the gate voltage is low.

As with the *n*-channel low side switch, there are several additions that must be made to the basic high side switch of Figure 6-21. A more practical version is shown in Figure 6-22. To drive the circuit with a logic level, add the 7406 open collector to the input. It requires a pull-up resistor, R1, connected to the high voltage supply. Since this supply is greater than the 20 V maximum V_{GS} that the IRF9530 can stand, a voltage divider must be included, R2. This voltage divider is different from the one used with the *n*-channel low side switch. When the output of the 7406 is open, no current flows through R1. It drops no voltage. So both the source and the gate of the MOSFET are at $+28V_{dc}$. When the output of the 7406 is driven to common, the gate must drop to

$$8\ V \le V_G \le 18\ V$$

IRF9530 key specifications

$V_{DSS} = -100\ V$
$I_{D\ max} = -12\ A$
$R_{on\ @\ 25°C} = 0.3\ \Omega$
$R_{on\ @\ 150°C} = 0.57\Omega$
$\Theta_{JC} = 1.7°C/W$

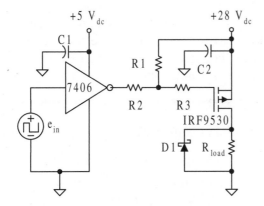

Figure 6-22 Practical *p*-channel high side switch

Since the source is at a constant $+28\ \mathrm{V_{dc}}$, these levels keep,

$$-20\ \mathrm{V} \le V_{GS} \le -10\ \mathrm{V}$$

enough to drive the MOSFET hard *on*, but not damage the gate.

To assure that the 7406 can drive the MOSFET *on* and *off* as quickly as possible, keep R1 + R2 as small as you can and still limit the sink current into the 7406 to less than 40 $\mathrm{mA_p}$. This allows the transistor's gate parasitic capacitance to charge and discharge through minimum resistance.

Resistor R3 is in the 10 Ω to 100 Ω range. As with the *n*-channel switch, it is in the circuit to dampen oscillations on the gate when it is hit with a step.

The decoupling capacitors, C1 and C2, should be placed as close to the IC or transistor pin as possible. Set C1 to about 0.1 μF. Select C2's value at about 100 μF for each ampere of load current.

The Schottky flyback diode is almost always necessary. If the leads to the load are short, place the diode directly across the load. If there is considerable inductance in the leads from the amplifier to the load, then place the diode close to the drain lead of the MOSFET, on the circuit board.

For simplicity, the two power supplies and the connections to common are shown as bubbles and triangles. But it is still important that there be a single point common, at the negative lead of the high voltage power supply. The logic supply common, the generator common, and the 7406 common may be bussed together and run there. The two decoupling capacitors (and perhaps the Schottky diode) should be connected to a separate bus and run to the single point common. Finally, the load should return to common in its own lead. These are the same procedures used with the *n*-channel low side switch in Figure 6-16.

Example 6-9

For the circuit in Figure 6-22:

1. Calculate the correct value for R1 and R2.

2. For $R_{load} = 12\ \Omega$, determine the voltage at each of the nodes when $e_{in} = 0.8\ \mathrm{V_p}$, and when $e_{in} = 2.4\ \mathrm{V_p}$.

3. When e_{in} is a 30 kHz, 80% duty cycle, calculate:

   P_{load}, P_Q, Θ_{SA} ($T_{J\ max} = 140°C$, $T_A = 50°C$ with a mica wafer)

4. Confirm steps 2 and 3 by simulation.

Solution

1. The 7406 sinks up to 40 mA$_p$ at an output voltage of 0.7 V$_p$.

$$R1 + R2 > \frac{28\,V_{dc} - 0.7\,V_p}{40\,mA_p} = 683\ \Omega$$

The two resistors must divide the +28 V$_{dc}$ setting V_G between 18 V$_p$ and 8 V$_p$. Set

$$R1 = 330\Omega \qquad R2 = 360\Omega$$

2. For $e_{in} = 0.8$ V,

The output of the 7406 is open. No voltage is dropped.
   output node of the 7406 = 28 V
   the junction of R1 and R2 = 28 V
   the gate = 28 V.
The transistor is *off*, and there is no load current so the load voltage is 0 V. The circuit is *noninverting*.

For $e_{in} = 2.4$ V

The output of the 7406 is shorted to common.

output node of the 7406 = 0.7 V (specification)

The junction of R1 and R2 is at

$$0.7\,V_p + \frac{360\Omega}{360\Omega + 330\Omega}\,28\,V_{dc} = 15.3\,V_p$$

$$v_G = 15.3\ V.$$

The difference between the gate and the source is

$$V_{GS} = 15.3\,V_p - 28\,V_{dc} = -12.7\,V_p$$

This is enough to turn the transistor hard *on* without damaging the gate.

The *on* resistance of the IRF9530 at 150°C is

$$R_{on\ IRF9530\ @\ 150°C} = 1.9 \times 0.3\ \Omega = 0.57\ \Omega.$$

$$v_{load} = 28\,V_{dc}\,\frac{12\Omega}{12\Omega + 0.57\Omega} = 26.7\,V_p$$

3. The load has an 80% duty cycle rectangular voltage. This produces an 80% rectangular current.

$$i_{\text{load}} = \frac{26.7\,\text{V}_{\text{p}}}{12\,\Omega} = 2.23\,\text{A}_{\text{p}}$$

From Table 3-2, this rectangular voltage and current cause

$$P_{\text{load}} = Dv_{\text{p}}i_{\text{p}}$$

$$P_{\text{load}} = 0.8 \times 26.7\,\text{V}_{\text{p}} \times 2.23\,\text{A}_{\text{p}} = 47.6\,\text{W}$$

For the transistor

$$v_{\text{Q}} = 2.23\,\text{A}_{\text{p}} \times 0.57\,\Omega = 1.27\,\text{V}_{\text{p}}$$

$$P_{\text{Q}} = 0.8 \times 1.27\,\text{V}_{\text{p}} \times 2.23\,\text{A}_{\text{p}} = 2.27\,\text{W}$$

The heat sink's maximum thermal resistance is

$$\Theta_{\text{SA max}} = \frac{T_{\text{J max}} - T_{\text{A}}}{P} - \Theta_{\text{JC}} - \Theta_{\text{CS}}$$

$$\Theta_{\text{SA max}} = \frac{150°\text{C} - 40°\text{C}}{2.27\,\text{W}} - 1.7\frac{°\text{C}}{\text{W}} - 2\frac{°\text{C}}{\text{W}} = 45\frac{°\text{C}}{\text{W}}$$

4. The waveforms with node voltages are shown in Figure 6-23. There is good agreement between the manually calculated values and those predicted by OrCAD. The differences are because the manual calculations use the worst case values for the IRF9530. The simulation uses typical values.

   The simulated load power is higher than the manual calculation. The simulation accounts for the time spent *while* the transistor is turning *on* and *off*. This time on the edges is considerable, even though the frequency is only 30 kHz. The *p*-channel is definitely slower than the *n*-channel.

   The simulated transistor power is lower than the manual calculations. This is because of the difference in the transistor's typical on resistance and its worst case.

Practice: Using the circuit from Example 6-9 and Figure 6-22, how many IRF9530's must be paralleled if you can tolerate only a 0.3 V_{p} drop across the transistors? What heat sink is needed?

Answer: Use 5 transistors. No heat sink is required.

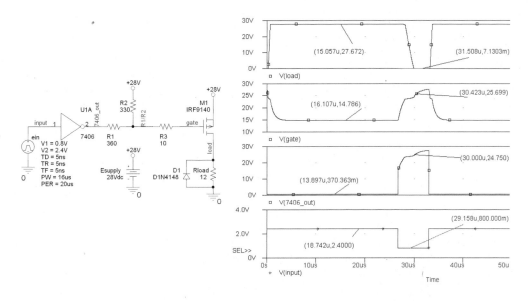

Figure 6-23 Simulation of Example 6-9 node voltages

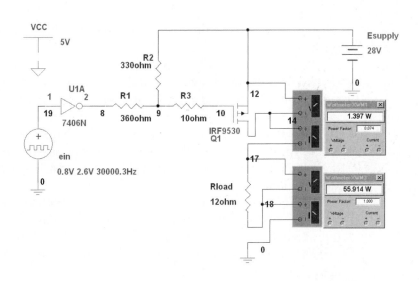

Figure 6-24 Simulation of Example 6-9 powers

PNP High Side Switch

If you need only 100 mA$_p$ of load current, or if switching speed is not critical, then you can use a *pnp* bipolar junction transistor or *pnp* Darlington pair configuration. This arrangement is particularly attractive if you are working with a single logic supply voltage. MOSFETs are a poor choice for a single +5 V system, since most require at least 10 V to turn *on*, regardless of what configuration you use. Look at Figure 6-25.

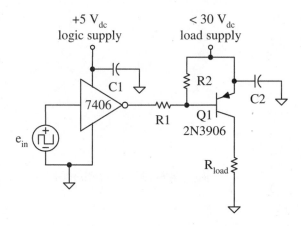

Figure 6-25 Bipolar junction transistor high side switch

As with the other switches in this chapter, the 7406 provides logic signal compatibility and sinks 40 mA$_p$ to drive the transistor *on*. When it is time to turn the transistor *off*, the 7406's open collector assures that the base is eventually pulled up to the load supply voltage, and is no longer forward biased. These same functions may be provided by the logic circuit or microprocessor creating the input, especially if the load supply voltage is also +5 V$_{dc}$.

When the output of the 7406 is switched to common, current flows from the load supply through the emitter-base junction of Q1, R1 and into the output of the 7406 to common. This saturates Q1 (if you have picked the parts correctly), and places almost all of the load supply across the load.

$$i_{\text{load}} = \frac{E_{\text{load supply}} - V_{\text{CE sat}}}{R_{\text{load}}}$$

The load current is created by the base current, as limited by the circuit. There must be enough to support the current drawn by the load.

$$i_{\text{base}} > \frac{i_{\text{load}}}{\beta}$$

When the transistor turns *on*, current also flows through R2.

$$i_{R2} = \frac{V_{\text{BE on}}}{R2}$$

These two currents combine and flow through R1. At this point in the design, there are several constraints.

$$i_{R1} < i_{\text{sink max G}} \quad (40 \text{ mA}_p \text{ for the 7406})$$

$$i_{R1} = i_{\text{base}} + i_{R2}$$

$$i_{R1} = \frac{\left(E_{\text{load supply}} - V_{\text{BE on}}\right) - V_{7406 \text{ on}}}{R1}$$

So, you have to juggle these relationships to determine values for R1 and R2. The lower their resistance, the more quickly parasitic capacitances can charge and discharge, so the load is driven with sharper edges. However, the smaller the resistors, the more current the transistor and the logic gate have to handle, the more power they must dissipate, and the more current that $E_{\text{load supply}}$ must provide.

When the logic gate (7406 in Figure 6-25) outputs an open, the transistor's base voltage must rise to $E_{\text{load supply}}$. This removes the 0.6 V emitter-base voltage and allows the transistor to turn *off*. But for this voltage to rise, parasitic base capacitance must be charged. It is charged through R2. The smaller R2 is, the more quickly the capacitance can charge and the sooner the transistor turns *off*. Remember, however, that R2 carries current when the gate outputs a low, turning the switch *on*. The smaller R2 is (for fast turn *off* time), the more current it must carry when the switch is turned *on*. Any current through R2 takes away from the base current, since R1 limits its current ($i_{R1} + i_{R2}$) to a fixed value.

So your balancing act must try to keep R2 small for fast turn *off*, but not so small that the transistor does not get enough base current when it is time to turn *on*.

6.5 MOSFET Switch Drivers

The higher the switching frequency, the smaller the capacitors, inductors, transformers, and other components needed, and the more rapidly the entire system can respond. The IRF530 n-channel MOSFET can go from fully *off* to fully *on* in less than 35 ns, and back *off* again in another 50 ns. The total cycle time, then, is less than 100 ns. Even if that represents a 10% duty cycle signal, the period of the entire wave would only be 1 µs, that is, 1MHz.

So, why then do the circuits in the previous section have trouble keeping up at 30 kHz to 50 kHz (3% to 5% of the MOSFET's potential)? The limitation is the transistor's input capacitance, C_{iss}. For the IRF530 it is typically 640 nF. To turn the transistor *on* or *off*, as much as 44 nC of charge must be moved onto or off of this parasitic capacitance. That takes time. How much time? That depends entirely on how much *current* you can provide. Current is defined as

$$I = \frac{\Delta Q}{\Delta t}$$

$$I_{\text{gate capacitance}} = \frac{Q_G}{t_{\text{rise or fall}}}$$

Substituting typical values for the IRF530,

$$I_{\text{gate capacitance}} = \frac{44\,\text{nC}}{25\,\text{ns}} = 1.76\,\text{A}_p$$

It seems that if you want to drive your high current switch *on* quickly, you have to provide lots of current to it. You need a fast, high current switch to drive the fast, high current switch. And what will drive that switch? It all seems rather self-defeating.

But, the 1.76 A_p to charge the gate capacitance only has to be available for 25 ns. The driver that has to produce this current only has to do it for a *very* short time. Once the gate capacitance is charged (or discharged) the current requirement drops to zero.

This certainly does *not* describe the 7406 used in all of the previous circuits. Charging the gate capacitance comes through a pull-up resistor. The 7406 cannot source any current. It can sink 40 mA$_p$, not 1.76 A$_p$. It can do that continuously, however, even though the charge should be over in less than 50 ns.

Example 6-10

The 7406-based driver circuit can provide only 40 mA$_p$ to move 44 nC (worst case) onto and off of the MOSFET's gate capacitance. With this drive current, what is the longest time required to turn the transistor *on* or *off*?

Solution

$$I = \frac{\Delta Q}{\Delta t}$$

$$\Delta t = \frac{\Delta Q}{I}$$

$$\Delta t = \frac{44\,\text{nC}}{40\,\text{mA}_p} = 1.1\,\mu\text{s}$$

Example 6-6 and Figure 6-13 found a rise time of 0.95 µs a consistent result.

Practice: A special driver IC can produce 6 A$_p$ into 2500 pF, causing the capacitor to charge with a rise time of 25 ns. Calculate the charge transferred to the capacitance.

Answer: $\Delta Q = 150$ nC

Low Side MOSFET Driver

Several manufacturers produce ICs specifically designed to drive a high current pulse of short duration into a capacitive load connected to common. The MAX4420 is typical. An application is shown in Figure 6-26.

The power supply voltage may be as low as 4.5 V$_{dc}$ or as large as 18 V$_{dc}$. Logically, it is noninverting. The 7406 is inverting. The output swings to within 25 mV of common when outputting a low and to within 25 mV of the positive supply when outputting a high. This means that you must provide a positive supply at least as large as the MOSFET's gate voltage requirement (typically 10 V$_p$). Even though the part works with a +5 V$_{dc}$ supply, that output is too small to turn most MOSFETs *on*.

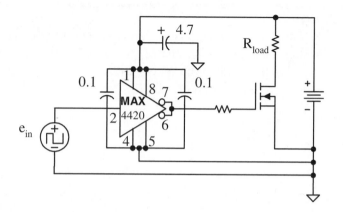

Figure 6-26 Low side switch and driver

The MAX4420 can source or sink 6 A_p while driving a 2500 pF capacitor, with rise and fall times of less than 60 ns (over the entire temperature range). However, there is also a delay time on each edge of no more than 100 ns. So, under worst case conditions, and into 2500 pF, a full cycle from low to high and back to low again takes less than 320 ns. If this is in response to a 10% duty cycle pulse, then the MAX4420 can drive its MOSFET switch up to at least 300 kHz. This is quite an improvement over the 7406. Considering that the IRF530's input capacitance is only 640 pF (25% of that at which the IC is tested), it is reasonable to expect even faster performance.

Such high current, high speed pulses require careful layout and rigorous decoupling. A ground plane is recommended. *Both* common pins (pin 4 and pin 5) must be connected together and to the circuit common (a ground plane is best). *Both* output pins (pin 6 and pin 7) should be connected together and then to the MOSFET gate. Power supply decoupling should include a single 4.7 μF high quality capacitor and *two* 0.1 μF film capacitors. One of the film capacitors must be immediately adjacent to pin 1 (to the ground plane or pin 4) and the other immediately adjacent to pin 8 (to the ground plane or pin 5). Keep the lead length of these decoupling capacitors as short as possible.

High and Low Side MOSFET Driver

The *p*-channel MOSFETs are generally much slower than comparable *n*-channel devices. *N*-channel transistors also cost less. So, the *n*-channel high side switch of Figure 6-20 is preferred. But to use an *n*-channel transistor as a high side switch, you must provide a gate voltage that is 10 V or more *above* the high voltage that is powering the switch and the load.

Linear Technology provides in a single IC, a low side driver (similar to the MAX4420), a high side driver, and a supply booster. This boosted supply assures that the gate voltage for the high side switch is 10 V above the high voltage that is powering that switch, up to 75 V_p. The use of the LT1160 as an *n*-channel high side switch driver is shown in Figure 6-27.

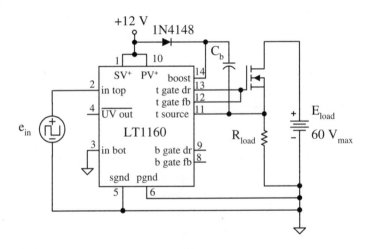

Figure 6-27 LT1160 driving an *n*-channel high side switch (*courtesy of Linear Technology*)

The SV$^+$ pin (pin 1) is the signal bias voltage. It must be at least as large as the MOSFET gate *on* voltage. The PV$^+$ pin (pin 10) is the positive supply voltage for the lower MOSFET driver. The sgnd pin (pin 5) is the bias common. It must be kept clean, and run to the circuit common with the input signal return. The pgnd pin (pin 6) is the return for the low side driver. It is not used in this example, but is normally connected directly to the low side switch source, which then goes to the supply common in its own, separate lead. Since the low side driver is not being used, the in bottom pin (pin 3) is tied to common. This assures that the low side driver outputs a steady low.

When power is first applied, before the high side driver becomes active, the 1N4148 diode forward biases, charging C_b through the low resistance R_{load} to almost +12 V_{dc}. That dc voltage (held on C_b) is presented to the boost pin (pin 14), which is the bias voltage for the high side driver.

Then the gate is driven up by a logic high at the in top pin (pin 2). The t gate dr pin (pin 13) initially goes up 12 V. That turns the MOSFET *on*. The source voltage begins to rise. Since the capacitor C_b cannot change its voltage instantaneously, as the load (and transistor source) voltage rise, the voltage on the boost pin rises too, staying 12 V above the source voltage. Even when the load reaches as much as 60 V, the boost voltage is 72 V, 12 V higher than the load. This is the voltage passed through the high side driver to the *n*-channel MOSFET's gate, keeping it on. The suspended supply amplifier of Chapter 1 used this same trick. As the output goes up, it drags its supply up with it.

Eventually, the input signal falls to 0 V. The gate voltage steps down as well. This turns the transistor *off*. With 0 V at the load, the diode turns on, replenishing the charge on C_b.

Run the LT1160 at 100 kHz.

The LT1160 is not as fast as the MAX4420. Driving 3000 pF, the rise time is 200 ns, the fall time is 140 ns, the turn-on delay is 500 ns and the turn-off delay is 600 ns. Application circuits provided by the manufacturer are all run at 100 kHz.

To achieve these speeds, the proper use of decoupling capacitors is important. These were omitted from Figure 6-27, but are necessary. Remember to use a ground plane and keep the capacitors' leads as short as possible. Place a 1000 μF capacitor at the drain of the high side switch. Add a 10 μF capacitor as close as possible between pins 1 and 5.

In the next section, you will see a circuit in which it is necessary to drive the load sometimes from the high side, and sometimes from the low side. The LT1160 is ideal for this application, since it provides both a high side and a low side driver. It features a lock-out circuit that allows *either* transistor to be on, but prevents *both* from being on at the same time.

6.6 H Bridge

You can control the amount of power that is applied to a load through a switch by altering the width of the pulse. A very low duty cycle delivers only a small amount of power each cycle. A high duty cycle means that the load is *on* almost all of the time. Since most loads are so much slower than the 50 kHz to 300 kHz at which you drive the switch, you never notice that the load is being turned hard *on* and hard *off* every few microseconds. At low duty cycles, the light glows dimly, or the motor barely turns. At high duty cycles, the heater blasts out the warmth, or the fan blows up a storm. Varying the duty cycle continuously changes the power delivered to the load. This is **pulse width modulation**. Proportional control is provided by a switch!

For the permanent magnet dc motor applying a low positive voltage turns the shaft slowly clockwise. Increase the duty cycle and the shaft turns more quickly. To turn counterclockwise, apply a negative voltage. Although this works, providing and controlling both positive and negative voltages at significant current levels is difficult. This is a particular problem in digital systems where everything should run from the same supply as the logic ICs.

Actually, if the motor (or other load) does not have to be tied to common, then the direction can be changed by forcing current to flow from the supply, but through the load in the reverse direction. This is accomplished with an **H bridge**. A simple H bridge is shown in Figure 6-28.

The load is placed across the center of an H, with high side switches on both ends *and* low side switches also at each end. The transistors are operated in diagonal pairs. Start by assuring all four transistors are *off*. To make the motor spin clockwise, turn Q1 and Q4 *on*. Current then flows down from the supply, through Q1, left to right through the load, and then down through Q4.

To reverse the direction of spin, turn all of the transistors *off*. Then turn Q2 and Q3 *on*. Current now flows down from the supply, through Q2, through the load from right to left, and then down to common through Q3.

Changing the duty cycle controls the speed of the motor. You can pulse width modulate the drive to the two transistors that are *on*, or you can turn one of the transistors in the pair *on* and pulse width modulate the other.

Schottky diodes are needed, as are drivers and some logic gates. These gates must make sure that the correct MOSFETs are *on* and that

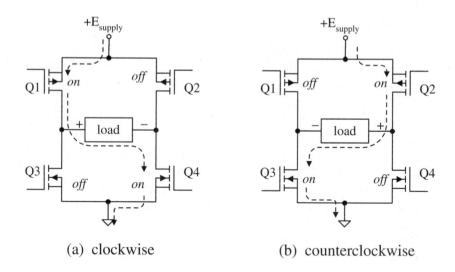

(a) clockwise   (b) counterclockwise

Figure 6-28 Simple H bridge operation

others are *off*. Usually the input signals to the logic are a pulse width modulation signal to indicate the speed, and a direction or polarity bit to indicate the direction the current should flow.

This logic also must contain circuitry to make absolutely sure that Q1 and Q3, or Q2 and Q4 are not on at the same time. If Q1 and Q3 (or Q2 and Q4) go on together, E_{supply} is shorted directly to common. Massive current flows until something dies. You must take into account the length of time required for the transistors to turn *off*. When the direction is to be changed, first *all* transistors are turned *off* and a lock-out delay is triggered. This pause gives time for each transistor to quit conducting and fully shut down. Only after this delay may you turn on the opposite pair of transistors.

And, be sure to remember that it takes a low to turn the p channel high side switches *on*. To turn them *off* their gates must be driven to within a few volts of E_{supply}. The low side switches are the opposite, requiring at least 10 V_p to turn *on*, and common to turn *off*.

Finally, in the previous section you learned that n channel MOSFETs make faster, cheaper high side switches than p channel transistors. But you then have to create a gate voltage that is 10 V above E_{supply}. And, this reverses the logic needed to drive the high side switches. Very high to turn them *on* and common to turn them *off*. Oh, but be sure that you do not exceed the MOSFETs' maximum gate-to-source voltage, which is easy to do when working with a supply above 20 V_{dc}.

All of these precautions must be rigorously observed, especially when providing several amps or more in 20 µs. Mistakes are spectacular! The result is a board full of logic gates, timers, drive transistors, load current sense and transistor temperature sense circuitry.

Or, you could buy it all in a single IC. Look at Figure 6-29.

FUNCTIONAL BLOCK DIAGRAM

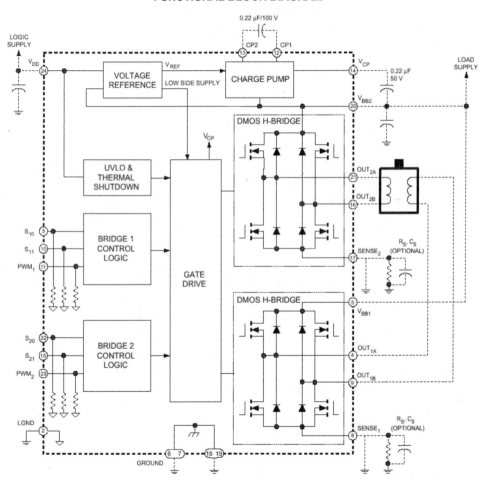

Figure 6-29 3971 H bridge integrated circuit (*courtesy of Allegro MicroSystems Inc.*)

At the heart of the Allegro 3971 are two independent H bridges, made with *n*-channel MOSFETs. Each transistor is protected with its own flyback diode. Like the LT1160, the load supply voltage is boosted, to provide the gate drive for the high side transistors. This IC uses a charge pump that requires two small, external capacitors. These are shown in Figure 6-29.

The logic supply voltage should be +5 V_{dc} to match TTL and CMOS input signals. The load supply may be as high as 50 V_{dc}. Each bridge can output up to 2.5 A_p. The application in Figure 6-29 shows two independent loads being driven. Out_{1A} and Out_{1B} power one set of windings of the inductive actuator. These output pins are controlled by the signals S_{10}, S_{11}, and PWM_1. An entirely separate set of controls and outputs, S_{20}, S_{21}, PWM_2, Out_{2A} and Out_{2B} drive the other winding.

To deliver twice the current to the load, you can wire the output pins together; S_{1A} to S_{2A}, and S_{1B} to S_{2B}. Of course you also have to be sure that each bridge receives the same commands. Wire S_{10} to S_{20}, S_{11} to S_{21}, and PWM_1 to PWM_2.

The three input control bits are defined in Table 6-1. To assure that no power is applied to the load enter 00 on the two control bits. Sending these two bits to 11 causes the two low side switches to be turned *on*. This also assures that no power is applied to the load. In addition, it places a short across the load. Any counter-emf created by turning an inductive load off quickly is shorted out. The inductive kicks, also called flyback, are eliminated. This is a good idea whenever you turn the load *off*.

The forward direction is defined as Out_A low (*on*) and Out_B driven down and up, inverted from the PWM signal. A 30% duty cycle at the PWM input applies voltage to the load 70% of the time.

The reverse direction changes the role of the Out_A and Out_B pins. Out_B is held *on*, and Out_A is driven opposite to the PWM input.

Table 6-1 3971 H bridge control truth table

S_{x0}	S_{x1}	Out_{xA}	Out_{xB}	function
0	0	open	open	off
0	1	low	$\overline{PWM_x}$	forward
1	0	$\overline{PWM_x}$	low	reverse
1	1	low	low	shorts load

With a 50 V_{dc} load supply, the output pin can be driven from low to off to high to off to low, completing a pulse, in about 1.3 µs. Therefore, typical applications provided by the manufacturer set the PWM signal's frequency at 100 kHz.

As with all of the other power switch circuits, layout, decoupling, and common routing are important. A ground plane board is needed. The IC should be soldered directly to the board, not put in a socket. The load supply connection should have 47 µF capacitors placed as close to pins 19 and 20, and pins 5 and 6 as possible. The main load current flows out pin 17 and out pin 8. So these two common pins must be routed back to the supply common in their own separate leads. The IC common (pins 6, 7, 18, and 19) and the logic common (pin 2) can be tied to the ground plane and then returned to the supply common in a different lead from the main load current returns (pins 17 and 8). There should also be a decoupling capacitor (0.1 µF film) tied immediately between the +5 V_{dc} logic supply (pin 24) and the ground plane.

There is built-in junction thermal monitoring. Should the junction temperature exceed 165°C, all output transistors are turned *off*, until their temperature falls to 150°C. Do not ignore power, thermal, and heat sink calculations. However, this thermal shutdown is intended as a fail safe, not as standard operation. Without proper heat sinking, the IC overheats far too soon, and shuts down even under nominal operation.

3971 specifications

$P_{IC\ max}$ = 2.2 W
$T_{J\ max}$ = 150°C
$R_{on\ high\ side\ max}$ = 0.375 Ω
$R_{on\ low\ side\ max}$ = 0.2 Ω

heat sink internally grounded

Example 6-11

The 3971 is to be used to drive a 15 Ω load from +28 V_{dc} with a duty cycle that may vary up to 85%. Calculate the maximum power delivered to the load, dissipated by the 3971, and the heat sink thermal resistance needed. Assume a maximum ambient temperature of 60°C.

Solution

When the load is turned on, current flows through a high side switch, the load, and then a low side switch.

$$i = \frac{28\,V_{dc}}{0.375\Omega + 15\Omega + 0.2\Omega} = 1.8\,A_p$$

This is below the 2.5 A_p maximum for the IC.

The load drops

$$v_{load} = 1.8\,A_p \times 15\,\Omega = 27.0\,V_p$$

This means that the power delivered to the load is

$$P = D v_p i_p$$

$$P_{load} = 0.85 \times 27.0\,V_p \times 1.8\,A_p = 41.3\,W$$

The high side switch drops

$$v_{high\,side} = 1.8\,A_p \times 0.375\Omega = 0.68\,V_p$$

This means that the high side switch must dissipate

$$P_{high\,side} = 0.85 \times 0.68\,V_p \times 1.8\,A_p = 1.0\,W$$

The low side switch drops

$$v_{low\,side} = 1.8\,A_p \times 0.2\Omega = 0.36\,V_p$$

This means that the low side switch must dissipate

$$P_{low\,side} = 0.85 \times 0.36\,V_p \times 1.8\,A_p = 0.56\,W$$

The IC dissipates

$$P_{IC} = 1.0\,W + 0.56\,W = 1.56\,W$$

This is below the maximum rated 2.2 W for the IC.
The required heat sink is

$$\Theta_{SA} = \frac{T_{J\,max} - T_A}{P} - \Theta_{JC} - \Theta_{CS}$$

The specifications do not list Θ_{JC}. But the MOSFETs you have used have $\Theta_{JC} < 2°C/W$. Since the heat sink is internally tied to circuit common, the heat sink may be connected directly to the package, setting $\Theta_{CS} = 0.2°C/W$.

$$\Theta_{SA} = \frac{150°C - 60°C}{1.56\,W} - 2\frac{°C}{W} - 0.2\frac{°C}{W} = 55.5\frac{°C}{W}$$

Practice: If the IC dissipates its maximum 2.2 W at $T_A = 85°C$, calculate the heat sink's thermal resistance, and the load peak current from a 36 V_{dc} supply with a 75% pulse width.

Answer: 27.3°C/W, 2.26 A_p

Summary

Diode switches made for line voltage rectification, such as the 1N400x series, have far too much reverse recovery time and too much forward voltage drop to be practical in high speed switching circuits. In contrast, Schottky diodes only drop 0.3 V when carrying several amps, and switch states in mere tens to hundreds of nanoseconds.

Bipolar junction transistors can be used as switches. The base current is produced by e_{in} and limited by R_{base}. It must be large enough to create more collector current than E_{supply} and R_{load} require. This collector current-to-base current relationship defines β. Unfortunately, at high collector current, β often falls below 20. To generate the needed base current, a second transistor can be added to drive the first: the Darlington pair. Though these transistors can provide amps of current, their switching speeds may be 5 µs or longer.

Driving the gate-to-source voltage of a MOSFET over 10 V_p sends its channel resistance to a few hundred *milli*ohms. Transitions in about 50 ns are common even when handling 10 A_p.

MOSFET *switches* may be wired directly in parallel. At the high V_{GS} levels used to force them to switch, the channel current goes *down* as the channel gets hotter. This negative temperature coefficient forces the paralleled transistors to evenly share the current. It is common to use several paralleled MOSFETs rather than one with a large heat sink.

A bipolar junction transistor can be effective as a high speed low side switch if you restrict its collector current to less than 100 mA_p. Doing that requires larger load resistors. This, in turn slows the circuit's ability to charge parasitic capacitance. This $R_{load} C_{parasitic}$ is the major limit in low side bipolar junction transistor circuits. An open collector gate (such as the 7406) ensures logic family compatibility.

In MOSFET switch circuit calculations, be sure to account for the rise in channel resistance as temperature rises. Both the load and the MOSFET power can be calculated as $P = D v_p i_p$. Junction temperature and heat sinks are determined just as they are for amplifiers.

At the high speeds and high currents present in MOSFET switching circuits, decoupling must occur *right at the pin*. Loads must be returned to the power supply common separately from the logic common. Inductive loads may create hundreds of flyback volts. Therefore, be sure to properly connect a Schottky diode across these inductances.

The switch can be moved to the high side of the load. To use a fast, inexpensive *n*-channel device, the gate voltage must be boosted 10 V above the supply voltage to be switched to the load. If this is impractical, a *p*-channel MOSFET can be used. Its drive logic is reversed, and it is slower and more expensive. For low current use a *pnp* transistor. Be sure to carefully consider the wide variety of design factors.

MOSFET gates have hundreds of picoFarads of capacitance. Before the transistor can switch, its gate charge must change. Moving charge in a short time requires a great deal of gate drive current. The MAX4420 is a low side driver capable of switching a MOSFET in a few hundred nanoseconds. The LT1160 can drive a low side switch as well. It also contains the boost circuit and driver needed to switch an *n*-channel high side switch.

You can send current either way through a load with a single power supply and four switches, arranged in an H. The load is connected across the center. You can build an H bridge from discrete transistors, logic gates, and sensors. But Allegro Microsystems offers a variety of fully integrated single and dual H bridge ICs. The decoding logic, hold-off timing, high side boost, drivers, and junction temperature shutdown are all built in.

The major *application* of high speed power switching circuits is in switch mode power supplies. Used in toys, pagers, cell phones, computers, automotive electronics, and avionics, they are now the standard way to provide reliable voltage to the electronics package. Linear regulators are used only when their ultra-clean output is demanded and their waste can be tolerated.

In the next chapter you will learn to use low side and high side switches as the foundation upon which the buck, boost, and off-line switching power supplies rely.

Problems

Switch Characteristics

6-1   Locate a data sheet for a 1N4004 diode. Determine its average forward current, maximum forward voltage, and peak reverse voltage. Calculate the maximum power it must dissipate.

6-2   Locate a data sheet for a 1N5819 Schottky diode. Determine its average forward current, maximum forward voltage, and peak reverse voltage. Calculate the maximum power it must dissipate.

6-3 Design a switch using a 2N3904 to drive a 100 Ω load powered from a 5 V_{dc} supply. Determine the collector current, base current, base resistor, power delivered to the load, and the transistor power dissipation for an 80% duty cycle pulse.

6-4 Design a switch using a Darlington pair to drive a 35 Ω load powered from a 24 V_{dc} supply. Determine the collector current, base current, base resistor, power delivered to the load, and the transistor power dissipation for a 75% duty cycle pulse.

6-5 Find a data sheet for a 2N2222. Determine the maximum collector current, maximum collector-emitter voltage, worst case β (or h_{FE}), worst case V_{CE}, worst case V_{BE}, and worst case switching time.

6-6 Find a bipolar junction transistor with a maximum collector current of at least 9 A. From its data sheet determine the maximum collector-emitter voltage, worst case β (or h_{FE}), worst case V_{CE}, worst case V_{BE}, and worst case switching time.

6-7 Design a switch using an IRF530 to drive a 35 Ω load powered from a 24 V_{dc} supply. Determine the drain current, required gate pulse amplitude, power delivered to the load, and the transistor power dissipation for a 75% duty cycle pulse, and the required heat sink maximum thermal resistance for an ambient temperature of 60°C.

6-8 Design a switch using an IRF630 to drive a 25 Ω load powered from a 150 V_{dc} supply. Determine the drain current, required gate pulse amplitude, power delivered to the load, and the transistor power dissipation for a 80% duty cycle pulse, and the required heat sink maximum thermal resistance for $T_A = 60$°C.

6-9 Locate a data sheet for an IRF630 MOSFET. Determine its maximum continuous drain current, maximum drain-to-source voltage, maximum gate-to-source voltage, junction-to-case thermal resistance, channel resistance at 100°C, gate-to-source voltage to drive it *on*, switching time, and gate capacitance.

6-10 Locate a data sheet for an IRFZ14 MOSFET. Determine its maximum continuous drain current, maximum drain-to-source voltage, maximum gate-to-source voltage, junction-to-case thermal resistance, channel resistance at 100°C, gate-to-source voltage to drive it *on*, switching time, and gate capacitance.

Parallel MOSFET Switches

6-11 It is decided to use four IRF530s in parallel to drive a 35 Ω load from a 24 V_{dc} supply with a 75% duty cycle. Calculate the voltage drop across the paralleled transistors, and the power *each* transistor must dissipate. Is a heat sink needed? Prove your answer with a calculation.

6-12 It is decided to use six IRF630s in parallel to drive a 25 Ω load from a 150 V_{dc} supply with an 80% duty cycle. Calculate the voltage drop across the paralleled transistors, and the power *each* transistor must dissipate. Is a heat sink needed? Prove your answer with a calculation.

Low Side Switches

6-13 A 2N3904 transistor switch is to drive a load consisting of a resistor tied to +15 V_{dc} and a 840 pF capacitor connected to common. Select the appropriate load resistor, and base resistor for a 2.4 V_p input, to provide a minimum rise time. Calculate that rise time.

6-14 Repeat problem 6-13 using a 7406.

6-15 For the circuit in Figure 6-30, calculate

R1, R2, R3, $t_{\text{rise of gate voltage}}$, $R_{on \ @ \ T = 140°C}$, P_{load}, P_Q, Θ_{SA}

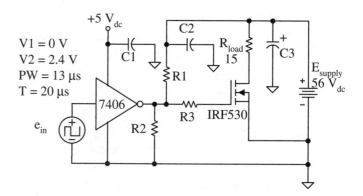

Figure 6-30 Schematic for Problems 6-15 and 6-16

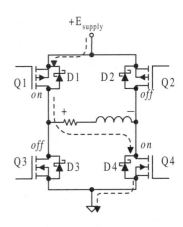

6-16 Repeat Problem 6-15 for a junction temperature of 160°C, $E_{supply} = 36$ V$_{dc}$, and a pulse width of 17 μs.

6-17 A load has 6 μH of inductance and carries 5 A$_p$ at the instant the switch driving it is turned *off*. Indicate the polarity of the voltage across the switch and calculate its peak value if the current through the switch has a fall time of 60 ns.

6-18 In the circuit of Figure 6-31, transistors Q1 and Q4 are *on* and current flows as indicated. The load has a significant inductance. Explain, in detail how, and which Schottky diodes prevent the flyback when Q1 and Q4 are turned *off*.

Figure 6-31 Schematic for Problem 6-18

High Side Switches

6-19 For the circuit in Figure 6-32, with a 30% input duty cycle, calculate R1, R2, $t_{rise\ of\ gate\ voltage}$, $R_{on\ @\ T\ =\ 140°C}$, P_{load}, P_Q, Θ_{SA}

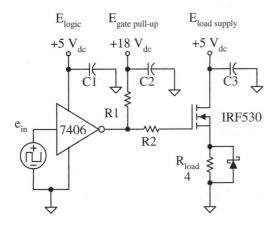

Figure 6-32 Schematic for Problem 6-19

6-20 If the load in Figure 6-32 is a 12 V, 2 A motor, redesign the circuit and repeat all of the calculations from Problem 6-19.

6-21 Design a *p*-channel high side switch to power a 5 V, 4 Ω load. For your design, assuming a 90% duty cycle to the load, calculate all components and $t_{rise\ of\ gate\ voltage}$, $R_{on\ @\ T\ =\ 140°C}$, P_{load}, P_Q, Θ_{SA}.

6-22 Design a *p*-channel high side switch to power a 12 V, 2 Ω load. For your design, assuming a 90% duty cycle to the load, calculate all components and $t_{\text{rise of gate voltage}}$, $R_{on \text{ @ T} = 140°C}$, P_{load}, P_Q, Θ_{SA}.

6-23 Design a *pnp* high side switch to power a 5 V, 50 mA$_p$ load. For your design, assuming an 80% duty cycle to the load, calculate all components, P_{load}, and P_Q.

6-24 Repeat Problem 6-23 for a 12 V, 100 mA$_p$ load.

MOSFET Switch Drivers

6-25 Find a low side MOSFET switch driver (other than the two presented in this chapter) capable of driving its switch at 60 kHz with a 10% duty cycle signal, from a +28 V$_{dc}$ supply. Include the specifications and the calculations you used to determine that the IC is appropriate.

6-26 Find a high side *n*-channel MOSFET switch driver (other than the two presented in this chapter) capable of driving its switch at 40 kHz with a 10% duty cycle signal, from a +28 V$_{dc}$ supply. Include the specifications and the calculations you used to determine that the IC is appropriate.

H Bridge

6-27 An H bridge built from bipolar transistors has the following speed specification:

$$t_{\text{high on delay}} = 5.3 \ \mu s, \quad t_{\text{high off delay}} = 2.4 \ \mu s, \quad t_{\text{low on delay}} = 1.8 \ \mu s,$$
$$t_{\text{low off delay}} = 3.3 \ \mu s, \quad t_{\text{lock-out delay}} = 8 \ \mu s.$$

Calculate the maximum PWM frequency for a 15% duty cycle signal, without changing direction.

6-28 The 3971 is to be used to drive a 12 Ω load from +18 V$_{dc}$ with a duty cycle that may vary up to 90%. Calculate the maximum power delivered to the load, dissipated by the 3971, and the heat sink thermal resistance needed. Assume a maximum ambient temperature of 50°C.

6-29 Find an H bridge IC capable of driving 2 A$_p$ from a 12 V$_{dc}$ supply with a PWM frequency of 80 kHz. The control must be entered serially from a microprocessor.

Transistor Switch Lab Exercise

A. Bipolar Junction Power Transistor

1. Build the circuit in Figure 6-33. Set the load resistor to 110 Ω, >10 W.

2. Set the signal from the generator to 0 V to 10 V, 50% duty cycle square wave at 5 kHz.

3. Verify that the output is correct by displaying both the input and the collector waveform with an oscilloscope. Look carefully at the collector low voltage level and the pulse widths.

4. Replace the load resistor with a motor. Keep the leads to the motor as short as possible.

5. Assure that the oscilloscope probe at the collector has a x10 attenuator. There may be some *big* transient spikes.

6. Record the wave shapes of the voltage at the input and across the transistor.

7. Measure the storage time, rise time, delay time, and the fall time of the voltage at the collector.

B. *N*-Channel Low Side Switch

1. Build the circuit in Figure 6-34. Take particular care to keep the leads as short as possible, and to provide the separate common returns as indicated. Connect the Schottky diode *directly* across the load terminals.

2. Increase the frequency of the signal from the generator to provide 0 V to 10 V, 50% duty cycle square wave at *50 kHz*.

3. Verify that the output is correct by displaying both the input and the drain waveform with an oscilloscope. Look carefully at the drain low voltage level and the pulse widths.

4. Replace the load resistor with a motor. Keep the leads to the motor as short as possible. Place the diode across the motor.

5. Assure that the oscilloscope probe at the drain has a x10 attenuator.

6. Record the wave shapes of the voltage at the input and across the transistor.

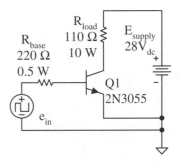

Figure 6-33 Bipolar junction transistor switch

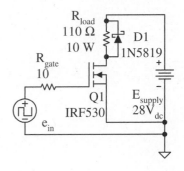

Figure 6-34 *N*-channel low side switch

7. Measure the storage time, rise time, delay time, and the fall time of the voltage at the drain.

C. Simple *N*-Channel High Side Switch
1. Build the circuit in Figure 6-35. Follow the schematic to assure separate common returns, decoupling, and the Schottky diode directly across the load.

2. Set the input signal from the generator to provide 0 V to 5 V, 50% duty cycle square wave at 50 kHz.

3. The motor should *not* run. Assure that your circuit is correct.

4. Record the wave shapes of the voltage at the input and across the load. Be sure to accurately record the low and the high level of the signal across the load.

5. Explain the waveform and why the motor does not run.

D. Boosted Supply *N*-Channel High Side Switch
1. Build the circuit in Figure 6-36. Although the schematic does not specifically show how, be sure to provide separate common returns for the logic, the motor, and the decoupling capacitors.

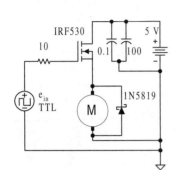

Figure 6-35 Simple *n*-channel high side switch

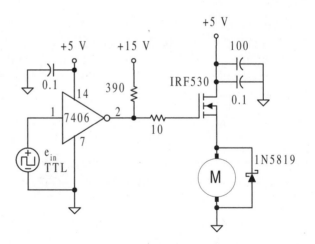

Figure 6-36 Correct *n*-channel high side switch

2. Set the input signal from the generator to provide 0 V to 5 V, 50% duty cycle square wave at 50 kHz.

3. The motor *should* run. Assure that your circuit is correct.

4. Record the wave shapes of the voltage at the input and across the load. Be sure to accurately record the low and the high level of the signal across the load.

E. *P*-Channel High Side Switch
   1. Build the circuit in Figure 6-37. Although the schematic does not specifically show how, be sure to provide separate common returns for the logic, the motor, and the decoupling capacitors.

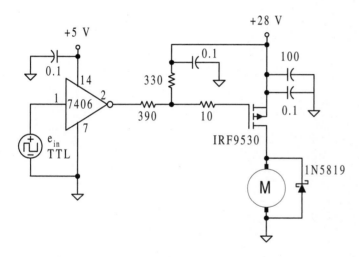

Figure 6-37 *P*-channel high side switch

2. Set the input signal from the generator to provide 0 V to 5 V, 50% duty cycle square wave at 50 kHz.

3. The motor *should* run. Assure that your circuit is correct.

4. Record the wave shapes of the voltage at the input and across the load. Be sure to accurately record the low and the high level of the signal across the load.

7

Switching Power Supplies

Introduction

The most common use of power switches is to provide steady, clean voltage to a circuit. Should the load demand more current, causing the filter components to discharge, the switch must be held on longer each cycle to compensate. If the input voltage goes up, causing the output to try to rise, the switch must be turned off sooner each cycle to keep the output voltage from rising.

You have already seen how pulse width modulation can produce an average output voltage that is lower than the input. This is a buck regulator. Often, though, the load voltage must be higher than the input. The flyback voltage across an inductor can be used to provide this boost.

A flyback converter provides a variety of voltages by chopping the input dc into a high frequency rectangular wave that can be passed through a pulse transformer to several secondary windings. This transformer also provides isolation, replacing the large, heavy, expensive line frequency transformer.

The high voltage sine is the most common waveform. This may be converted immediately to dc and then stepped to the variety of regulated voltages provided by a flyback converter. You must be careful: careful to avoid the damage such high input voltages and current levels can produce, and careful to assure that the high speed transients from your switches do not contaminate everything else connected to the utility line.

Objectives

By the end of this chapter, you will be able to:

- For the buck regulator, boost regulator, and flyback converter:

    analyze a given circuit to determine all waveforms, voltages, currents, and power dissipation.

    design a circuit to meet specifications.

    illustrate appropriate circuit layout.

    discuss advantages and limitations.

- List precautions for operation directly from the line voltage.

7.1 Buck Regulators

Basics

Voltage can be provided to a load through a high side switch, and its average value may be altered by changing how long each cycle the switch is on. This is shown in Figure 7-1.

$$V_{dc} = De_{p\,in}$$

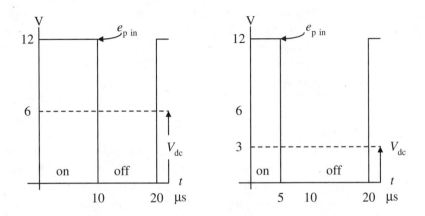

Figure 7-1 Average value varies with the duty cycle

Example 7-1

Calculate the average value for the second signal in Figure 7-1.

Solution

$$T = \text{period} = 20 \;\mu s$$

$$D = \text{duty cycle} = \frac{t_{on}}{T} = \frac{5\,\mu s}{20\,\mu s} = 0.25$$

$$V_{dc} = 0.25 \times 12 \, V_p = 3 \, V_{dc}$$

Practice: What must be done to keep the output at 3 V_{dc} if the amplitude changes to 7 V_p?

Answer: Raise the duty cycle to 43% by increasing the pulse width to 8.6 μs.

You can alter the voltage delivered to the load by changing the duty cycle of the high side switch. You can also keep the average voltage across the load constant by correcting the duty cycle in response to changes in the input's level.

This takes more than a simple high side switch and a driver. Look at Figure 7-2. The input is delivered from some source of raw, unregulated voltage, such as the utility line, a battery, a generator, or a solar array. The reference provides a stable voltage against which the output can be compared. This comparison is done by the pulse width modulator (**pwm**), usually an IC. Should the output drop, the switch is driven on longer. The pulse width modulator responds to increases at the output by reducing the duty cycle.

The output of the switch is a rectangle, such as in Figure 7-1. Filtering is needed to convert this into useful dc. The inductor opposes a change in current. It stores energy when the switch is *on* and generates current when the switch is *off.* The capacitor opposes a change in voltage. It also charges when the switch is *on.* When the switch goes *off* and the voltage tries to drop, the capacitor discharges, sending current to the load to hold the voltage up.

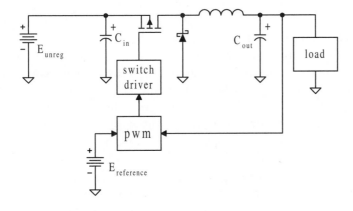

Figure 7-2 Basic buck regulator

The relationship of the switch to the charging and discharging of the inductor and capacitor and the operation of the Schottky diode requires a closer examination. Figure 7-3 is extracted from the basic buck regulator you just saw in Figure 7-2.

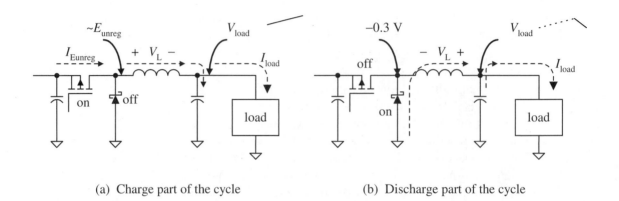

(a) Charge part of the cycle   (b) Discharge part of the cycle

Figure 7-3 Buck regulator operation

When the switch is turned *on* the charge part of the cycle begins. Since the transistor switch has very little resistance, the voltage at its drain is very close to the voltage at its source, E_{unreg}. This positive voltage on the cathode of the diode turns it *off*. The capacitor is holding a charge. Since this is directly across the load, the voltage across the capacitor is V_{load}. The buck regulator outputs a voltage that is below its input. So $E_{unreg} > V_{load}$. The voltage across the inductor, V_L, is higher on the left than on the right. So current flows through the inductor, left to right. Most of this current goes to the load, I_{load}, but some goes to charge the capacitor. This current flowing through the inductor causes energy to be stored in its magnetic field. The field lines expand. As the capacitor charges, its voltage, and the voltage across the load, increase.

The switch goes off, this starts the discharge part of the cycle. The magnetic field begins to collapse. As the lines of flux fall back they cut the wire of the inductor. This generates a voltage that is the opposite polarity of the voltage impressed across the inductor as the magnetic field grew. So the polarity flips as the inductor converts from a load to a generator. This places a negative at the cathode of the diode. It turns *on*, clamping the voltage at its cathode at about −0.3 V. The diode completes the discharge path for the inductor, connecting its left end to common.

As the magnetic field lines fly back, current is generated by the inductor and sent to the load. When this current falls, the load voltage falls. The capacitor discharges, also sending current into the load.

From these two cycles you can obtain the specifications for the diode, inductor, and capacitor. The switching frequency is well above the audio range, perhaps as high as 1 MHz. So the diode must be fast. However, it also handles all of the current through the inductor during the discharge. This is also the current through the load. So pick a Schottky diode with

$$I_{\text{diode continuous}} > 1.3 I_{\text{dc load max}}$$

The diode is turned *off* when the switch turns *on*, placing E_{unreg} on its cathode. Be sure that the diode you use has a reverse voltage rating of

Diode ratings

$$PIV > 1.3 E_{\text{max unreg}}$$

The primary relationship for an inductor is

$$v_{\text{L}} = L\frac{di}{dt}$$

During the charge cycle, this equation becomes

Inductance calculations

$$v_{\text{L}} = L\frac{\Delta i}{\Delta t}$$

The voltage across the inductor is

$$v_{\text{L}} = E_{\text{unreg}} - V_{\text{load}}$$

During the very short time that the inductor is charging, neither, E_{unreg} nor V_{load} change noticeably. So it is reasonable to treat V_{L} as a constant.

The time that the switch is *on* is Δt. This comes from the duty cycle and the frequency (and period, T) at which the switch is being driven.

$$on\ time = \Delta t = D \times T$$

Remember, the duty cycle is set by the relationship of input to output (Figure 7-1).

$$D = \frac{V_{\text{load}}}{E_{\text{unreg}}}$$

The purpose of the inductor (L) is to oppose a change in current (Δi). The larger you make the inductor, the less its current changes. This changing current flows into the low reactance of the capacitor, charging it. Very little of the *variation* in current actually goes to the load. So there is no reason to restrict Δi severely. Typically, if you are designing the supply and have to pick a value, pick

$$\Delta i \approx 0.25 I_{\text{dc load}} \text{ to } 0.7 I_{\text{dc load}}$$

At or below 200 mA$_{\text{dc}}$ you may allow the larger **ripple** current. Above 1 A$_{\text{dc}}$, restrict the ripple current to about 25% $I_{\text{load dc}}$. But if you are analyzing a given circuit, the value of this ripple current is set by the inductor.

Example 7-2

A buck regulator has an input voltage of 12 V$_{\text{dc}}$, and produces a load voltage of 5 V$_{\text{dc}}$ into a 10 Ω load. Its switch is driven at 100 kHz. The inductor is 220 µH. Calculate the following: duty cycle, on time, ripple current, average inductor current, peak inductor current

Solution

The duty cycle must by set to control the load voltage.

$$D = \frac{V_{\text{load}}}{E_{\text{unreg}}} = \frac{5 \, V_{\text{dc}}}{12 \, V_{\text{dc}}} = 42\%$$

The *on time* also depends on the switching frequency or period.

$$T = \frac{1}{f} = \frac{1}{100 \, \text{kHz}} = 10 \, \mu s$$

$$\Delta t = 0.42 \times 10 \, \mu s = 4.2 \, \mu s$$

The *ripple current* depends on the inductor.

$$V_{\text{L}} = L \frac{\Delta i}{\Delta t}$$

Solve for Δi.

$$\Delta i = \frac{V_{\text{L}} \Delta t}{L} = \frac{(12 \, V - 5 \, V) 4.2 \, \mu s}{220 \, \mu H} = 134 \, \text{mA}_{\text{pp}}$$

The load current is

$$I_{dc\,L} = I_{dc\,load} = \frac{5\,V_{dc}}{10\,\Omega} = 500\,mA_{dc}$$

As a check $\quad \dfrac{\Delta i}{I_{dc\,load}} = \dfrac{134\,mA_{pp}}{500\,mA_{dc}} = 0.27$

This is very close to 25%. This looks reasonable.

The peak inductor and load currents are

$$i_{p\,inductor} = I_{dc\,load} + \frac{\Delta i}{2} = 500\,mA_{dc} + \frac{134\,mA_{pp}}{2} = 567\,mA_p$$

Practice: Calculate the rms value of the current that flows through the capacitor. (Hint: it is a triangle.)

Answers: 156 mA$_{rms}$ Look in Table 3-1.

In addition to its inductance, the inductor has a maximum current rating. At current levels above this specification, the core saturates. Once in saturation, no additional magnetic field lines are produced. The inductor can store no more energy. It appears that the inductance has suddenly dropped to zero. There is only the small resistance of the wires to drop V_L. The inductor can no longer limit the current. The transistor or the diode may be damaged. Be sure to select an inductor whose peak current rating is above

$$i_{p\,inductor} = I_{dc\,load} + \frac{\Delta i}{2}$$

Inductor current spec

The construction technique also strongly affects the inductor's performance. The bobbin-core inductors look like a small spool of wire. They are inexpensive. But since the core is just a vertical piece of metal, the magnetic field spreads out from both ends, badly contaminating any conductors it intercepts. This type of inductor also tends to heat up and shift its value.

Types of inductors

Inductors may be made by wrapping wire around a toroid, a ring made of a variety of magnetic materials. In this configuration, most of the magnetic field is constrained to the core, so electromagnetic interference is much lower. Because the wiring is more spread out around the toroid, the inductor does not heat up as much as the bobbin-core type. However, toroids are more expensive and take up much more space.

The input capacitor, C_{in}, must be located as close to the transistor switch as possible. It provides a pool of charge to the circuit at the instant that the switch turns on. That initial current does not have to flow from E_{unreg}, through the connector's and wiring's inductance and resistance.

A key parameter for C_{in} is its **equivalent series resistance, or *ESR***. Large capacitors, C_{in} and C_{out}, may have several hundred milliohms of internal resistance. Current flowing off and onto these capacitors produces a voltage drop and forces the capacitor to dissipate power (producing heat). Therefore, a low *ESR* rating is important. Some capacitors are designed for use in switching power supplies. Their *ESR* at high frequency is small, and is listed in the specifications. Examples of through-hole, aluminum electrolytic high frequency capacitors are the PL series from Nichicon, and the HFQ series from Panasonic. Surface-mount applications usually use tantalum capacitors, such as the AXV TPS series or the 595D series from Sprague.

Secondly, C_{in}'s current handling must be considered. Select a capacitor that has

Input capacitor ratings

$$I_{rms\ Cin} > 0.5I_{dc\ load}$$

Finally, the capacitor's voltage rating must be high enough that it can survive any excursions at the input. Capacitors with a higher voltage rating have lower *ESR*. So when in doubt, select a capacitor with the largest voltage rating that the available space and budget can tolerate.

$$V_{Cin} > 1.5E_{max\ unreg}$$

Nothing has been said about the capacitance of C_{in}. Its actual value is not critical. Select C_{in} based on *ESR, I_{rms}, V* ratings, and availability. Any value above 100 μF should work.

$$C_{in} \geq 100\mu F$$

At the output, the reactance of that capacitor is far smaller than the load's resistance. So, virtually all of the ripple current from the inductor (Δi) flows through the capacitor. It develops a triangularly shaped voltage across the load, depending on the output capacitor's *ESR*. The lower C_{out}'s *ESR*, the lower the ripple voltage sent to the load.

Output capacitor's specifications

$$\Delta v_{pp\ load} = \Delta i_{pp\ inductor} \times ESR_{Cout}$$

Be sure that the output capacitor can handle the ripple current sent to it. Current ratings are in A$_{rms}$, but the ripple current is a triangle.

$$i_{rms\ Cout} = \frac{\Delta i_{pp\ inductor}}{\sqrt{3}}$$

A capacitor opposes a change in load voltage. The larger the output capacitor, the more charge is available to hold up the load while the switch is *off*. A large capacitor also charges slowly. Therefore, when the pulse width modulator tries to adjust the output, the capacitor causes the entire system to lag. The manufacturer of the pwm IC often offers tables to guide in your selection. Without this guidance, set

$$C \approx 500 \frac{\mu F}{A} \times I_{dc\ load}$$

Ideally the voltage passed to the load is dc, smooth and steady. In reality, there are two wave forms present. One is the triangle caused by the ripple current from the inductor flowing through the output capacitor's *ESR*. This can be reduced by using a larger inductor to lower Δi, or by finding a capacitor with a lower *ESR*.

There are also spikes present across the load. These are caused by the flyback from parasitic inductance in the main current path. Every wire and every trace that the current must run through has inductance. Keep the main path, from the input capacitor to the switch to the inductor to the output capacitor and to the diode back to the inductor, as short as possible. Typically, surface mount technology is extensively used. A ground plane is *necessary*. It should cover the entire area under and around the switching regulator, *except* beneath the inductor (where it would intercept and distort the magnetic field). There should be several common returns, one for the pwm, control, and ground plane, and a separate one for the load. As with all of the circuits you have seen, these are kept apart, and joined only at the circuit's single point common, at the terminal of the primary source of energy.

The purpose of this circuit is to deliver voltage to a load at significant current levels, while wasting as little power as possible. You must keep track of the currents, voltage drops, and power dissipated throughout. Look back at Figure 7-3. When the transistor is *on*, the primary, series path is from E_{unreg} through the transistor, the inductor, and to the load. So, during the charge part of the cycle, these components carry the same current. The average value of that current is $I_{dc\ load}$. The inductor

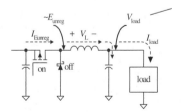

Figure 7-3 Buck regulator charge cycle (again)

causes it to ripple evenly above and below that level by a total amount, Δi_{pp}. This is shown in Figure 7-4 (a).

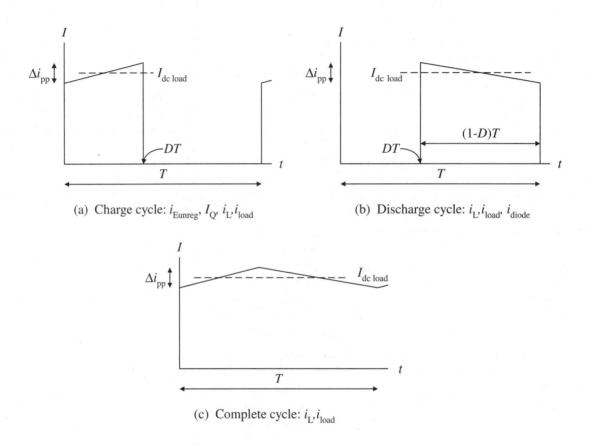

(a) Charge cycle: i_{Eunreg}, I_Q, i_L, i_{load} (b) Discharge cycle: i_L, i_{load}, i_{diode}

(c) Complete cycle: i_L, i_{load}

Figure 7-4 Currents in a buck regulator

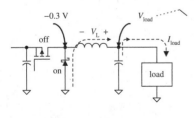

Figure 7-3 Buck regulator
charge cycle (again)

When the transistor is turned *off*, the current from the supply and through the transistor falls to zero. So Figure 7-4 (a) is correct for these two components across the entire operation of the regulator. During the discharge cycle, the inductor's field collapses, sourcing current through the load, and turning the diode *on* to complete the path. This is shown in Figure 7-4 (b). Since the diode does not carry current during the charge cycle, Figure 7-4 (b) accurately depicts that current across the entire operation.

The inductor and the load carry current during both the charge and the discharge cycles. Their combined current waveform is shown in Figure 7-4 (c).

Example 7-3

Calculate the rms value for the current through the transistor given $I_{dc\ load} = 500$ mA$_{dc}$ and $\Delta i_{pp} = 134$ mA$_{pp}$. D = 0.42. Complete the calculation twice, first using the trapezoid equation from Table 3-1, and secondly assuming that the signal is a simple rectangle with $i_p = I_{dc\ load}$. Compare the results.

Solution

For the trapezoid,

$$I_{rms} = \sqrt{\frac{D}{3}\left(I_p^2 + I_p I_{min} + I_{min}^2\right)}$$

where

$$I_p = I_{dc} + \frac{\Delta i_{pp}}{2} \quad \text{and} \quad I_{min} = I_{dc} - \frac{\Delta i_{pp}}{2}$$

Substituting values for I_{dc} and Δi_{pp} gives:

$$I_p = 567 \text{ mA}_p \quad \text{and} \quad I_{min} = 433 \text{ mA}_p$$

$$I_{rms} = \sqrt{\frac{0.42}{3}\left((567 \text{ mA}_p)^2 + (567 \text{ mA}_p)(433 \text{ mA}_p) + (433 \text{ mA}_p)^2\right)}$$

$$I_{rms} = 325.0 \text{ mA}_{rms}$$

Result using exact trapezoid equation

For a rectangle with a flat top of 500 mA$_p$,

$$I_{rms} = \sqrt{D}I_p = \sqrt{0.42} \times 500 \text{ mA}_p = 324.0 \text{ mA}_{rms}$$

Result using a retangular approximation

As long as the ripple peak-to-peak is a relatively small fraction of the average value, the simpler rectangular equation gives good results.

Practice: Calculate the *average* value of the same wave using the trapezoid and the rectangular equations.

Answers: $I_{dc} = 210$ mA$_{dc}$ from *both* equations.

So, using the rectangular approximations you can now calculate the power dissipated by the transistor, and the Schottky diode. This must be done to assure that these parts do not overheat. Knowing how much power they dissipate also allows you to estimate the efficiency of the regulator.

Example 7-4

1. Calculate the power dissipated by the IRF9530 for the circuit in Examples 7-2 and 7-3..

2. Find the power dissipated by the diode, assuming a 0.3 V_p drop when it is *on*.

Solution

Example 7-3 demonstrated that the results with the rectangular wave equations are as good as those using the trapezoid equations. So it may be assumed that the current is a rectangular pulse with a peak amplitude of 500 mA_p and a duty cycle of 42%.

In Chapter 6 you saw that the resistance of the IRF9530 at 150°C is 0.6 Ω. So, when the transistor is *on* it drops

$$v_Q = 500\,mA_p \times 0.6\,\Omega = 0.3\,V_p$$

This is a rectangular voltage pulse, produced by a rectangular current pulse, both with a 42% duty cycle. The power they dissipate is

$$P_Q = D v_p i_p$$

$$P_Q = 0.42 \times 0.3\,V_p \times 500\,mA_p = 63\,mW$$

This is not enough to cause the transistor to heat up, even without a heat sink.

By coincidence, the diode also drops 0.3 V_p when carrying 500 mA_p with a duty cycle of 42%. So it too must dissipate 63 mW.

Practice: Assuming that these are the only losses in the regulator, calculate the power delivered to the load, and the efficiency, η.

Answers: $P_{load} = 2.5$ W, $\eta = 95\%$

It is more reasonable to assume an efficiency of 90%, since there are losses *as* the transistor and diode change states, and losses in the resistance of the inductor and the capacitors. So this means that

$$P_{load} = 0.9 P_{input}$$

Since both are dc, $V_{dc\ load} \times I_{dc\ load} = 0.9 E_{dc\ unreg} \times I_{dc\ in}$

LM2595 Simple Switcher® Buck Regulator IC

The easy part of building a switching regulator is selecting the diode, inductor, and capacitor. Designing the high side switch, its driver, a voltage reference, and the pulse width modulator is much more demanding. National Semiconductor makes the Simple Switcher® series of ICs. Each IC contains all of these parts and more. A buck regulator using the LM2595 is shown in Figure 7-5.

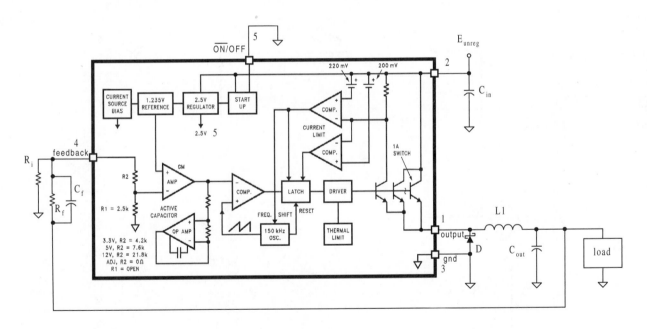

Figure 7-5 LM2595 used in a buck regulator (*courtesy of National Semiconductor*)

The high side switch as well as its driver and pulse width modulator are contained within the IC. These transistors can handle 1 A_{dc} and an input voltage E_{unreg} as high as 40 V. Buried in the center of the diagram is a 1.23 V_{dc} reference, and a fixed 150 kHz oscillator. You do not have to add any external resistors or capacitors to establish either of these parameters.

There are additional control and monitoring circuits. A logic low at the $\overline{\text{ON}}$ / OFF pin (pin 5) allows the IC to operate, providing a regulated voltage to the load. Taking this pin high turns the switch *off* sending the output voltage and current to zero. This provides handy logic control over the power supply. There is also a junction temperature monitor circuit. Should the junction temperature exceed 150°C, the IC turns itself *off*, no current passes through the switches, the load floats down toward 0 V_{dc} as the inductor and capacitor discharge, and the IC cools. Once it has cooled enough, it turns back *on* automatically.

Look at the IC, diode, inductor, and capacitor as a powerful, sophisticated op amp. This is shown in Figure 7-6.

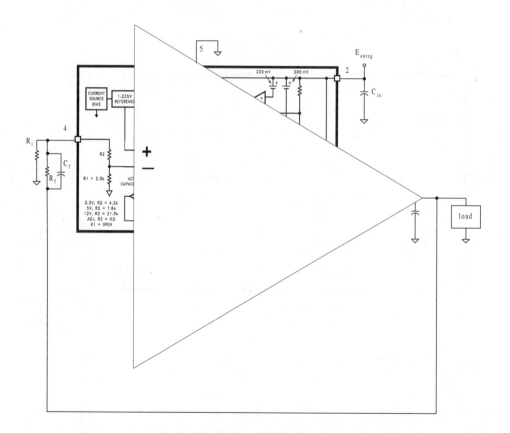

Figure 7-6 Buck regulator seen as a negative feedback amplifier

The internal 1.23 V_{dc} reference is the noninverting input to this "amplifier". The load voltage is sampled by R_f and R_i and sent to the IC as negative feedback. The IC changes its output in any way necessary to force the voltage fed back to equal the 1.23 V_{dc} input (reference). So, if the entire regulator circuit is working correctly, the voltage at the negative feedback pin (pin 4) is 1.23 V_{dc}. The load voltage, then, must be

$$1.23\,V_{dc} = \frac{R_i}{R_f + R_i} V_{dc\ load}$$

Solving this for the load voltage $V_{dc\ load}$,

$$V_{dc\ load} = \frac{R_f + R_i}{R_i} 1.23\,V_{dc}$$

$$V_{dc\ load} = 1.23\,V_{dc}\left(1 + \frac{R_f}{R_i}\right)$$

You can control the load voltage by changing the feedback resistor R_f. It could be a front panel potentiometer, a PCB trimmer, or even a digitally controlled IC that is set by a microprocessor.

The capacitor C_f across R_f provides high frequency stability. The load voltage should be steady dc. Any transients or other noise at the load are undesirable, and their effects should be reduced. At dc, C_f looks like an open, it has no effect on the dc operation. But, at 150 kHz, the capacitor looks like a short. The noise is not attenuated as it is fed back to the IC. Remember this is *negative* feedback. So, the larger the feedback, the lower the resulting output. The system is more stable.

If the feedback capacitor is too small, it does not adequately pass the higher frequency variation. If it is too large, it interferes with the dc operation of the circuit. The manufacturer suggests a film capacitor.

Table 7-1 contains the recommended values for the feedback capacitor and the output capacitor.

Allowing the LM2595 to repeatedly overheat, and turn itself *off*, cool down, turn itself *on*, just to overheat again is a bad idea. It subjects the IC to excessive temperatures, shortening its life. It also cycles the output power to the load *on* and *off*, over and over. The on-board temperature sensing and shutdown is intended as a fail-safe, not for normal operation. Under most circumstances, the LM2595 is so efficient that a heat sink is *not* needed. But in any design, it is wise to verify that before you build and test the circuit. Most of the heat within the IC is generated by the main pass transistors. Look back at Figure 7-4a. The current through the transistor is a pulse with a peak amplitude of $I_{dc\ load}$ and a duty cycle of D. During that time, the transistors drop a small voltage. It is listed in the specifications as V_{sat}.

Table 7-1 LM2595 based regulator capacitors (*courtesy of National Semiconductor*)

Output Voltage (V)	Through Hole Electrolytic Output Capacitor			Surface Mount Tantalum Output Capacitor		
	Panasonic HFQ Series (µF/V)	Nichicon PL Series (µF/V)	Feedforward Capacitor	AVX TPS Series (µF/V)	Sprague 595D Series (µF/V)	Feedforward Capacitor
1.2	330/50	330/50	0	330/6.3	330/6.3	0
4	220/25	220/25	4.7 nF	220/10	220/10	4.7 nF
6	220/25	220/25	3.3 nF	220/10	220/10	3.3 nF
9	180/25	180/25	1.5 nF	100/16	180/16	1.5 nF
1 2	120/25	120/25	1.5 nF	68/20	120/20	1.5 nF
1 5	120/25	120/25	1.5 nF	68/20	100/20	1.5 nF
2 4	82/35	82/35	1 nF	33/25	33/35	220 pF
2 8	82/50	82/50	1 nF	10/35	33/35	220 pF

So the power that the switch transistors must dissipate is

$$P_Q = DI_{dc\ load}V_{sat}$$

The IC contains quite a bit of circuitry. This circuitry draws current continuously while the IC is operating, from the E_{unreg} input supply. That current is called the **quiescent current** and is a specification. The resulting power that the support electronics must dissipate is

$$P_{IC} = I_{quies}E_{unreg}$$

All together, then, the LM2595 must dissipate

$$P_{LM2595} = I_{quies}E_{unreg} + DI_{dc\ load}V_{sat\ max}$$

The standard thermal equation that relates junction and ambient temperature, power, and thermal resistance applies to the LM2595. Under most circumstances, the IC can be used bare case, *without* a heat sink.

$$T_J = T_A + P(\Theta_{JC} + \Theta_{CS} + \Theta_{SA})$$

LM2595 specifications

$V_{sat\ max} = 1\ V_p$
$I_{quies\ max} = 10\ mA_p$

$\Theta_{JC} = 2°C/W$
$\Theta_{JA\ bare} = 50°C/W$

Or, bare case, $$T_J = T_A + P\Theta_{JA}$$

Where you place the components, and *how* you wire them together have a significant effect on the performance of your circuit. It is strongly recommended that you include a ground plane, to cover as much of both sides of the circuit board as practical. However, you must be sure to remove the ground plane from beneath the inductor. The ground plane reduces the inductance of the inductor.

Place all of the components as close to the IC as you practically can. This lowers the effect of parasitic inductance and the radiation and reception of interference along the traces. Certainly a surface mount layout should be quieter than a through-hole design. Keep the leads in the main path components as wide and short as possible. This lowers their resistance and inductance. These components include C_{in}, the IC common (pin 3), the IC $\overline{ON}/OFF$ (pin 5), D1, and C_{out}. However, make the connection to the IC output pin (pin 1) as narrow and short as is consistent with the current it must carry. This lowers the opportunity for coupling noise from this power switching trace to adjacent sensitive pins.

The feedback resistors R_f and R_i and the capacitor C_f should be placed as close to the feedback pin (pin 4) as possible. This way the low level signal ($1.23V_{dc}$) is connected directly to the electronics within the IC, and does not run through a long lead where it can pick up interference.

Be sure also to look carefully at the possibility that the magnetic field from the inductor may intercept a conductor, and induce an unwanted signal. This is particularly a problem along the feedback path (R_f and R_i), the IC's common pin (pin 3), and the output capacitor. Try to place the inductor so that magnetic coupling into these components is minimized.

Example 7-5

Design a buck regulator using a LM2595 to convert an unregulated input voltage of 12 V_{dc} to 5 V_{dc} into 10 Ω. Select all components. Assuming that $T_A = 80°C$, determine the heat sink needed.

Solution

This example uses the same parameters as in the previous discrete buck regulator examples.

$$D = \frac{V_{load}}{E_{unreg}} = \frac{5\,V}{12\,V} = 42\%$$

The period is $T = \dfrac{1}{f} = \dfrac{1}{150\,\text{kHz}} = 6.7\,\mu s$

The transistor is *on* for

$$\Delta t = 0.42 \times 6.7\,\mu s = 2.8\,\mu s$$

The load current is

$$I_{\text{dc load}} = \frac{V_{\text{load}}}{R_{\text{load}}} = \frac{5\,V_{\text{dc}}}{10\,\Omega} = 500\,\text{mA}_{\text{dc}}$$

Pick the inductor ripple current at about

$$\Delta i_{\text{pp}} = 0.25 \times 500\,\text{mA} = 125\,\text{mA}_{\text{pp}}$$

The voltage across the inductor is

$$v_{\text{L}} = E_{\text{unreg}} - V_{\text{load}} = 12\,V_{\text{dc}} - 5\,V_{\text{dc}} = 7\,V_{\text{dc}}$$

The fundamental relationship for an inductor is

$$v_{\text{L}} = L\frac{\Delta i}{\Delta t}$$

Solving this for the value of the inductor gives

$$L = v_{\text{L}}\frac{\Delta t}{\Delta i} = 7\,\text{V}\frac{2.8\,\mu s}{125\,\text{mA}} = 157\,\mu H$$

A standard value above this minimum is 220 μH. You must also specify the continuous and peak current through the inductor.

$$I_{\text{dc L}} = I_{\text{dc load}} = 500\ \text{mA}_{\text{dc}}$$

$$I_{\text{p L}} = 500\,\text{mA}_{\text{dc}} + \frac{125\,\text{mA}_{\text{pp}}}{2} = 563\,\text{mA}_{\text{p}}$$

From Table 7-1, C_{out} = 220 μF/10 V tantalum.

The input capacitor should be

$$C_{\text{in}} > 100\ \mu F$$

Its current rating should be

$$I_{\text{Cin}} > 0.5 I_{\text{dc load}} = 250\,\text{mA}_{\text{rms}}$$

The voltage rating of the input capacitor is

$$V_{\text{Cin}} > 1.5 E_{\text{max unreg}} = 18\,\text{V}$$

The output voltage is set by the reference voltage of 1.23 V and the feedback resistors.

$$V_{\text{load}} = 1.23\,\text{V}\left(1 + \frac{R_f}{R_i}\right)$$

Rearranging this a little and entering the input voltage gives

$$\frac{R_f}{R_i} = \frac{5\,V_{dc}}{1.23\,V_{dc}} - 1 = 3.07$$

A few minutes of trying different combinations of standard values lead to $R_i = 3.3\,\text{k}\Omega$ and $R_f = 10\,\text{k}\Omega$.

The LM2595 must dissipate

$$P_{\text{LM2595}} = I_{\text{quies}} E_{\text{unreg}} + DI_{\text{dc load}} V_{\text{sat max}}$$

$$P_{\text{LM2595}} = 10\,\text{mA} \times 12\,\text{V} + 0.42 \times 500\,\text{mA} \times 1\,\text{V} = 0.33\,\text{W}$$

To keep the IC from overheating, select a heat sink that keeps the junction temperature below 150°C.

$$\Theta_{\text{SA}} < \frac{T_{\text{Jmax}} - T_{\text{A}}}{P} - \Theta_{\text{JC}} - \Theta_{\text{CS}}$$

Since the tab is tied to common, within the IC, the heat sink can be bolted directly to the IC without an electrical insulator.

$$\Theta_{\text{CS}} = 0.2\frac{°\text{C}}{\text{W}}$$

$$\Theta_{\text{SA}} < \frac{150°\text{C} - 80°\text{C}}{0.33\,\text{W}} - 2\frac{°\text{C}}{\text{W}} - 0.2\frac{°\text{C}}{\text{W}} = 210\frac{°\text{C}}{\text{W}}$$

Since the bare case thermal resistance of the IC is 50°C/W, far below the calculated maximum, no heat sink is needed.

Practice: Select the inductor, output capacitor, R_i R_f, and heat sink for a LM2595 buck regulator with an input of 28 V_{dc}, a regulated output of 15 V_{dc} at 800 mA$_{dc}$.

Answers: $L \approx 220\,\mu\text{H}$ @ 800 mA$_{dc}$, 900 mA$_p$, $C_{\text{out}} = 68\,\mu\text{F}/20$ V tantalum, $R_i = 1\,\text{k}\Omega$, $R_f = 11.2\,\text{k}\Omega$, $\Theta_{\text{SA}} < 3.4\ °\text{C/W}$

7.2 Boost Regulators

Basics

By rearranging the components of the buck regulator, you can build a circuit that steps the output *up*, producing an output that is higher than the input. Look at Figure 7-7. The inductor, transistor and diode have been rotated. Also, a low side switch is used.

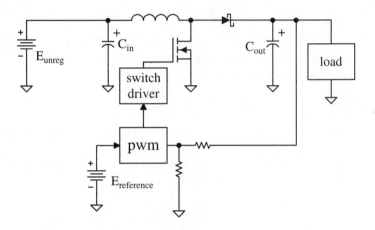

Figure 7-7 Boost regulator block diagram

When the pulse width modulator turns the transistor *on*, the node between the inductor and the diode is connected to circuit common. Since the capacitor, C_{out}, is charged to V_{load}, the cathode of the diode is held at a potential that is more positive than that of its anode. This reverse biases the diode and it turns *off*.

With the transistor turned hard *on*, the right end of the inductor is at common and the left end is at E_{unreg}. So the entire input voltage is across the inductor. It charges, storing energy in its magnetic field.

$$v_{inductor} = E_{unreg} = L\frac{\Delta i}{\Delta t}$$

Δi = the peak-to-peak variation in the current through the inductor
Δt = the time that the switch is *on*.

This relationship allows you to determine how much the current through the inductor varies, if you are given a circuit to analyze.

$$\Delta i = \frac{E_{\text{unreg}} \Delta t}{L}$$

Alternatively, if you are designing a boost regulator, you can set the amount of ripple current allowed through the inductor by selecting the value of the inductor. Generally the ripple current is set to about 25% of the average *input* current.

$$L = \frac{E_{\text{unreg}} \Delta t}{\Delta i}$$

Inductor specifications

The inductance is only one of three specifications needed to select an inductor. You must also know its average and its peak current. It is reasonable to expect that the boost regulator is about 90% efficient. That means

$$P_{\text{load}} = 0.9 P_{\text{in}}$$

Since both the input and load are, more or less, dc,

$$V_{\text{dc load}} I_{\text{dc load}} = 0.9 E_{\text{unreg}} I_{\text{in}}$$

The inductor is connected directly to the input source, so the input average current is also the inductor's average current, Figure 7-8.

$$I_{\text{ave L}} = \frac{V_{\text{dc load}} I_{\text{dc load}}}{0.9 E_{\text{unreg}}}$$

The peak inductor current is

$$I_{\text{p L}} = I_{\text{dc L}} + \frac{\Delta i}{2}$$

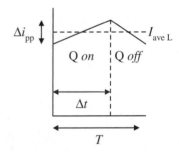

Figure 7-8 Inductor current

When the transistor turns *on*, current flows from the input source, through the inductor, and then through the transistor. Of course, when the transistor turns *off*, there is no current flow though it. The length of time the transistor is *on* is defined as the duty cycle, *D*. This wave shape is shown in Figure 7-9.

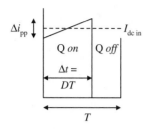

Figure 7-9 Transistor current

In the previous section, a calculation of the rms value of a trapezoidal wave with a 25% ripple was compared to the approximation that the wave is a simple rectangular pulse with a flat top. There is very little difference. So for simplicity, assume that the current waveform through the transistor is a simple rectangular pulse whose height is $I_{dc\ in}$ and whose duration is DT. This means that

$$I_{ave\ Q} = DI_{dc\ in}$$

$$I_{rms\ Q} = \sqrt{D}\,I_{dc\ in}$$

When the transistor goes *off*, the junction of the inductor, transistor, and diode is no longer held at circuit common. The magnetic field begins to fall, cutting the inductor. This generates a voltage across the inductor that is *opposed* to the polarity induced when the field expanded. That is, the voltage polarity across the inductor reverses polarity. This places a + at the node, driving the anode of the Schottky diode up above its cathode. The diode goes *on* when the transistor is *off*.

Transistor specifications

The diode's current is shown in Figure 7-10. Notice that the diode is *on* for $(1-D)T$. This is the rest of the cycle. Using the same rectangular pulse approximation that was used for the transistor gives

Diode specifications

$$I_{ave\ diode} = (1-D)I_{dc\ in}$$

$$I_{rms\ diode} = \sqrt{(1-D)}\,I_{dc\ in}$$

The boost regulator outputs a voltage that is higher than its input. Where does this extra voltage come from? Look at Figure 7-11(a). When the transistor turns *on*, the right end of the inductor is connected to circuit common. The current through the inductor ramps up as energy is stored in its magnetic field. These field lines build up. The inductor is acting as a load. The voltage polarity is + to −.

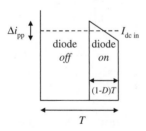

Figure 7-10 Diode current

When the transistor turns *off*, the magnetic field begins to collapse. As it falls, it cuts the wire of the inductor, *generating* a voltage. Since the direction of the field movement is now opposite to its direction when the field was building, the induced voltage reverses direction, − to +.

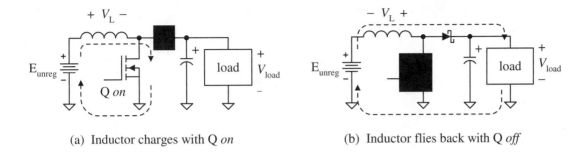

(a) Inductor charges with Q *on*   (b) Inductor flies back with Q *off*

Figure 7-11 Inductor charge and flyback generates the boost voltage

The magnitude of this induced voltage is

$$v_L = L\frac{\Delta i}{\Delta t} = L\frac{\Delta i}{(1-D)T}$$

where Δi is the peak-to-peak ripple current through the inductor

$\Delta t = (1 - D)T$, the time the inductor discharges

This voltage across the inductor is in series with the input unregulated voltage. These two voltages add.

$$v_{load} = E_{unreg} + v_L$$

The input voltage is increased by the flyback voltage induced across the inductor. Using these relationships, you can determine the output voltage. When first talking about the inductor, its value was derived.

$$L = \frac{E_{unreg}\Delta t}{\Delta i} = \frac{E_{unreg}DT}{\Delta i}$$

where $\Delta t = DT$, the time that the inductor charges.

Substituting this value of the inductor into the equation for the voltage across the inductor gives

$$v_L = \frac{E_{unreg}DT}{\Delta i}\frac{\Delta i}{(1-D)T}$$

This simplifies to $v_L = \dfrac{DE_{unreg}}{(1-D)}$

The output voltage is

$$v_{\text{load}} = E_{\text{unreg}} + v_{\text{L}} = E_{\text{unreg}} + \frac{DE_{\text{unreg}}}{(1-D)}$$

Obtaining a common denominator

$$v_{\text{load}} = \frac{(1-D)+D}{(1-D)} E_{\text{unreg}}$$

Load voltage   Finally
$$v_{\text{load}} = \frac{E_{\text{unreg}}}{(1-D)}$$

Since $D < 1$, the denominator is also less than 1. Dividing by a number smaller than 1 results in an answer larger than the numerator. The load voltage is larger than the input. That's right, it's a *boost* regulator.

Switching regulators are very efficient. Typically

$$P_{\text{load}} = 0.9 P_{\text{in}}$$

$$V_{\text{dc load}} I_{\text{dc load}} = 0.9 E_{\text{unreg}} I_{\text{dc in}}$$

Input current

$$I_{\text{dc in}} = \frac{V_{\text{dc load}} I_{\text{dc load}}}{0.9 E_{\text{unreg}}}$$

The load voltage is larger than the input voltage, so the input current is larger than the load current. You "pay" for the step up in voltage by providing more current from the input than the load requires.

When the transistor turns *off*, and the inductor flies back, the diode turns *on*. Current is sent to power the load and to charge the capacitor. But when the transistor turns *on*, the diode's anode is connected to common while its cathode is held at $V_{\text{dc out}}$ by the charge on the capacitor. The diode turns *off*. During the entire time that the transistor is *on*, the output voltage is provided by the capacitor. It discharges into the load resistance. Given long enough, the output voltage would follow the familiar RC discharge equation

$$v_{\text{load}} = V_{\text{dc load}} e^{-\frac{t}{RC}}$$

But the capacitor has to support the load for only a short while, before the diode turns *on* again, allowing the capacitor to recharge. So, the discharge is linear.

$$i_{\text{load}} = C_{\text{out}} \frac{dv_{\text{load}}}{dt}$$

$$I_{\text{dc load}} = C_{\text{out}} \frac{\Delta v_{\text{load}}}{\Delta t}$$

Solving this for C_{out}

$$C_{\text{out}} = \frac{DTI_{\text{dc load}}}{\Delta v_{\text{load}}}$$

Output capacitor rating

where D is the duty cycle
T is the period of the switching signal
Δv_{load} is the amount of variation in the load voltage

Although your load may be able to tolerate several hundred millivolts of variation, this voltage is also fed back to the pulse width modulator and controller. Keep $\Delta v_{\text{load}} < 20 \ \text{mV}_{\text{pp}}$. Otherwise, the controller may oscillate.

Also remember to select a capacitor that is rated to handle the load current at the frequency produced by the pulse width modulator. A cheap, electrolytic capacitor designed as a filter for a line frequency power supply produces disappointing results.

Example 7-6

Design a boost regulator that converts an input voltage of 12 V_{dc} to a load voltage of 22 V_{dc} at 0.7 A_{dc}. Use a controller frequency of 100 kHz. Assume that the transistor has an *on* resistance of 0.4 Ω and that the diode drops 0.5 V_p when it is conducting.

Calculate the power and current that must be provided to the regulator, the inductor's inductance and current ratings, the power that the transistor and the diode must dissipate, and the value of the output capacitor.

Solution

$$P_{\text{load}} = V_{\text{dc load}} I_{\text{dc load}}$$

$$P_{\text{load}} = 22 \, V_{\text{dc}} \times 0.7 \, A_{\text{dc}} = 15.4 \, \text{W}$$

$$P_{\text{load}} = 0.9 P_{\text{supply}}$$

$$P_{\text{supply}} = \frac{P_{\text{load}}}{0.9} = 17.1 \, \text{W}$$

$$P_{\text{supply}} = E_{\text{unreg}} I_{\text{in}}$$

$$I_{\text{dc in}} = \frac{P_{\text{supply}}}{E_{\text{unreg}}} = \frac{17.1\,\text{W}}{12\,\text{V}_{\text{dc}}} = 1.4\,\text{A}_{\text{dc}}$$

For the inductor, select

$$\Delta i_{\text{pp}} = 0.25 I_{\text{dc in}} = 350\,\text{mA}_{\text{pp}}$$

$$I_{\text{dc L}} = I_{\text{dc in}} = 1.4\,\text{A}_{\text{dc}}$$

$$I_{\text{p L}} = I_{\text{dc L}} + \frac{\Delta i_{\text{pp}}}{2}$$

$$I_{\text{p L}} = 1.4\,\text{A}_{\text{dc}} + \frac{0.35\,\text{A}_{\text{pp}}}{2} = 1.6\,\text{A}_{\text{p}}$$

The duty cycle of the switch is

$$D = 1 - \frac{E_{\text{unreg}}}{V_{\text{load}}}$$

$$D = 1 - \frac{12\,\text{V}_{\text{dc}}}{22\,\text{V}_{\text{dc}}} = 0.45$$

The switch is *on* for Δt.

$$\Delta t = DT = 0.45 \times 10\,\mu\text{s} = 4.5\,\mu\text{s}$$

For the inductor,

$$v_{\text{L}} = \text{L}\frac{di}{dt}$$

$$\text{L} = v_{\text{L}}\frac{\Delta t}{\Delta i}$$

$$\text{L} = 12\,\text{V}_{\text{dc}}\frac{4.5\,\mu\text{s}}{0.35\,\text{A}_{\text{pp}}} = 154\,\mu\text{H}$$

For the transistor, the current is (more or less) a rectangle with a peak equal to the dc from the input supply. This current produces a rectangular voltage across the resistance of the turned *on* transistor.

$$V_{pQ} = 1.4 \, A_p \times 0.4 \, \Omega = 0.56 \, V_p$$

With a rectangular voltage across it and a rectangular current through it, the transistor must dissipate

$$P_Q = DV_p I_p$$

$$P_Q = 0.45 \times 0.56 \, V_p \times 1.4 \, A_p = 353 \, mW$$

The diode also must dissipate

$$P_{diode} = DV_p I_p$$

Remember that the diode is *on* when the transistor is *off.*

$$P_{diode} = (1 - 0.45) \times 0.5 \, V_p \times 1.4 \, A_p = 390 \, mW$$

The output capacitor discharges through the load during the time that the transistor is *on.*

$$i_C = C \frac{dv}{dt}$$

$$C = I_{dc \ load} \frac{\Delta t}{\Delta v}$$

$$C = 0.7 \, A \frac{4.5 \, \mu s}{20 \, mV} = 158 \, \mu F$$

Practice: Repeat these calculations for a second boost regulator that could convert the 22 V_{dc} at 0.7 A_{dc} to 56 V_{dc}. Notice that this problem specifies the *supply* current.

Answers: I_{load} = 248 mA$_{dc}$, D = 0.61, L = 767 μH, P_Q = 120 mW, P_{diode} = 137 mW, C = 76 μF

LM2585 Simple Switcher® Flyback Regulator IC

National Semiconductor integrates the electronics necessary to build a boost switching regulator into a single IC, the LM2585. Figure 7-12 combines this IC with the other components to implement the boost regulator.

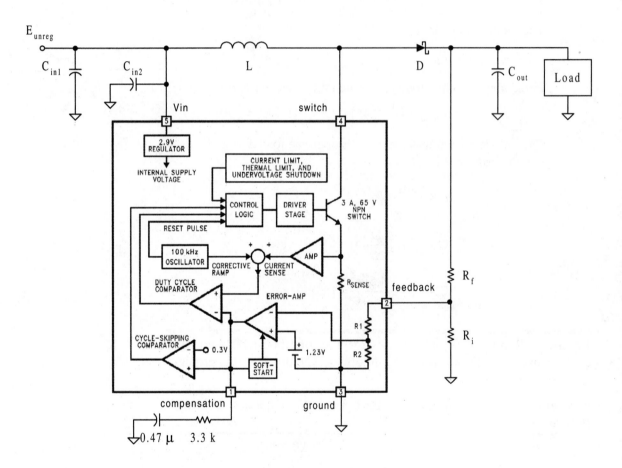

Figure 7-12 Simple Switcher® boost regulator (*courtesy of National Semiconductor*)

There is an internal, 100 kHz oscillator.   You need to connect *no* external components to create this signal.

Four versions are available.   The 3.3 V, 5 V, and 12 V use fixed internal resistors at the feedback pin to set the regulated load voltage.   The ADJ, adjustable version, removes the resistors and connects the feedback pin directly to the inverting input of the error amp.   You then supply R_f and R_i externally to set the load voltage to whatever level you need.   The key restriction is

$$E_{unreg} < V_{load} < 65\,V_{dc}$$

Keep $$4\,V_{dc} < E_{unreg} < 40\,V_{dc}$$

The feedback voltage is compared to the internal reference, 1.23 V_{dc}. As with the buck regulator and any noninverting, negative feedback circuit,

$$V_{dc\ load} = 1.23\,V_{dc}\left(1 + \frac{R_f}{R_i}\right)$$

The manufacturer recommends that

$$1\,k\Omega < R_i < 5\,k\Omega$$

The current through the internal switch is limited to 3 A_p.   This is the same current calculated for the transistor at the beginning of the basic boost regulator, and shown in Figure 7-9.   Assuming 90% efficiency,

$$I_{p\ Q} = I_{dc\ in} = I_{ave\ L} = \frac{V_{dc\ load}I_{dc\ load}}{0.9E_{unreg}} < 3\,A_p$$

However, if the load shorts, none of the inductor current flows through the internal switch.   It all goes to the shorted load.   The R_{sense} shown in Figure 7-12 never receives any current.   So the control does not know that there is an external short.   The inductor saturates very quickly, and looses its ability to limit the load current.   Be sure to include some form of external short circuit load protection.

When power is first applied, the inductor and the capacitor hold no charge.   The load voltage is zero.   Sensing this, the controller turns its internal switch on, and keeps it on, sending the duty cycle to 100%.   Increasing the duty cycle *should* increase the output voltage.   But if the transistor is kept *on*, it never turns *off*, so the inductor can never fly back, producing the boosted voltage, turning on the diode and building up the output voltage. The output voltage cannot be generated because the switch won't turn *off*. But the switch won't turn *off* because the output voltage is too low. The current ramps up and the inductor and IC may be damaged.

The soft-start circuit forces the duty cycle to be small when power is first applied. Over many cycles the compensation resistor and capacitor allow the duty cycle to gradually grow, gently increasing the inductor and switch current and allowing the load voltage to grow. Depending on input and output voltages, you may need to alter C_C above or below the 0.47 µF to provide a soft-start and stable steady-state operation.

The temperature of the silicon is monitored. Should it become too high, the IC is turned *off* until the chip cools. This thermal shutdown is a fail-safe. Do not use it as a regular way to limit load current during nominal operation. Routinely exposing any silicon device to repeated temperature cycles seriously degrades the part's performance, and shortens its life. Select a heat sink that keeps the junction temperature below 110°C.

The major element that determines the power that the IC must dissipate is the power of the main switch.

$$P_{\text{switch}} = \left(I_{\text{rms Q}}\right)^2 R_{\text{on Q}}$$

$$I_{\text{rms Q}} = \sqrt{D}\, I_{\text{dc in}}$$

The manufacturer specifies that

$$R_{\text{on Q}} = 0.15\,\Omega$$

The other electronics within the IC also use a small amount of current. It is about 2% of the current from E_{unreg}.

$$P_{\text{IC bias}} = \frac{1}{50} D I_{\text{dc in}} E_{\text{unreg}}$$

So the total power that the IC turns into heat is

$$P_{\text{IC total}} = P_{\text{switch}} + P_{\text{IC bias}}$$

There are two final precautions you must observe to build a practical circuit with the LM2585. First, two input capacitors are needed. The larger, C_{in1}, provides a pool of charge for the switch when it turns *on*. It should be at least 100 µF, able to work at the 100 kHz of the internal oscillator, and supply the current that the IC requires. Capacitor C_{in2}

LM2585 thermal parameters
$T_{J \text{ upper limit}} = 110°C$
$\Theta_{JA \text{ bare case}} = 65°C/W$
$\Theta_{JC} = 2°C/W$

should be a 1 µF ceramic. This type capacitor is much faster than the 100 µF electrolytic, allowing it to respond to the turn *on* transients.

The other practical issue is the physical layout of the circuit. All of the precautions presented for the LM2595 buck regulator are doubly important for this boost regulator, because the voltages and the transients are so much larger. Use a ground plane, except under the inductor. Keep the leads to the capacitors and between the capacitors and the ground pin of the IC as short as possible. Bring R_f and R_i as close to the feedback pin as possible. Assure a separate short return from the load to the unregulated input supply's common.

Example 7-7

Design a boost regulator that converts an input voltage of 12 V_{dc} to a load voltage of 22 V_{dc} at 0.7 A_{dc} using the LM2585.

Calculate the feedback resistors, power, and current that must be provided to the regulator, the power that the IC must dissipate, and the junction temperature assuming $T_A = 80°C$ and that there is *no* heat sink.

This is a continuation of Example 7-6.

Solution

$$V_{dc\ load} = 1.23\,V_{dc}\left(1+\frac{R_f}{R_i}\right)$$

$$\frac{R_f}{R_i}=\frac{V_{dc\ load}}{1.23\,V_{dc}}-1$$

$$\frac{R_f}{R_i}=\frac{22\,V_{dc}}{1.23\,V_{dc}}-1=16.9$$

$$R_f = 16.9\times R_i$$

Pick $R_i = 1.3$ kΩ $R_f = 22$ kΩ

From the beginning of Example 7-6

$$I_{dc\ in}=\frac{P_{supply}}{E_{unreg}}=\frac{17.1W}{12\,V_{dc}}=1.4\,A_{dc}$$

This is well below the 3A_p rating of the IC switch.

$$D = 1 - \frac{E_{\text{unreg}}}{V_{\text{load}}}$$

$$D = 1 - \frac{E_{\text{unreg}}}{V_{\text{load}}} = 1 - \frac{12\,\text{V}_{\text{dc}}}{22\,\text{V}_{\text{dc}}} = 0.45$$

For the switch $I_{\text{rms Q}} = \sqrt{D}\,I_{\text{dc in}}$

$$I_{\text{rms Q}} = \sqrt{0.45} \times 1.4\,\text{A}_{\text{dc}} = 0.94\,\text{A}_{\text{rms}}$$

The main switch must dissipate

$$P_{\text{switch}} = \left(I_{\text{rms Q}}\right)^2 R_{\text{Q on}}$$

$$P_{\text{switch}} = \left(0.94\,\text{A}_{\text{rms}}\right)^2 \times 0.15\,\Omega = 132\ \text{mW}$$

The rest of the IC dissipates

$$P_{\text{IC bias}} = \frac{1}{50} D I_{\text{dc in}} E_{\text{unreg}}$$

$$P_{\text{IC bias}} = \frac{1}{50} \times 0.45 \times 1.4\,\text{A}_{\text{dc}} \times 12\,\text{V}_{\text{dc}} = 151\ \text{mW}$$

$$P_{\text{total IC}} = 132\,\text{mW} + 151\,\text{mW} = 283\,\text{mW}$$

The IC temperature is

$$T_{\text{J}} = 80°\text{C} + 0.28\,\text{W} \times 65\frac{°\text{C}}{\text{W}} = 98.4\,°\text{C}$$

A heat sink is *not* needed.

Practice: Repeat these calculations for a second boost regulator that could convert the 22 V_{dc} at 0.7 A_{dc} to 56 V_{dc}. Notice that this problem specifies the *supply* current.

Answers: R_i = 2.7 kΩ, R_f = 120 kΩ, D = 0.61, P_{IC} = 233 mW, T_{J} = 95°C

7.3 Line Voltage Flyback Converter

The buck regulator prooduces a single voltage, lower than its input. The boost's single output is always above its input. If the input voltage begins above the output voltage, but falls below the output voltage (such as with a battery discharging), neither alone works. If you need several different output voltages, you must have several different buck and/or boost regulators, each complete with its own IC, inductor, diode, capacitors, and feedback.

The flyback converter replaces the inductor of the boost with a transformer primary. There may be as many secondaries as you need. And for each you need only a diode and capacitor. There are no inductors.

Now that there is a transformer in the system, the large, heavy, expensive low frequency line transformer can be replaced. The large line voltage can be *directly* rectified and filtered, sending a high voltage dc to the flyback transformer. This transformer provides the isolation between the line and your regulated voltages. Since the flyback converter is running above audio frequencies, that transformer is small, light, available as a surface mount component, and less expensive.

The Flyback Converter

A flyback converter using the LM2585 is shown in Figure 7-13. The inductor of the boost has been replaced with a transformer. It may have as many secondaries as you need. The output voltages and their polarities are determined by the transformer's turns ratio, the duty cycle of the switch, the input voltage, and how you decide to connect the rectifiers and commons.

Each secondary is rectified with a Schottky diode and then filtered with a capacitor. The secondary provides the inductance for each output filter. Separate inductors are not needed.

The voltage from the most critical output is fed back to the controller. This is called the **master** output. At the controller this voltage is compared to the IC's internal reference. If the feedback voltage is too low, the switch is driven *on* longer each cycle. If the voltage is too high, the *on* time is reduced. Altering the *on* time changes the output voltage of the outputs, bringing the master output back to its correct value.

The other outputs are referred to as **slaves**. This scheme provides good regulation in the slave outputs for changes in the unregulated input voltage, and about 8% variation for changes in the slave outputs due to changes in their loads.

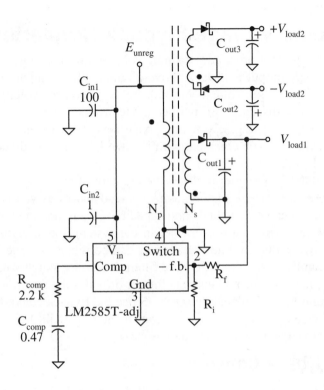

Figure 7-13 Simple flyback converter

The key to a flyback regulator's operation is that the primary and the secondaries are *out of phase*. Notice the polarity dots on the transformer in Figure 7-13. When the switch within the IC turns *on*, the bottom of the primary is tied to common. This makes the top of the primary, the dot, positive. On the secondaries, the dots are induced positive, the non-dotted ends negative. All of the diodes are *off*.

Since no current flows in any of the secondaries, the primary acts as a simple inductor. Current ramps up in it, just as it did in the boost regulator. When the switch turns *off*, the field in the core reverses direction, and begins to fall. This reversal of direction changes the polarities of the voltage at all of the secondary windings. The dots are induced positive. The diodes turn *on* and energy is transferred to the output capacitors and loads. The loads are powered on the **flyback** part of each cycle, thus the converter's name.

For signals that vary with time, whether they are sinusoids or some other shape, voltages and currents are induced in the winding of the transformer based on the ratio of the number of turns on the primary compared to the number of turns of wire on each secondary. You have seen these relationships before when studying ac circuits. Remember, the voltage follows the turns ratio, but the current is the inverse of that ratio.

$$\frac{v_{pri}}{v_{sec}} = \frac{N_{pri}}{N_{sec}} \qquad \frac{i_{pri}}{i_{sec}} = \frac{N_{sec}}{N_{pri}}$$

At the beginning of the cycle, the transistor turns *on*. This connects the lower end of the primary to common, and places the unregulated input voltage, E_{unreg}, across the primary. All of the rectifiers on the secondaries go *off*. This continues for the *on* time, t_{on}.

When the transistor turns *off*, the secondary current steps to the peak coupled from the primary. It then ramps down, keeping V_{load} across the secondary. This voltage is coupled back into the primary by the turns ratio N_{pri} / N_{sec}. Look at Figure 7-14.

When the transistor turns *off*, the collapsing field induces the voltage

$$V_{load} \frac{N_{pri}}{N_{sec}}$$

across the primary. Its polarity is opposite to that when the transistor was *on*. That is, the non-dotted end is positive. The drain of the transistor is driven up by this voltage. It is in series aiding with the supply voltage, E_{unreg}. This is the maximum voltage the transistor must tolerate.

$$V_{Q\,off} = E_{unreg} + V_{load} \frac{N_{pri}}{N_{sec}}$$

When the transistor is *on*, there is E_{unreg} across the primary. Current ramps up in the primary, inducing a secondary voltage of

$$V_{sec} = E_{unreg} \frac{N_{sec}}{N_{pri}}$$

The polarity is such as to reverse bias the diode. The output capacitor holds V_{load} on the cathode. So their maximum reverse voltage is

$$V_{diode\,off} = E_{unreg} \frac{N_{sec}}{N_{pri}} + V_{load}$$

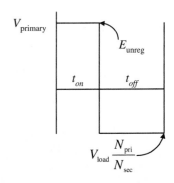

Figure 7-14 Primary voltage

Maximum transistor voltage

Maximum diode *off* voltage

Since the primary has very little dc resistance, it is critical that the average voltage across the primary be zero. The area above the curve must be made to equal the area below the curve. Otherwise there is a net dc voltage across the transformer primary causing large dc primary currents, saturation of the magnetics, and overheating of the transformer. The turns ratio, input and load voltages, and the period are all fixed parameters of the design. But you (or the controller IC) have some control over the *on* time.

For the average value to be zero

$$area\ above\ x\ axis = area\ below\ x\ axis$$

$$E_{unreg} t_{on} = V_{load} \frac{N_{pri}}{N_{sec}} t_{off}$$

$$t_{off} = T - t_{on}$$

Substitute this for t_{off} and solve for V_{load}.

$$V_{load} = \frac{E_{unreg} t_{on}}{\dfrac{N_{pri}}{N_{sec}} (T - t_{on})}$$

As with the buck and the boost regulator, the duty cycle is defined as

$$D = \frac{t_{on}}{T} \quad \text{or} \quad t_{on} = DT$$

Substitute DT for t_{on} and simplify.

Load voltage dc

$$V_{load} = \frac{E_{unreg} D}{\dfrac{N_{pri}}{N_{sec}} (1 - D)}$$

A few steps of rearranging this equation isolate the duty cycle.

Duty cycle

$$D = \frac{\dfrac{N_{pri}}{N_{sec}} V_{load}}{E_{unreg} + \dfrac{N_{pri}}{N_{sec}} V_{load}}$$

The power delivered to the load is

$$P_{load} = V_{dc\ load} I_{dc\ load}$$

Both the load voltage and load current are specifications of the circuit.

Switching regulators are very efficient, sending to the load approximately 90% of the power provided by the unregulated source.

$$P_{\text{load}} = 0.9 P_{\text{in}}$$

Or

$$P_{\text{in}} = \frac{P_{\text{load}}}{0.9}$$

The input power is provided from E_{unreg} and the average value of the current sent into the primary.

$$E_{\text{unreg}} I_{\text{ave pri}} = P_{\text{in}}$$

$$I_{\text{ave pri}} = \frac{P_{\text{in}}}{E_{\text{unreg}}}$$

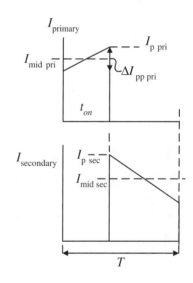

The primary and secondary currents are shown in Figure 7-15. When the transistor turns *on*, current from the collapsing secondary field induces a step in the primary current. With a constant voltage across the primary, its current then begins to ramp up. This induces the secondary voltage that turns the diodes *off*. So the primary is acting as a simple inductor.

$$v_{\text{L}} = L \frac{di}{dt}$$

The voltage across the inductor is the constant unregulated input, and the *dt* is the *on* time, $t_{on} = DT$. So the current ramps up.

$$\Delta i_{\text{pp pri}} = \frac{E_{\text{unreg}} DT}{L_{\text{pri}}}$$

Or

$$L_{\text{pri}} = \frac{E_{\text{unreg}} DT}{\Delta i_{\text{pp pri}}}$$

Figure 7-15 Flyback transformer currents

In previous sections of this chapter it was decided that the average value and rms value of a trapezoid was practically the same as for a rectangular wave with a peak value equal to the middle of the ramp, $I_{\text{mid pri}}$.

$$I_{\text{dc pri}} = I_{\text{mid pri}} D$$

$$I_{\text{rms pri}} = I_{\text{mid pri}} \sqrt{D}$$

Primary currents

$$I_{p\ pri} = I_{mid\ pri} + \frac{\Delta i_{pp\ pri}}{2}$$

When the transistor turns *off*, the primary current is coupled into the secondary as the diodes are biased *on*. Since current does not change instantaneously through an inductor, the peak primary current is coupled to the secondary.

$$I_{p\ sec} = \frac{N_{pri}}{N_{sec}} I_{p\ pri}$$

The average of the secondary current is the load dc current, usually a specification. Remember that the secondary ramp is happening during the *off* time, $(1 - D)T$.

$$I_{dc\ sec} = I_{dc\ load} = I_{mid\ sec}(1 - D)$$

Secondary currents

$$I_{rms\ sec} = I_{mid\ sec}\sqrt{1 - D}$$

During the entire time that the transistor is *on*, the diode is *off* and the output voltage is provided by the capacitor. It discharges into the load resistance. The capacitor has to support the load for only a short while before the diode turns *on* again, allowing the capacitor to recharge. So, the discharge is virtually linear.

$$i_{load} = C_{out}\frac{dv_{load}}{dt}$$

$$I_{dc\ load} = C_{out}\frac{\Delta v_{load}}{\Delta t}$$

Solving this for C_{out}

Output capacitor rating

$$C_{out} = \frac{t_{on}I_{dc\ load}}{\Delta v_{load}}$$

Although your load may be able to tolerate several hundred millivolts of variation, this voltage is also fed back to the pulse width modulator and controller. Keep $\Delta v_{load} < 20\ \text{mV}_{pp}$. Otherwise, the controller may oscillate.

Example 7-8

Design a flyback converter to meet the following specifications.

$$E_{unreg} = 12 \text{ V}_{dc} \qquad V_{load} = 5 \text{ V}_{dc} \qquad I_{load} = 1 \text{ A}_{dc} \qquad T = 10 \text{ µs}$$

Solution

There are only a few off-the-shelf switching transformers commonly available So, select an available transformer and then confirm that it will work in the circuit.

From Pulse Engineering, the PE-68421 transformer is designed to work with the LM2585

$$\frac{N_{pri}}{N_{sec}} = \frac{1}{0.8} = 1.25 \text{ (characteristic of PE-68421)}$$

$$L_{pri} = 85 \text{ µH} \qquad \text{(characteristic of PE-68421)}$$

The maximum voltage across the transistor when it is *off* is

$$V_{Q\ off} = E_{unreg} + V_{load} \frac{N_{pri}}{N_{sec}}$$

$$V_{Q\ off} = 12 \text{ V}_{dc} + 5 \text{ V}_{dc} \times 1.25 = 18.3 \text{ V}_{dc}$$

Transistor (IC) maximum voltage

The maximum voltage across the diode when it is *off* is

$$V_{diode\ off} = E_{unreg} \frac{N_{sec}}{N_{pri}} + V_{load}$$

$$V_{diode\ off} = 12 \text{ V}_{dc} \frac{0.8}{1} + 5 \text{ V} = 14.6 \text{ V}$$

Diode maximum voltage

$$D = \frac{\dfrac{N_{pri}}{N_{sec}} V_{load}}{E_{unreg} + \dfrac{N_{pri}}{N_{sec}} V_{load}}$$

$$D = \frac{1.25 \times 5 \text{ V}_{dc}}{12 \text{ V}_{dc} + 1.25 \times 5 \text{ V}_{dc}} = 0.34$$

The power delivered to the load is

$$P_o = 5 \text{ V}_{dc} \times 1 \text{ A}_{dc} = 5 \text{ W}$$

So the unregulated supply must provide

$$P_i = \frac{P_o}{0.9} = \frac{5\,\text{W}}{0.9} = 5.6\,\text{W}$$

The average current from the unregulated supply is

**Transistor (IC)
average current**

$$I_{\text{dc in}} = \frac{5.6\,\text{W}}{12\,\text{V}_{\text{dc}}} = 467\,\text{mA}_{\text{dc}}$$

This is also the average primary current.

$$I_{\text{dc pri}} = I_{\text{mid pri}} D$$

So

$$I_{\text{mid pri}} = \frac{I_{\text{dc pri}}}{D} = \frac{467\,\text{mA}_{\text{dc}}}{0.34} = 1.37\,\text{A}_p$$

The primary and transistor rms current is

$$I_{\text{rms pri}} = I_{\text{mid pri}} \sqrt{D}$$

$$I_{\text{rms pri}} = I_{\text{rms Q}} = 1.37\,\text{A}_p \sqrt{0.34} = 800\,\text{mA}_{\text{rms}}$$

**IC power and
heat sink**

You can now apply the LM2585 special power equation, from the previous section, to select the IC's heat sink.

The ramp's amplitude is controlled by the primary's inductance.

$$\Delta I_{\text{pp pri}} = \frac{E_{\text{unreg}} DT}{L_{\text{pri}}}$$

$$\Delta I_{\text{pp pri}} = \frac{12\,\text{V}_{\text{dc}} \times 0.34 \times 10\,\mu\text{s}}{85\,\mu\text{H}} = 480\,\text{mA}_{\text{pp}}$$

The ramp rises a total of 480 mA, from 240 mA below $I_{\text{mid pri}}$ to a peak of 240 mA above $I_{\text{mid pri}}$.

**Transistor (IC)
maximum current**

$$I_{\text{p pri}} = 1.37\,\text{A} + 0.24\,\text{A} = 1.61\,\text{A}_p$$

Be sure that the transistor can handle this peak current.

$$I_{\text{min pri}} = 1.37\,\text{A} - 0.24\,\text{A} = 1.13\,\text{A}$$

It is important to verify that the minimum is above 0 A. Otherwise the transformer's magnetic field would completely collapse every cycle. This discontinuous operation leads to much larger peak currents, and may confuse the IC enough to cause it to lose control of the circuit.

When the transistor turns *off*, this peak primary current induces into the secondary

$$I_{p\ sec} = \frac{N_{pri}}{N_{sec}} I_{p\ pri}$$

$$I_{p\ sec} = 1.25 \times 1.61\,A_p = 2.86\,A_p$$ **Diode maximum current**

Be sure that the diode can handle this peak current.

The secondary rms current flows into the output capacitor.

$$I_{dc\ sec} = I_{dc\ load} = I_{mid\ sec}(1-D)$$

$$I_{mid\ sec} = \frac{I_{dc\ load}}{(1-D)}$$

$$I_{mid\ sec} = \frac{1\,A_{dc}}{(1-0.34)} = 1.5\,A$$

$$I_{rms\ sec} = I_{mid\ sec}\sqrt{1-D}$$

$$I_{rms\ sec} = 1.5\,A\sqrt{1-0.34} = 1.2\,A_{rms}$$

Be sure that the capacitor you use can handle at least this much current at 100 kHz.

The output capacitor holds the load voltage steady.

$$C_{out} = \frac{t_{on}I_{dc\ load}}{\Delta v_{load}}$$ **Capacitor rating**

$$C_{out} = \frac{0.34 \times 10\mu s \times 1\,A_{dc}}{20\,mV_{pp}} = 170\mu F$$

Practice: Can the same transformer be used with an input of 5 V_{dc} to output 12 V_{dc} at 400 mA$_{dc}$? Calculate D, $I_{p\ pri}$ and $I_{min\ pri}$.

Answer: $D = 0.75$, $I_{p\ pri} = 1.6\,A_p$, and $I_{min\ pri} = 1.2\,A$
This transformer can be used with the LM2585 to set the input voltage up or down.

Line Voltage Input

All of the regulators you have seen so far in this chapter are dc to dc converters. But most electrical power is provided as a rather large ac. This signal must be efficiently reduced in amplitude. Then something must convert the sinusoid into a steady dc. And, for the safety of the equipment and the people using it, normally the dc should be isolated from the commercial power line.

To do all of this, a traditional power supply begins with a low frequency (50–60 Hz) step down transformer to reduce the voltage and provide isolation from the line. This is followed by a rectifier and line frequency filter capacitor. The largest, heaviest, most expensive component is the 50–60 Hz transformer.

But the switching regulator already has a transformer. It is small, light, and inexpensive because it operates at 150 kHz. You can use it to provide the required step down in voltage and the isolation. Just move the rectifier and the filter in front of the transformer. Convert the line voltage *first* to a large dc. Feed this large dc signal into the transformer of the switching regulator. Look at Figure 7-16.

But before you can work directly from the line, there are several concerns. First, and always, exercise extreme caution. **Line voltages are lethal.** This is not the low level analog or 5 V digital circuits you are too comfortable with.

> **Line voltages are lethal.
> Use extreme caution.**

- You must *not* move anything, or even touch *anything* while power is turned on.

- Use only one hand when working in the circuit. This keeps you from getting *across* a large voltage, completing the circuit.

- Be sure to wear shoes, and stand on an insulated pad.

- Always remove all conductive jewelry.

- Wear eye protection. A mistake or a faulty component may cause something to explode. An exploding part can easily end up in your eye.

- Work with a partner. If one of you becomes "hung" on the high voltage, the other can turn off the power and call help.

- Be sure that you have rapid access to call for emergency medical technicians.

- Become certified in Cardiopulmonary Resuscitation.

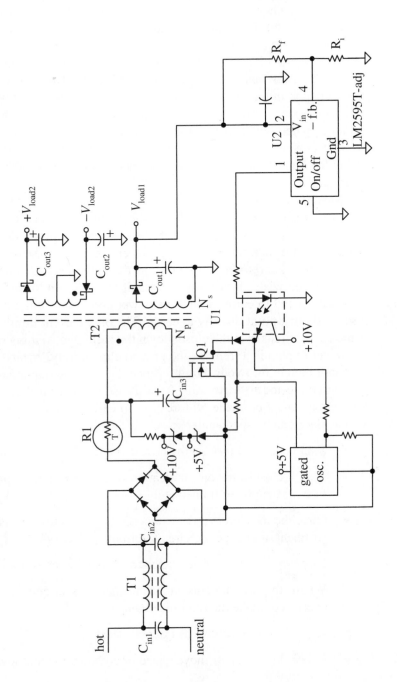

Figure 7-16 Line powered flyback switching regulator

There are two other safety precautions. The large filter capacitor, C_{in3}, holds its charge *long* after the power is turned off. So, even though the power switch is *off* and the plug is clearly pulled out, lying on the bench top, there are still lethal voltages within the circuit. Before doing *anything* to the circuit, discharge the filter capacitor through a resistor. Select a resistor that allows the charge to bleed off in a second or two, but limits the initial current to the same value as normally flows through the transformer. While prototyping, it may even be wise to temporarily install a switch with this resistor directly across C_{in3}. This allows you to remove power, then flip the switch to discharge the capacitor without trying to connect the discharge resistor across the charged capacitor with your hands.

Most oscilloscopes are powered from the same line voltage to which this circuit is connected. For safety, the oscilloscope's case and ground clip are connected to **earth** through the third, round prong on the power plug. But the neutral of the line power system is also tied to earth somewhere. Therefore, you cannot just clip the oscilloscope's ground wherever you wish. If you use it at all, it *must* be tied to earth. Since it is already wired to earth through its power cord, connecting it anywhere else shorts that point directly to earth. The only current limiting element is the circuit breaker in the closet down the hall. Arc welding occurs.

A better plan is to remove the ground clip entirely, or tape it back so it cannot dangle into your circuit. Since the oscilloscope already has its ground connected to earth, all of the measurements on the line side of your circuit must be made with respect to earth. Placing the clip in the wrong place can be spectacular.

- Stand back and turn *on* the power.

- Watch closely for any indication of an error and be prepared to turn *off* the power immediately.

- Once the measurements are recorded and it is time to move an instrument or change the circuit, turn the power *off*.

- Discharge the main filter capacitor through a resistor and switch.

- When the instruments indicate that the circuit is fully de-energized, make the required changes.

- Repeat the steps above as often as necessary.

- *Never* skip a step or move a lead or component with power *on*.

Not shown in Figure 7-16 are several input elements. You will want to include some form of power switch, a pilot indicator lamp, and a fuse.

Following these, and at the immediate left of the schematic is an *LC* π filter. The 150 kHz switching noise generated by the MOSFET often completely bypasses the main filter capacitor, C_{in3}. This electrolytic capacitor is designed to handle large voltages and high currents at line frequencies. It may have enough internal equivalent series resistance and inductance to look like an open to the 150 kHz produced by the transistor. So this high frequency noise passes through the bridge rectifier and out onto the commercial power line. There it can contaminate any other equipment connected to that branch of the distribution system. Depending on the application of the power supply, there are strict regulations governing the noise that your power supply can send back out onto the line.

Capacitors C_{in1} and C_{in2} and the inductors within T1 form a fourth order low pass filter. The cut-off is well above the line frequency, so the power passes in without attenuation. But the filter's cut-off is far below the switching frequency. So the 150 kHz is drastically attenuated as it attempts to leave the circuit.

The value of the two capacitors is not critical; 1 μF is typical. But be sure that they are *not* polarized, and can handle the line peak voltage. The little film capacitors that you have used extensively usually *cannot* tolerate the line voltage.

The transformer, T1, is a coupled inductor. Each side couples an identical signal (of the 150 kHz noise) into the opposite coil. But since the opposite coil is on the other line, the noise there is of the reverse polarity. By transformer action, the noise on the opposite side is canceled. This works both ways, the top coupling to and canceling noise in the bottom winding, and noise at the bottom winding coupling to and canceling noise at the top winding. Be sure to select a coupled inductor that can pass the line frequency current and has an isolation voltage between the two windings greater than the line peak voltage.

The full-wave bridge rectifier comes next. Full-wave rectification is chosen over a single diode half-wave rectifier. The bridge doubles the output frequency. This then doubles the effectiveness of the main filter capacitor. The peak inverse voltage rating of the bridge diodes is the line peak voltage. Be sure to look at both the diode's average forward current and repetitive peak ratings. Both of these must be larger than that required by the transformer primary.

When power is first applied, the main filter capacitor, C_{in3}, is discharged. It appears as a short circuit to the line voltage. At the peak of

Input *LC* high frequency filter

C_{in1} and C_{in2}

Coupled inductor T1

Bridge rectifier

the first cycle, the only thing to limit the current through the bridge diodes and into C_{in3} is the diodes' and capacitor's internal resistance. These are intentionally very small. So that initial charge current would be huge. Either the input fuse (not shown) blows, or the line filter or bridge is damaged.

Thermistor

The resistor, R1, is a **thermistor**. It has a negative temperature coefficient. When it is cool, when power is first applied, its resistance is high. As it warms up, though, its resistance drops. When the capacitor is discharged, the thermistor has a high resistance. It limits the current through the circuit. After many cycles, the capacitor reaches its full charge. Then, only enough current is drawn from the line to replace the charge sent to the transformer. The charging current falls into the region of the current to the primary. This is an appropriate level that the filter and bridge can handle. So the limiting of the thermistor is no longer needed. By this time, the thermistor has heated up enough to lower its resistance so far that it, indeed, no longer has any significant effect.

When you select a thermistor you have to consider three items. Its resistance at ambient temperature must be just high enough to assure that the initial charge current into the discharged C_{in3} is below the peak current ratings of the line filter, diodes, and C_{in3}. Its hot resistance, however, must be low enough to keep it from having any noticeable effect once the capacitor has reached its operating charge. Finally, select a wattage that allows the thermistor to heat up enough to lower its resistance and remove itself from the circuit. The thermistor must be able to operate continuously at this warm, low resistance level, though.

Main filter capacitor, C_{in3}

Capacitor C_{in3} is the main line frequency filter. It must hold all of the charge needed to keep the voltage into the switching transformer above the switching regulator's minimum specification. A common guideline is to set the value of the capacitor above 1000 µF for each ampere of peak primary current. Since this is a *very* nonlinear circuit, even the simplest analysis results in a set of transcendental equations (no closed form solution). The best simulators have difficulty with the high switching rate, long time constant, and multiple feedback loops of the circuit. The practical solution is to select a reasonable value for C_{in3} and then rigorously test it under maximum load and minimum line voltage. Be sure that the capacitor's smallest output voltage is just comfortably above that level needed by the switching regulator. Also, remember that the capacitor is across the line. Its voltage rating must exceed the peak line voltage.

Zener regulator

The resistor and the two zener diodes are used to provide a low, regulated voltage to the MOSFET driver electronics. The +10 V is used to provide an adequate gate voltage to turn the MOSFET hard *on*. The + 5 V powers the logic IC within the gated oscillator.

The transformer is just like those discussed for the low voltage versions of the flyback regulator, but it must be able to handle the high voltage presented by the line. This also means that for typical output voltages a turns ratio of about 10:1 or higher is appropriate.

Switching transformer, T2

The main switch, Q1, also must be sized to withstand much higher voltages. Remember, the maximum *off* state voltage of the switch is E_{unreg} plus the voltage coupled back to the primary from the secondary as the fields collapse. This should be at least twice the peak line voltage. Also, keep in mind that there are transient spikes throughout the switching part of this supply. Double check your calculations and select a transistor that can withstand considerably more voltage than you expect in the worse case. The gated oscillator drives the switch as the supply powers up. Be sure, then, that the transistor you choose can be driven hard *on* by the logic level signal from the gated oscillator.

Switching transistor, Q1

It is convenient to place the controller IC on the secondary. There it can directly sense the load voltage. The secondary voltages are also more appropriate to the IC's maximum rating. But this presents a problem. The controller IC turns on Q1, that couples energy to the secondary to power the controller IC. When power is first applied, Q1 is not switching, so no energy is sent to the secondary, so the IC has no voltage, so it does not oscillate, driving Q1. The circuit never starts.

Gated oscillator

The gated oscillator solves this. As soon as the charge on C_{in3} is large enough for the zeners to present power to the gated oscillator, it begins to run, driving Q1 *on* and *off*. This couples energy to the secondary. When the controller IC powers up and begins to oscillate, its signal is sent through the opto-coupler back to the switching transistor. That signal is sensed by the gated oscillator, and it turns *off*. The controller IC now is powered and has taken control of the switch.

The switching signal from the controller IC must be coupled from the secondary side of the circuit to the line voltage side. You cannot just wire the IC to the MOSFET because that would defeat the isolation we have worked so hard to obtain. The opto-coupler passes the signal and maintains the isolation. Watch the phase relationship. When the IC's switch turns *on*, the LED is lit, turning the opto-transistor *on*, placing 10 V_p on the gate of the MOSFET, turning it *on*.

This off-line switching power supply is much more complicated and more difficult to work with than a simple line frequency transformer, rectifier, filter, and linear regulator. However, it is smaller, lighter, *far* more efficient, and less expensive.

Summary

A switching transistor, combined with an inductor, a diode, a capacitor, and perhaps a transformer, can efficiently step a dc voltage down (buck regulator), up (boost regulator), or both (flyback converter).

The output voltage for the step-down buck regulator is controlled by the duty cycle, D.

$$V_{load} = DE_{unreg}$$

The switch is in series with the input voltage and the inductor. There is a capacitor in parallel with the load. When the switch turns *on*, energy is coupled to the inductor, the capacitor, and the load. When the switch turns *off*, the voltage across the inductor reverses polarity, turning the Schottky diode *on*. The inductor and capacitor then discharge, powering the load until the switch comes *on* again.

The type of inductor and capacitor are as important as their values. Select components that do not contaminate the area with interference, and can work efficiently at the high switching frequency of the supply.

The LM2595-adj provides the feedback controller, 1 A switch, current and temperature limiting, the oscillator (150 kHz), and a 1.23 V_{dc} reference. It adjusts its switch's duty cycle to assure that the negative feedback voltage is 1.23 V_{dc}.

$$V_{load} = \left(1 + \frac{R_f}{R_i}\right)1.23\,V_{dc}$$

To boost the load voltage above the input, the inductor, switch and diode must be rearranged. The inductor is powered directly from the input supply, and the switch completes this circuit to ground. When the switch goes *on*, current ramps up in the inductor, storing energy. When the switch turns *off*, the polarity of the inductor's voltage reverses, adding to the supply voltage and coupling energy to the capacitor and load.

$$V_{load} = \frac{E_{unreg}}{1 - D}$$

While the inductor is charging, the diode is *off*. The capacitor must hold up the load voltage.

The LM2585-adj is a boost IC. It contains a 3 A switch, current and temperature limiting, the oscillator (100 kHz), a 1.23 V_{dc} reference, and a soft-start circuit. It too adjusts the duty cycle to keep the feedback voltage at 1.23 V_{dc}. This can be set by changing the feedback voltage divider.

In the flyback converter, the inductor is replaced with the primary of a transformer. There may be multiple secondaries, allowing a variety of output voltages. The most critical one is fed back to the LM2585 controller IC. The output voltage from that secondary is

$$V_{load} = \frac{D}{(1-D)\dfrac{N_{pri}}{N_{sec}}} E_{unreg}$$

When the switch is *on*, the current through the primary's inductance ramps up. All of the secondaries are *off*. The voltage to the loads is provided by each output capacitor. When the transistor turns *off*, the secondary polarities reverse. Current is now coupled into the load and the output capacitor.

Replacing the traditional line frequency step-down transformer with the high frequency flyback transformer provides considerable reduction in size, weight, and cost. But to work directly on the commercial power line requires several precautions. Be sure to follow all of the other safety precautions outlined.

You must also protect the circuit from the extreme voltages and the commercial line from the noise of the switcher. Use an *LC* π filter to keep the switching interference from spreading through the power grid. Add a thermistor to force the main filter capacitor to acquire its initial charge slowly. The transformer provides isolation for the secondary voltages, but you must also provide isolation for the feedback.

The switching circuits you have seen in this chapter provide up to 100 W to the load. In the next chapter you will read about thyristors. They are switches that work directly from the commercial power line. They easily pass hundreds to thousands of watts to the load at over 98% efficiency

Just like the off-line flyback switcher, you must exercise care with thyristors. They may generate electromagnetic interference, and may be damaged by switching the high line voltage and current. Isolation of the control circuit from the line is very important. There is a family of opto-isolators specifically designed to work with thyristors.

Time proportioning sends power to the load in half cycle packets, firing at every zero crossing. Phase-angle firing turns the thyristor *on* for an adjustable part of every cycle.

Problems

Buck Regulator Basics

7-1　Analyze the circuit in Figure 7-17. Calculate the following.
given:

$$E_{unreg} = 9 \text{ V}_{dc}, \quad V_{load} = 5 \text{ V}_{dc}, \quad I_{dc\ load} = 0.8 \text{ A}_{dc}, \quad L = 100 \text{ μH},$$
$$C = 470 \text{ μF}, \quad ESR = 0.1 \text{ Ω}, \quad f_{osc} = 150 \text{ kHz}, \quad Q = \text{IRF530}$$

find:

$$D, \quad t_{on}, \quad \Delta i_{pp\ L}, \quad i_{ave\ L}, \quad i_{p\ L}, \quad \Delta v_{pp\ load}, \quad i_{rms\ C}, \quad P_Q, \quad P_{diode}, \quad I_{dc\ in}$$

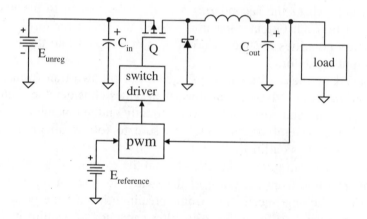

Figure 7-17　Schematic for Problems 7-1 through 7-4

7-2　Analyze the circuit in Figure 7-19, calculate the following,
given:

$$E_{unreg} = 28 \text{ V}_{dc}, \quad V_{load} = 18 \text{ V}_{dc}, \quad I_{dc\ load} = 0.3 \text{ A}_{dc}, \quad L = 1 \text{ mH},$$
$$C = 220 \text{ μF}, \quad ESR = 0.2 \text{ Ω}, \quad f_{osc} = 52 \text{ kHz}, \quad Q = \text{IRF } 530$$

find:

$$D, \quad t_{on}, \quad \Delta i_{pp\ L}, \quad i_{ave\ L}, \quad i_{p\ L}, \quad \Delta v_{pp\ load}, \quad i_{rms\ C}, \quad P_Q, \quad P_{diode}, \quad I_{dc\ in}$$

7-3　Design the circuit in Figure 7-17. Calculate the following.
given:

$$E_{unreg} = 9 \text{ V}_{dc}, \quad V_{load} = 5 \text{ V}_{dc}, \quad I_{dc\ load} = 0.8 \text{ A}_{dc}, f_{osc} = 150 \text{ kHz},$$
$$\Delta v_{pp\ load} = 0.1 \text{ V}_{pp}$$

find:

$$L, \quad C_{out}, \quad ESR_{Cout}, \quad I_{dcin\ dc}, \quad P_{Eunreg}$$

7-4 Design the circuit in Figure 7-17. Calculate the following, given:

$E_{unreg} = 28\ V_{dc}$, $V_{load} = 18\ V_{dc}$, $I_{dc\ load} = 0.3\ A_{dc}$, $f_{osc} = 52\ kHz$,
$\Delta v_{pp\ load} = 0.03\ V_{pp}$

find:

L, C_{out}, ESR_{Cout}, $I_{dcin\ dc}$, P_{Eunreg}

LM2595 Simple Switcher Buck Regulator IC

7-5 An LM 2595 controller IC is added to Problem 7-1. Calculate the following, given:

$R_f = 11\ k\Omega$, $R_i = 3.6\ k\Omega$, $T_A = 50°C$

find:

$V_{dc\ load}$, P_{LM2595}, Θ_{SA} for the heat sink

7-6 An LM 2595 controller IC is added to Problem 7-1. Calculate the following, given:

$R_f = 1.1\ k\Omega$, $R_i = 1.8\ k\Omega$, $T_A = 65°C$, $I_{dc\ load} = 0.6\ A_{dc}$

find:

$V_{dc\ load}$, P_{LM2595}, Θ_{SA} for the heat sink

7-7 Select R_f and R_i using standard 5% resistors for an LM 2595 that inputs 28 V_{dc} and outputs 18 V_{dc} at 0.3 A_{dc}.

7-8 Select R_f and R_i using standard 5% resistors for an LM 2595 that inputs 18 V_{dc} and outputs 9 V_{dc} at 0.5 A_{dc}.

Boost Regulators

7-9 Analyze the circuit in Figure 7-18. Calculate the following, given:

$E_{unreg} = 5\ V_{dc}$, $V_{load} = 9\ V_{dc}$, $I_{dc\ load} = 0.8\ A_{dc}$, $L = 100\ \mu H$,
$C = 470\ \mu F$, $f_{osc} = 100\ kHz$, $Q = IRF530$

find:

D, t_{on}, $\Delta i_{pp\ L}$, $i_{ave\ L}$, $i_{p\ L}$, $\Delta v_{pp\ load}$, $i_{rms\ C}$, P_Q, P_{diode}, $I_{dc\ in}$

7-10 Analyze the circuit in Figure 7-18. Calculate the following, given:

$E_{unreg} = 18\ V_{dc}$, $V_{load} = 28\ V_{dc}$, $I_{dc\ load} = 0.5\ A_{dc}$, $L = 100\ \mu H$,
$C = 470\ \mu F$, $f_{osc} = 100\ kHz$, $Q = IRF530$

find:

D, t_{on}, $\Delta i_{pp\ L}$, $i_{ave\ L}$, $i_{p\ L}$, $\Delta v_{pp\ load}$, $i_{rms\ C}$, P_Q, P_{diode}, $I_{dc\ in}$

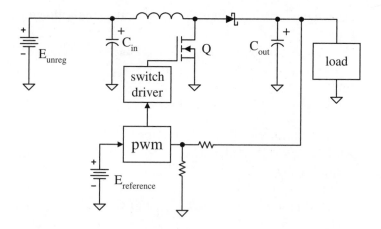

Figure 7-18 Schematic for Problems 7-9 through 7-12

7-11   Design the circuit in Figure 7-18. Calculate the following, given:

$E_{unreg} = 5$ V$_{dc}$, $V_{load} = 9$ V$_{dc}$, $I_{dc\ load} = 0.8$ A$_{dc}$, $f_{osc} = 100$ kHz, $\Delta v_{pp\ load} = 0.1$ V$_{pp}$

find:

L, C_{out}, $I_{dc\ in}$, P_{Eunreg}

7-12   Design the circuit in Figure 7-18. Calculate the following, given:

$E_{unreg} = 18$ V$_{dc}$, $V_{load} = 28$ V$_{dc}$, $I_{dc\ load} = 0.5$ A$_{dc}$, $f_{osc} = 100$ kHz, $\Delta v_{pp\ load} = 0.03$ V$_{pp}$

find:

L, C_{out}, $I_{dc\ in}$, P_{Eunreg}

LM2585 Simple Switcher Flyback Regulator IC

7-13   An LM 2585 controller IC is added to Problem 7-11. Calculate the following,

given:

$R_f = 7.5$ kΩ, $R_i = 1.1$ kΩ, $T_A = 50°C$

find:

$V_{dc\ load}$, P_{LM2585}, Θ_{SA} for the heat sink

7-14 An LM 2585 controller IC is added to Problem 7-12. Calculate
the following,
given:
$R_f = 22$ kΩ, $R_i = 1$ kΩ, $T_A = 65°C$
find:
$V_{dc\ load}$, P_{LM2585}, Θ_{SA} for the heat sink

7-15 Select R_f and R_i using standard 5% resistors for an LM 2585 that
inputs 5 V_{dc} and outputs 15 V_{dc} at 0.3 A_{dc}.

7-16 Select R_f and R_i using standard 5% resistors for an LM 2585 that
inputs 9 V_{dc} and outputs 18 V_{dc} at 0.5 A_{dc}.

Flyback Regulator Basics

7-17 Analyze the circuit in Figure 7-19. Calculate the following,
given:
$E_{unreg} = 9$ V_{dc}, $I_{dc\ load} = 0.8$ A_{dc}, $L_{primary} = 22$ μH,
$N_{pri}/N_{sec} = 1:1$, $R_f = 11$ kΩ, $R_i = 3.6$ kΩ, $C_{out} = 470$ μF
find:
$V_{dc\ load}$, D, $I_{dc\ pri}$, $I_{p\ pri}$, $I_{dc\ sec}$, $I_{p\ sec}$, $\Delta v_{pp\ load}$, P_{IC}, $I_{dc\ in}$

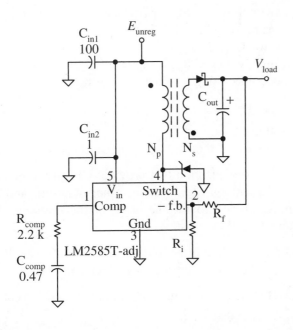

Figure 7-19 Schematic for Problems 7-17 through 7-20

7-18   **a.**   Repeat Problem 7-17 for $R_f = 49$ kΩ, $I_{dc\ load} = 0.3$ A$_{dc}$.

     b.   Discuss the advantages and limitations of using a flyback converter compared to the buck or the boost regulators.

7-19   Design the circuit in Figure 7-19. Calculate the following, given:

$$E_{unreg} = 18\ V_{dc},\ \ V_{load} = 28\ V_{dc},\ I_{dc\ load} = 0.5\ A_{dc},$$
$$\Delta v_{pp\ load} = 0.03\ V_{pp}$$

find:

a switching transformer (indicate manufacturer, part number, turns ratio and primary inductance)

$R_f,\ R_i,\ C_{out},\ I_{dc\ in}$

IC max voltage, IC average current, IC max current

diode max voltage, diode max current

P_{IC}

7-20   Design the circuit in Figure 7-19. Calculate the following, given:

$$E_{unreg} = 18\ V_{dc},\ \ V_{load} = 9\ V_{dc},\ I_{dc\ load} = 0.5\ A_{dc},$$
$$\Delta v_{pp\ load} = 0.05\ V_{pp}$$

find:

a switching transformer (indicate manufacturer, part number, turns ratio and primary inductance)

$R_f,\ R_i,\ C_{out},\ I_{dc\ in}$

IC max voltage, IC average current, IC max current

diode max voltage, diode max current

P_{IC}

Line Voltage Input

7-21   Explain why each of the safety precautions is important.

7-22   Locate an organization that provides CPR training and certification. Report on the location, time, duration, and cost of the training. Also provide the name of the trainer and the phone number to call to register.

7-23   Find the $LC\ \pi$ line filter components for use on a 120 V_{rms} line at 0.5 A$_{rms}$. Indicate source, part numbers, and cost.

7-24   Find a thermistor that limits the initial charge current to 2 A$_p$.

7-25   Locate a MOSFET with $I_{max\ DS} > 3$ A$_p$ and $V_{max\ DS} \geq 300$ V$_p$.

7-26   Design the gated oscillator.

Buck Regulator Lab Exercise

A. Initial Performance

   1. Build the circuit shown in Figure 7-20.

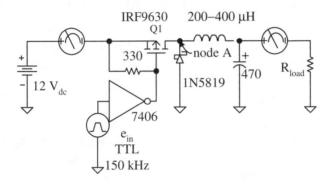

Figure 7-20 Manually adjusted pulse width buck regulator

   2. Layout is critical. Use a protoboard mounted on a ground plane and no busses. Keep the layout tight, and the leads as short as possible. Be sure to return the load to the 12 V_{dc} supply in its own, separate lead. Set $R_{load} = 30\ \Omega$, 5 W.

   3. Adjust the function generator to provide a 150 kHz, 0 V_{dc} to 5 V_{dc} variable duty cycle rectangular pulse. Set the duty cycle at the function generator to 40%.

   4. Connect one channel of the oscilloscope at the gate of Q1, and the other channel across the load. Also connect the digital multimeter, configured as a dc voltmeter across the load.

   5. Turn the power supplies on. Carefully monitor the current. Be sure that it does not significantly exceed 300 mA_{dc}.

   6. Enable the function generator output. Verify that the voltage across the load is ≈5 V_{dc}. Alter the *duty cycle* of the function generator slightly to set the output voltage to 5.00 V_{dc}.

7. Record the actual duty cycle, the theoretical duty cycle, and the percent error.

8. Record the input and load currents. Compare these to the theoretical values.

9. Calculate the power provided to the circuit from the supply and the power delivered to the load. Also calculate the circuit's efficiency.

10. Move channel 2 of the oscilloscope from the load to the junction of the main transistor switch, the diode, and the inductor, node A. Save that trace to include in your report.

B. Load Regulation

1. Alter the load resistance to set the load current to 200 mA$_{dc}$, then the duty cycle as necessary to keep the load voltage at 5.00 V$_{dc}$. You may need to slightly adjust the load resistance and the duty cycle several times to obtain 5.00 V$_{dc}$ at 200 mA$_{dc}$.

2. Record the duty cycle.

3. Change the oscilloscope input channel coupling to ac and the volts/div to allow you to view the *triangular ripple* across the load. Ignore the transient spikes.

4. Record the triangular peak-to-peak ripple.

5. Repeat steps B1 through B4 for load currents of 200 mA$_{dc}$, 300 mA$_{dc}$, 400 mA$_{dc}$, and 500 mA$_{dc}$.

C. Line Voltage Regulation

1. Alter the voltage from the input dc power supply to 12 V$_{dc}$.

2. Adjust the duty cycle to keep the load V_{load} = 5.00 V$_{dc}$.

3. Assure that the load is adjusted to give I_{load} = 500 mA$_{dc}$.

4. Record the actual duty cycle of Q1, the theoretical duty cycle, and the percentage error.

5. Measure the current from the input power supply.

6. Calculate the power delivered to that load and the power provided by the input power supply.

7. Calculate the circuit's efficiency.

8. Repeat steps C1 through C7 for input voltages of 8 V_{dc}, 10 V_{dc}, and 14 V_{dc}.

D. Load Voltage Adjustment

1. Return $V_{dc\ in} = 12\ V_{dc}$ and $I_{load} = 500\ mA_{dc}$.

2. Adjust the duty cycle to produce $V_{load} = 4.0\ V_{dc}$ at $I_{load} = 500\ mA_{dc}$.

3. Record the actual duty cycle of Q1, the theoretical duty cycle, and the percentage error.

4. Measure the current from the input power supply.

5. Calculate the power delivered to that load and the power provided by the input power supply.

6. Calculate the circuit's efficiency.

7. Repeat steps D1 through D6 for load voltages of 4 V_{dc}, 6 V_{dc}, and 9 V_{dc}.

Boost Regulator Lab Exercise

A. Initial Performance

1. Build the circuit shown in Figure 7-21.

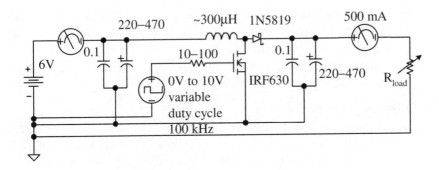

Figure 7-21 Boost regulator

2. Layout is critical. Use no busses on your protoboard. Keep the layout tight, and the leads as short as possible. Be sure to provide the *three* ground returns and proper layout on a ground plane board. Set $R_{load} = 80 \, \Omega$.

3. Adjust the function generator to provide a 100 kHz, 0 V_{dc} to 5 V_{dc} variable duty cycle rectangular pulse. Set the duty cycle at the function generator to 50%.

4. Connect one channel of the oscilloscope across the function generator, and the other channel across the load. Also connect the digital multimeter, configured as a dc voltmeter across the load.

5. Turn the power supply on. Carefully monitor its current. Be sure that it does not exceed 200 mA_{dc}.

6. Enable the function generator output. Verify that the voltage across the load is $\approx 12 \, V_{dc}$. Alter the *duty cycle* of the function generator slightly to set the output voltage to 12.0 V_{dc}.

7. Record the actual duty cycle, the theoretical duty cycle, and the percent error.

8. Record the input and load currents. Compare these to the theoretical values.

9. Calculate the power provided to the circuit from the supply, and the power delivered to the load. Also calculate the circuit's efficiency.

10. Capture the waveform of the voltage at the transistor's drain, and the voltage across the load. Use ac coupling for the voltage across the load. Adjust the settings until the transients are well centered and fill a significant part of the screen. Save these traces to include in your report.

B. Load Regulation
  1. Adjust the load resistance to set the load current to 50 mA_{dc}, then the duty cycle as necessary to keep the load voltage at 12.0 V_{dc}. You may need to slightly adjust the load resistance and the duty cycle several times to obtain 12.0 V_{dc} at 50 mA_{dc}.

  2. Record the actual duty cycle, the theoretical duty cycle, and the current from the input supply.

3. Record the input and load currents. Compare these to the theoretical values.

4. Calculate the power provided to the circuit from the supply and the power delivered to the load. Also calculate the circuit's efficiency.

5. Repeat steps B1 through B4 for load currents of 100 mA_{dc}, 150 mA_{dc}, 200 mA_{dc}, and 250 mA_{dc}.

C. Line Voltage Regulation
   1. Alter the voltage from the input dc power supply to 3 V_{dc}.

   2. Adjust the duty cycle to keep the load $V_{load} = 12.0\ V_{dc}$.

   3. Assure that the load is adjusted to give $I_{load} = 150\ mA_{dc}$.

   4. Record the actual duty cycle of Q1, the theoretical duty cycle, and the percentage error.

   5. Measure the current from the input power supply.

   6. Calculate the power delivered to that load and the power provided by the input power supply.

   7. Calculate the circuit's efficiency.

   8. Repeat steps C1 through C7 for input voltages of 4 V_{dc}, 5 V_{dc}, 6 V_{dc}, and 7 V_{dc}.

D. Load Voltage Adjustment
   1. Return $V_{dc\ in} = 6\ V_{dc}$ and $I_{load} = 150\ mA_{dc}$.

   2. Adjust the duty cycle to produce $V_{load} = 10.0\ V_{dc}$ at $I_{load} = 150\ mA_{dc}$.

   3. Record the actual duty cycle of Q1, the theoretical duty cycle, and the percentage error.

   4. Measure the current from the input power supply.

   5. Calculate the power delivered to that load and the power provided by the input power supply.

   6. Calculate the circuit's efficiency.

   7. Repeat steps D1 through D6 for load voltages of 15 V_{dc} and 20 V_{dc}.

8

Thyristors

Introduction

The switches of Chapter 7 can deliver up to a few hundred watts to a load. But to work from the commercial power line requires exceptional procedures and parts. Thyristors are semiconductor switches specifically designed to control that line voltage, delivering *thousands* of watts to the load at over 98% efficiency. They are constructed differently from the transistor switches of Chapters 6 and 7. Like the off-line switching power supply, you must assure that they are not misfired or damaged by, or that they do not contaminate, the power line. The circuits that protect the thyristors and the line are called snubbers.

To control the operation of the thyristor, you must be able to accurately trigger it, providing the correct amplitude and duration pulse. You must limit that pulse so that the gate is not burned out. Coupling from an isolated circuit into the power line is critical.

Adjusting power to a load proportionally by controlling when a switch turns *on* takes some special effort. Time proportioning circuits are particularly quiet and gentle on the components. Phase angle fired circuits are popular because they provide power more continuously.

Objectives

By the end of this chapter, you will be able to:

- Describe the SCR and the triac, properly using their key parameters.

- Calculate all of the snubbing components needed for a thryistor and explain the purpose of each component.

- Analyze and design each of the following trigger circuits calculating all component values and/or voltages, currents, and time delays:
  >*RC* lag driving a diac
  >opto-coupler; instantaneous and zero crossing

- Analyze and design each of the following proportional power interfaces, calculating all component values and/or voltages, currents, time delays, angles, and powers:
  >time proportioning
  >phase angle fired, analog and digital controlled

8.1 Thyristor Device Characteristics

The bipolar transistor is made of three layers of semiconductor material, *npn* or *pnp*. Enhancement mode MOSFETs also appear to have three layers, but when biased, they turn into a single conducting channel.

Thyristors have *four* alternating semiconductor layers, *pnpn*. As such, they are always inherently *off*. With either polarity voltage across the thyristor, at least one of these junctions is reverse biased. But when a gating pulse is applied, the thyristor turns *on*. Once well *on*, the thyristor holds itself *on*, even when the gating pulse is removed. Rated at hundreds to thousands of volts and tens to hundreds of amperes, these line switches can be turned *on* with a single, 100 μs low level pulse. How and why this happens requires a closer look inside the thyristor.

Silicon Controlled Rectifier

The silicon controlled rectifier (SCR) is a four layer device. Its top *p* material is called the **anode**. The bottom *n* material is the **cathode**. There is a **gate** connection to the center *p* material. This construction and the schematic symbol are shown in Figure 8-1.

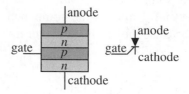

Figure 8-1 SCR layers and symbol

The schematic symbol suggests that the SCR is a rectifier, allowing current to pass only from the anode to the cathode. This is also supported by the first and bottom *pn* layers. Only if the anode is positive with respect to the cathode can those layers be forward biased, and conduct. When the anode is more negative than the cathode, these two junctions reverse bias and the SCR turns *off*.

But what about the second junction? It has *n* on top and *p* below. With a positive voltage on the anode this junction is reverse biased. Without a voltage applied to the gate, the SCR is indeed always *off*, when the anode is negative with respect to the cathode (as a simple diode) and when the anode is positive with respect to the cathode.

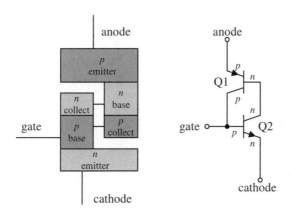

Figure 8-2 Bipolar junction transistor of a silicon controlled rectifier

The gate is the key. Partition the SCR into two parts and redraw it as bipolar junction transistors. This is done in Figure 8-2. The upper *p* layer is the emitter of a *pnp* transistor switch. The upper *n* layer is both the base of that same *pnp* transistor and the collector of an *npn* transistor switch. The lower *p* region completes the collector of the top *pnp* transistor. But it is also the base of the lower *npn* transistor. This is where the gate is tied. Finally, the bottom *n* layer is the emitter of the lower *npn* switching transistor.

When the gate is open, even with a positive anode-to-cathode voltage, Q1's base-to-collector junction is reverse biased. There is no base current for Q2, and it is *off* too. This means that there is no base current for Q1. Both transistors are *off*. This is shown in Figure 8-3 (a).

When *off* the SCR stays *off*.

But when a positive V_{GK} is applied, the base-to-emitter junction of Q2 is forward biased, Figure 8-3 (b). Base current into Q2 produces Q2 collector current. Q2's collector current is Q1's base current. Q1 now turns *on*. The collector current from Q1 flows into the base of Q2. The positive V_{GK} signal turns *on* Q2. Its collector current turns Q1 *on*. And, Q1's collector current becomes the base current of Q2.

+V_{GK} turns the SCR *on* as long as V_{AK} is positive.

Once Q1 is *on*, driving current into Q2's base, the gating signal can be removed. This is shown in Figure 8-3 (c). Even though the base current from the gate supply voltage is now gone, transistor Q1's collector

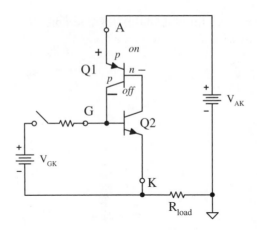

(a) SCR is *off* with the gate open

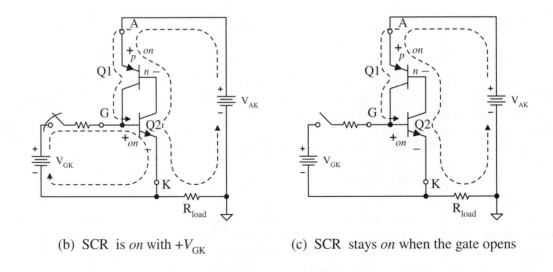

(b) SCR is *on* with +V_{GK} (c) SCR stays *on* when the gate opens

Figure 8-3 SCR model behavior

Once *on* the SCR stays *on* even when the gate signal is removed.

current provides the base current to keep Q2 *on*. And Q2's collector current is Q1's base current, keeping Q1 *on*. Q2 keeps Q1 *on*. Q1 keeps Q2 *on*. This **regenerative feedback** takes less than 100 μs to establish. Once *on* the SCR stays *on* even when the gating signal is removed.

So, how is an SCR that is *on* turned *off*? The internal, regenerative feedback must be broken. Reduce the anode current below a certain level, and there is not be enough collector current from Q1 to hold Q2 *on*. It goes *off*, sending the base current of Q1 to zero. Transistor Q1 also turns *off*. Then you are right back where you started, at Figure 8-3a. This key anode current level is called the **holding current** and is a manufacturer's specification.

The simplest way to send the anode current below the holding current is to wait. Since most SCRs are used to switch the commercial ac line voltage to a load, once every cycle the line voltage falls to zero, and then goes negative. The SCR turns *off*.

Example 8-1

Experiment with the 2N1599 shown in Figure 8-4. Verify that

a.   if the switch is *open* the SCR does not turn *on*.

b.   once turned *on* the SCR stays *on* for the rest of that cycle.

c.   the SCR turns *off* every negative line cycle and must be triggered back *on*.

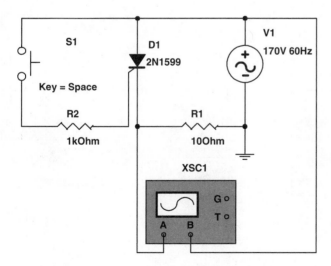

Figure 8-4 Simulation schematic for Example 8-1

Solution

You can toggle the switch while the simulation is running by selecting the switch, and then pressing the space bar. Expanding the oscilloscope gives you a good view of the results.

Figure 8-5 (a) shows that when the switch is closed, the SCR fires very early in each positive half-cycle, and stays *on* that entire half cycle. That voltage is passed to the load. At the zero crossing, when the line current goes negative, the SCR turns *off*, and stays *off* for the entire negative half-cycle. During that time, no voltage is passed to the load.

Figure 8-5 (b) shows the effect of closing the switch, and firing the SCR at various times *during* the positive half-cycle. As soon as the switch is closed, the SCR fires, passing the rest of that cycle to the load. It goes *off* at the current zero crossing and waits to be triggered back *on* sometime during the next positive half-cycle. The sooner each cycle the SCR is triggered, the more voltage is passed to the load.

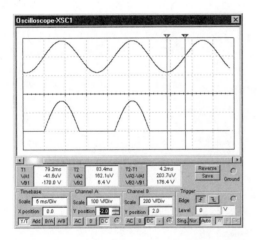 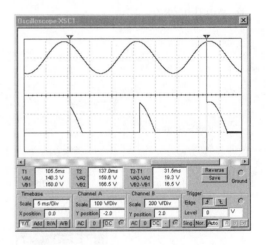

(a) SCR fired at the beginning of every cycle   (b) SCR fired earlier every cycle

Figure 8-5 Results of simulations with the 2N1599 SCR

There are several important specifications that allow you to match the SCR to your circuit's requirements. Figure 8-6 gives these specifications for the 2N1595 model from OrCAD.

✑**OrCAD**▰

```
.SUBCKT 2N1595              anode gate cathode



* "Typical" parameters
X1 anode gate cathode Scr PARAMS:
+ Vdrm=50v    Vrrm=50v    Ih=5ma       Vtm=1.1v     Itm=1
+ dVdt=1e9    Igt=2ma     Vgt=.7v      Ton=0.8u     Toff=10u
+ Idrm=10u
* 90-5-18     Morotola    DL137, Rev 2, 3/89
.ENDS
```

Figure 8-6 OrCAD SCR model and parameters

- V_{DRM} *Repetitive forward blocking voltage* This is the maximum positive anode voltage that the SCR can withstand before it turns *on* without a gating pulse. Be sure that the SCR's V_{DRM} is larger than the line peak voltage. The 2N1595 is *not* appropriate for a 120 V_{rms} line application, since it turns *on* when the anode voltage reaches 50 V, even though no gating signal is applied.

- I_H *Holding current* The anode current must be lowered below the holding current to cause the SCR to turn *off*.

- V_{TM} *Forward voltage drop* Since there are four layers and two junctions between the anode and the cathode, when the SCR is *on*, V_{AK} is greater than 0.6 V. Although this parameter seems small compared to the line voltage passed to the load, it is central in calculating the power that the SCR must dissipate.

- I_{rms} *Conducting rms current* This is the maximum continuous forward current that the SCR can carry. Remember, since the SCR is inherently a rectifier, only a half-cycle of current flows through it.

So $I_{rms} = I_p / 2$. The familiar square root of 2 only applies to full waves.

- I_{TSM} *Surge current* This is the largest repetitive peak current that the SCR can tolerate. Because of inductive kicks and capacitive discharges, this may be much, much greater than the rms current. It is not simply the load's half-cycle peak current.

- I_{gt} *Gate trigger current* Your trigger circuit must drive at least this much current into the gate to assure that the SCR turns *on*.

- V_{gt} *Gate trigger voltage* The gate-to-cathode junction must be forward biased. Since the SCR is silicon, this is typically +0.7 V.

- P_G *Maximum gate power* Your trigger circuit must not require that the gate dissipate more than this much power, or the gate overheats and fails. This is not shown in the simulation model, because the simulated part does not die. Usually to ensure a strong, accurate trigger, a large amount of current is driven into the gate, but only for 100 µs or so. For the remainder of the line cycle, the gate pulse is zero, and the gate can cool. On the average, the power is kept small by keeping the gating pulse width narrow. This is a key characteristic of all thyristor control circuits.

Figure 8-7 gives a sample of two data sheets. The top is from Teccor, and the lower illustrates two SCRs from International Rectifier. These were selected to show the wide variety of options you have. The first SCR is the S051. It can tolerate only 50 V before it turns *on* without a gating pulse. When it is *on* it passes only 1 A$_{rms}$. It only requires 10 mA of gate current to turn it *on*. It comes in a small plastic TO-92 package.

At the other extreme is the ST303C04 from International Rectifier. It is housed in a E-PUK TO-200 package that looks a lot like a hockey puck. It is capable of handling 400 V$_p$ at 1180 A$_{rms}$. There are versions that are rated at 800 V. To get this industrial strength switch *on*, you must provide only a 200 mA pulse!

With a little investigation you can find a silicon controlled rectifier that matches your load.

TYPE	Part Number		IT	VDRM & VRRM	IGT	IDRM & IRRM			VTM	VGT	
	Isolated	Non-Isolated	Maximum On-State Current (1) (2)	Repetitive Peak Off-State Forward & Reverse Voltage	DC Gate Trigger Current V_D = 12VDC R_L = 60Ω (4)	Peak Off-State Forward & Reverse Current at V_{DRM} & V_{RRM} (13)			Peak On-State Voltage at Max Rated RMS Current T_C = 25°C (3)	DC Gate-Trigger Voltage V_D = 12VDC R_L = 60Ω (8)	
	TO-92 / TO-220AB / TO-202AB / TO-220AB		Amps			mAmps				Volts	
	See "Package Dimensions" section for variations.		$I_{T(RMS)}$ / IT(AV)	Volts	mAmps	T_C=25°C / T_C=100°C / T_C=125°C			Volts	T_C=25°C / T_C=125°C	
			MAX / MAX	MIN	MIN / MAX	MAX			MAX	MAX / MIN	
1.0 Amp S051E			1.0 / 0.64	50	1 / 10	.01	0.2	0.5	1.6	1.5	0.2
S101E			1.0 / 0.64	100	1 / 10	.01	0.2	0.5	1.6	1.5	0.2
S201E			1.0 / 0.64	200	1 / 10	.01	0.2	0.5	1.6	1.5	0.2

(12) results			parameter:	VDRM	IT (RMS)	IT(av) comp. (a)	@ TC	ITSM (50Hz)	ITSM (60Hz)	Vgt	Igt	VTM comp. (a)	@ ITM	Tq	Rth (JC)
			units:	Volts	Amps	Amps	°C	Amps	Amps	Volts	mA	Volts	Amps	us	°C/W
			limit:	MIN	MAX	MAX	---	MAX	MAX	Max	Max	MAX	---	MAX	MAX
Part #	**Package**	**Circuit**	condition:	---	---	---	---	---	---	---	---	---	---	---	---
ST303C04CFL0	E-PUK (TO-200AB)	DISCRETE		400	1180	620	55	6690	7000	3	200	2.16	1255	10	.04
ST303C04CFN0	E-PUK (TO-	DISCRETE		400	1180	620	55	6690	7000	3	200	2.16	1255	15	.04

Figure 8-7 Sample of silicon controlled rectifers (*courtesy of Teccor and International Rectifier*)

Triac

The triac is an ac switch. When *off*, it blocks current in both directions (as long as the voltage does not exceed V_{DRM}). It may be triggered *on* with a pulse of either polarity and passes current in both directions. Like the SCR, once the triac is *on*, it continues to conduct even when the gating pulse is removed. To turn the triac *off*, the main path current must be reduced below the holding current, I_H.

The schematic symbol is shown in Figure 8-8. There are two main path terminals, MT1 and MT2, and a gate. Although current can flow in either direction, do *not* confuse the two terminals. MT2 is connected to the line. MT1 is tied to the load. The gate is fired with respect to MT1. Reversing MT1 and MT2 does not damage the triac, but it does not properly trigger *on*.

MT2

gate MT1

Figure 8-8 Triac symbol

Example 8-2

a. Verify by simulation that the triac is *off*, blocking voltage from the load when no gate pulse is applied.

b. Confirm that the triac can be triggered *on* and passes both the positive and negative line cycles.

Solution

The schematics, model, and Probe outputs for the 2N5444 are shown in Figure 8-10. With no gating pulse, there is no significant voltage passed to the load. With the gate tied through a resistor to MT2 the triac sends the entire line voltage to the load.

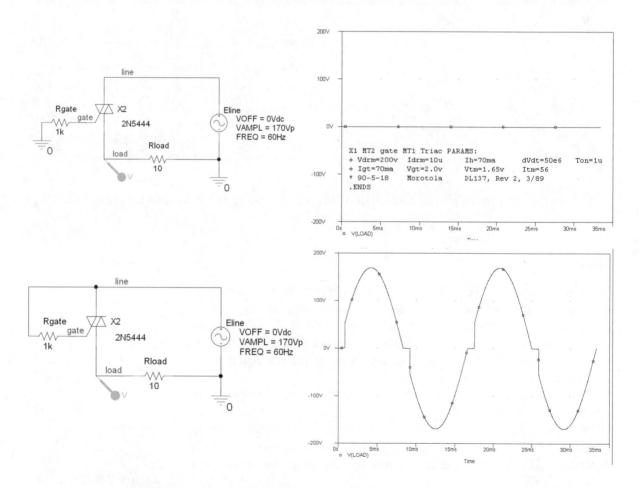

Figure 8-9 Simulations of a triac *off* and *on* for Example 8-2

Low Side versus High Side Switch

The arrangement of the line voltage, the thyristor switch, and the load has a major impact on the safety of the equipment you design and build. A mistake or a misunderstanding at these voltages can be fatal. In the simple world of low voltage electronics, the position of the voltage source, the switch, and the load makes little difference. After all, it is a series circuit. Elements in a series circuit can be rearranged without changing the circuit's operation. Current is still the same, regardless of which part it passes thorough first.

Look at Figure 8-10. One end of the line source is **neutral**. Somewhere in the wiring of the power distribution system this side is tied to **earth**, perhaps at a cold water pipe or other metal rod that actually runs deep into the soil. This end is safe. There is little difference in potential between this end and the chassis of the equipment you operate, or the concrete or metal floor on which you stand.

The other end of the source is **hot**. It is lethal. It is over 100 V above earth, with massive current carrying ability. Touching this point while standing on earth may result in a serious shock, sending enough current through you to kill.

The switch in Figure 8-10 is tied to the **low side**, neutral and earth. When it is closed, neutral is applied to one side of the load and hot to the other. The lamp lights, the heater warms, the motor spins. All is well.

But opening the switch creates a hazard. True, with the switch open, no current flows, and the load is deactivated. With no current flowing through the load, it drops no voltage. So both ends of the load are at the same potential. They are *both* hot. The line voltage now appears across the open switch. Anyone seeing that the load is deactivated could reasonably assume that power had been removed, and that the load is safe to work on. But both ends are hot. Touching either end of the deactivated load, when it is controlled by a low side switch, connects you to the line!

The proper arrangement is shown in Figure 8-11. The switch has been moved to the **high side**, and the load is now tied to neutral. With the switch closed, hot powers the load. But when the switch is opened, with no current through the load, the load drops no voltage. Both ends are at the same potential, neutral. With the load obviously deactivated, it is now safe to touch.

Using the thyristor as a high side switch is the preferred configuration. However, it does present two difficulties. Should it be necessary

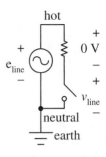

Figure 8-10 Low side switch

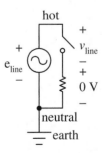

Figure 8-11 High side switch

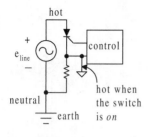

Figure 8-12 High side switch control problem

to work on the thryristor switch, you must remember that *both* ends are connected to the line hot, even when it is *off* and the load is deactivated.

The second problem is illustrated in Figure 8-12. The electronics driving the firing pulse into the gate must be connected to the lower end of the switch (the cathode or MT1). When the switch is *off*, that end is at neutral. So the common of the control electronics is tied to earth. But when the switch is fired, practically all of the line voltage is passed through the thyristor to the load (and the cathode or MT1) and to the common of your electronics. The common to the op amps or microprocessor or personal computer has just been stepped up over 100 V. At a minimum that circuitry malfunctions. It probably is damaged.

Use the thyristor as a high side switch, but be sure to provide **isolation** between the controlling electronics and the switch. This is usually done with either an opto-coupler or a pulse transformer.

8.2 Snubbing

Thyristors turn *on* in 10 μs to 100 μs. The voltage across them falls from as much as the line peak of several hundred volts to a volt or two. The current through the switch rises tens to hundreds of amperes in that same time. These extremely large, very fast edges create massive amount of trouble. The thyristor may actually burns itself out. The electromagnetic interference contaminates the power distribution system.

But thyristors are designed to switch the commercial line this fast. But switching the line creates major problems. It sounds like an impossible situation. The solution is to add a few components to **snub** these transients, slowing them or diverting them while protecting the power system from the interference.

Critical Rate of Rise of Current

In comparison to other semiconductor devices, the thyristor is *huge*, as it must be to carry large amounts of current. When first triggered *on*, a path between the main terminals is found that has the lowest resistance. Current causes the crystal to heat up along that path. The hotter the semiconductor gets, the more free current carriers are created. A lot of free current carriers means lower resistance. So, the path that is favored because it has lower resistance supports the current, that in turn lowers *that* path's resistance. As the current increases, it flows through only that path, creating more heat, more current carriers, and lower resistance, and taking more current. This can happen so fast that *all* of the main path current tries to flow through a very small part of the semiconductor. There is so much localized heating that the thyristor burns itself open.

The solution is to slow the **rate of rise of** this **current**. Forcing the main path current to rise more slowly gives the heat time to spread across the crystal. As the heat spreads out, so does the current. Inductors oppose a change of current, so place an inductor in series with the thyristor, as shown in Figure 8-13.

$$v_L = L\frac{di}{dt}$$

Solve this for L.

$$L = \frac{v_L}{\dfrac{di}{dt}}$$

The largest current occurs when the thyristor is triggered *on* at the peak of the line voltage. So v_L is the line peak voltage. *di /dt* is a specification given by the manufacturer. It assures proper operation if you assure that the current does not rise any faster than that.

$$L > \frac{V_{p\,line}}{\dfrac{di}{dt}_{spec}}$$

hot

L

load

neutral

Figure 8-13 Snubbing inductor limits *di/dt*

The snubbing inductor controls the rate of rise of current.

Example 8-3

The Q4004L triac from Teccor is operating under the following conditions:

$V_{line} = 220\ V_{rms}$, $f_{line} = 50\ Hz$, $R_{load} = 80\ \Omega$ (resistive),
$di/dt = 50\ A/\mu s$

a. Calculate the snubbing inductor needed.

b. Calculate the effect on the load voltage the drop across this inductor has when the triac is *on*.

Solution

$$L = \frac{220\,V_{rms} \times \sqrt{2}}{50\,\dfrac{A}{\mu s}} = 6.2\,\mu H$$

At 50 Hz this inductor has an impedance of

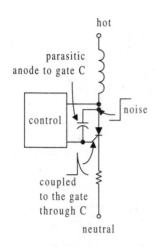

Figure 8-14 Critical rate of rise of voltage gates the SCR *on*

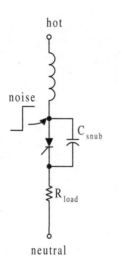

Figure 8-15 Snubbing capacitor

$$\overline{Z_{\text{L}}} = 2\pi f L \angle 90°$$

$$\overline{Z_{\text{L}}} = 2\pi \times 50\text{Hz} \times 6.2\,\mu\text{H} \angle 90° = 0.00195\,\Omega \angle 90°$$

Compared to the load impedance of 80 $\Omega \angle 0°$ the impedance of the inductor, 0.00195 $\Omega \angle 90°$, has no effect.

Practice: Repeat this problem for a Teccor S8070W SCR operating from a 440 V_{rms}, 60 Hz line, and driving a 7 Ω load.

Answers: *di/dt* = 200 A/µs, L = 3.1 µH, $\overline{Z_{\text{L}}}$ = 1.17 mΩ∠90°

The values from Example 8-3 show that the inductance needed is quite small, though you can always use more. Thyristors often drive motors, relays, and solenoids. These all have enough inductance that a separate snubbing inductor is not needed.

Critical Rate of Rise of Voltage

Between the main terminal of a thyristor and its gate is a considerable parasitic capacitance. Any noise on the line is coupled into the thyristor, and through this capacitance to the gate. It only takes a few volts at the gate to turn the thyristor *on*. It is quite possible that the switch can be fired by the noise even without a gating pulse from the control circuit. This is shown in Figure 8-14.

In consumer electronics, having a random occasional operation is annoying. But when you are controlling industrial loads, having the SCR turn *on* when it is supposed to be *off* may power the motor, driving the crane over the workmen loading it. This thyristor problem does not destroy the thyristor, it may kill people.

The solution is to short the noise around the thyristor in a capacitor large enough to divert the energy from the parasitic anode-to-gate capacitance. The manufacturer specifies the fastest rise of voltage on the main terminal the thyristor can tolerate. You then must select a capacitor that assures that the response to a step is slowed below that critical rate of rise of voltage. The circuit is shown in Figure 8-15.

The voltage across the snubbing capacitor cannot change instantaneously. Instead, it rises exponentially, as C_{snub} charges through R_{load}.

$$v_{\text{C}} = V_{\text{p noise}}\left(1 - e^{-\frac{t}{R_{\text{load}}C_{\text{snub}}}}\right)$$

The reason for including this snubbing capacitor is to limit the rate of rise of voltage, dv/dt. So, take the derivative of both sides of this equation.

$$\frac{dv_{C_{snub}}}{dt} = \frac{V_{p\,noise}}{R_{load}C_{snub}} e^{\frac{t}{R_{load}C_{snub}}}$$

The worst case for this is at $t = 0$. That is, the capacitor charges quickest at the beginning.

$$\left.\frac{dv_{C_{snub}}}{dt}\right|_{worst\,case} = \left.\frac{V_{p\,noise}}{R_{load}C_{snub}}\right|_{t=0}$$

Solve this for the value of the snubbing capacitor.

$$C_{snub} = \frac{V_{p\,noise}}{R_{load} \times \dfrac{dv}{dt}}$$

For $V_{p\,noise}$ choose the V_{DRM} of the thyristor you are using. Any signal larger than this causes the switch to turn *on* without a gating signal. So there is no reason to protect the circuit from noise any larger than the thyristor can tolerate on its main terminal.

The manufacturer specifies the highest rate of change of voltage that the part can handle, dv/dt.

The snubbing capacitor controls the rate of rise of voltage.

$$C_{snub} > \frac{V_{DRM}}{R_{load} \times \dfrac{dv}{dt}_{specification}}$$

Example 8-4

The Q4004L triac from Teccor is operating under the following conditions:

$V_{line} = 220\ V_{rms}$, $f_{line} = 50\ Hz$, $R_{load} = 80\ \Omega$ (resistive),
$dv/dt = 75\ V/\mu s$, $V_{DRM} = 400\ V_p$

a. Calculate the snubbing capacitor needed.

b. Calculate the current through this capacitor when the triac is *off*.

Solution

$$C_{snub} > \frac{V_{DRM}}{R_{load} \times \dfrac{dv}{dt}_{specification}}$$

$$C_{snub} > \frac{400\,V}{80\,\Omega \times 75\dfrac{V}{\mu s}} = 0.067\,\mu F$$

Choose a 0.1 μF capacitor.

At 50 Hz this capacitor has an impedance of

$$\overline{Z_C} = \frac{1}{2\pi f C} \angle -90°$$

$$\overline{Z_C} = \frac{1}{2\pi \times 50\,Hz \times 0.1\,\mu F} \angle -90° = 32\,k\Omega\angle -90°$$

When the triac is *off*, the capacitor is in series with the load resistor (ignoring the snubbing inductor for simplicity). The current that flows through this series circuit, bypassing the triac, is

$$\overline{I_{load}} = \frac{\overline{V_{line}}}{\overline{Z_C} + \overline{Z_{R\,load}}}$$

$$\overline{I_{load}} = \frac{220\,V_{rms}\angle 0°}{32\,k\Omega\angle -90° + 80\,\Omega\angle 0°} = 6.9\,mA_{rms}\angle 90°$$

This **leakage** current causes the load to drop

$$\overline{V_{load}} = 6.9\,mA_{rms}\angle 90° \times 80\,\Omega\angle 0° = 0.5\,V_{rms}\angle 90°$$

This causes the load to dissipate 3 mW.

Practice: Repeat this problem for a Teccor S8070W SCR operating from a 440 V_{rms}, 60 Hz line, and driving a 7 Ω load.

Answers: $dv/dt = 475$ V/μs, $V_{DRM} = 800$ V, $C_{snub} > 0.24\mu F$, $I_{leakage} < 40$ mA$_{rms}$, $P_{load} = 11$ mW

This example shows that the snubbing capacitor limits the rate of rise of the voltage across the thyristor without affecting the load when the switch is *off*. However, there is a serious problem. The charge on the capacitor follows the line voltage. If the switch is fired at the peak of the line voltage, the capacitor has also charged to that peak voltage. The capacitor is *directly* across the thyristor, with only its *on* resistance and the resistance of the wiring to limit the discharge current. That current may be huge, much larger than the switch could stand. If the thyristor is fired at its peak, the snubbing capacitor dumps this discharge current through the switch. The thyristor may be damaged by one of the parts designed to protect it.

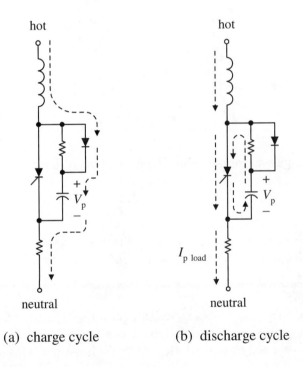

(a) charge cycle   (b) discharge cycle

Figure 8-16 Snubbing network resistor and diode operation

The solution is to add a resistor and diode as shown in Figure 8-16. During the charge cycle, and for any positive noise steps, the diode turns *on*. This places the capacitor in parallel with the SCR. It then limits the rate of rise of voltage at the anode to prevent false triggering. So the diode must be as fast as any of the noise pulses you expect to encounter.

If the SCR is fired at its peak, the capacitor discharges through the resistor, since the diode is reverse biased. The discharge current through the resistor is

$$I_{discharge} = \frac{V_{p\,line}}{R_{snubbing}}$$

At the anode, this combines with the load current to pass through the thyristor. You must select the snubbing resistor to assure that these two together are smaller than the maximum repetitive peak current that the SCR can handle.

$$I_{SCR} = I_{p\,load} + I_{discharge} < I_{TSM}$$

$$I_{TSM} > I_{p\,load} + I_{discharge}$$

Substituting for the discharge current gives

The snubbing resistor limits the capacitor's discharge current.

$$I_{TSM} > I_{p\,load} + \frac{V_{p\,load}}{R_{snubbing}}$$

Solve this equation for the value of the snubbing resistor.

hot

$$R_{snubbing} > \frac{V_{p\,line}}{I_{TSM} - I_{p\,load}}$$

The triac controls current in both directions, so you should include two snubbing networks, one for each direction, as shown in Figure 8-17.

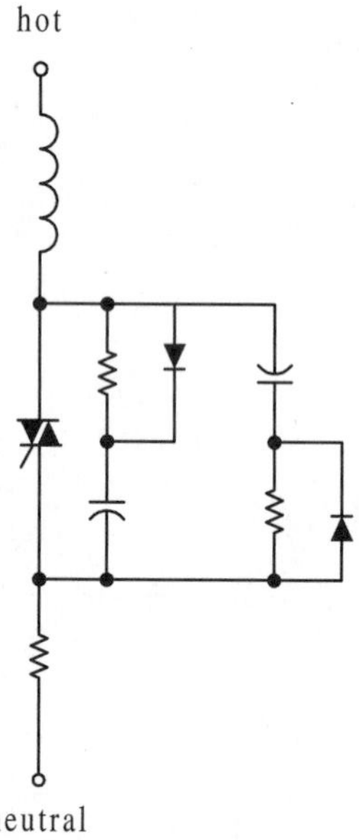

Example 8-5

Calculate the snubbing resistor for a Q4004L triac from Teccor is operating under the following conditions:

$V_{line} = 220\ V_{rms},$ $R_{load} = 80\ \Omega$ (resistive), $I_{TSM} = 46\ A_p$

Solution

$$R_{snubbing} > \frac{220\ V_{rms}\sqrt{2}}{46\ A_p - \dfrac{220\ V_{rms}\sqrt{2}}{80\ \Omega}} = 7.4\,\Omega$$

---

neutral

Figure 8-17 Triac snubbing

8.3 Triggers

A good thyristor trigger is able to pass enough current to the gate to be able to turn the switch *on*, but not so much that the gate is damaged. This gate current must be provided at a large enough voltage to forward bias the gate. It must be present long enough for the internal regenerative feedback to be established, so that the thyristor stays *on* when the gate pulse is removed. Once the switch is *on*, though, the trigger should remove the pulse. This assures that the gate is not required to dissipate so much power that it overheats and burns open.

Table 8-1 Gate requirements of two thyristors

part	type	V_{DRM} V_{rms}	I_{TSM} A_p	V_{gt} V_p min	V_p max	I_{gt} A_p min	A_p max	t_{gt} μs
Q4004L	triac	400	4	0.2	2.5	0.025	1.2	3.0
S8070W	SCR	800	70	0.2	2 .0	0.050	5.0	2.5

Shockley Diode

The Shockley diode is an SCR without a gate. It is normally *off*. Increasing the forward voltage across the diode has little effect. But, like the SCR's V_{DRM}, once the voltage reaches a specific point, the diode turns *on*. This point is the **breakover** voltage. Once *on* the Shockley diode stays *on*, passing large amounts of current, and dropping very little voltage, until the current through the diode falls below its specified holding current. At that point the diode turns back *off*.

The diode is *off* until $V_{AK} > V_{BO}$

Once *on* the diode stays *on* until $I_{AK} < I_H$

Both the breakover voltage and the holding current are specifications.

The characteristic curve of a Shockley diode is shown in Figure 8-18. Initially the switch is *off*. Increasing the anode-to-cathode voltage has little effect, until the breakover voltage is reached. At that point it turns *on*. The voltage across the diode drops to about a volt, and the current increases significantly. In the reverse direction the diode is *off*.

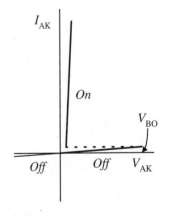

Figure 8-18 Shockley diode characteristic curve

Diac

Typical diac specifications

V_{BO}	10 V to 40 V
I_{TSM}	2A$_p$
I_H	0.5 mA

The diac is much more commonly available. Its schematic symbol and characteristic curve are given in Figure 8-19.

It behaves like two back-to-back Shockley diodes, or a triac without a gate, which is what the schematic symbol resembles. In either direction, the diac is *off* until the voltage across it exceeds its breakover voltage. It then goes *on*, providing large current with low voltage drop across it. When the current through the switch falls below the holding current, the diac goes *off*. This part is also called a **bilateral switch**.

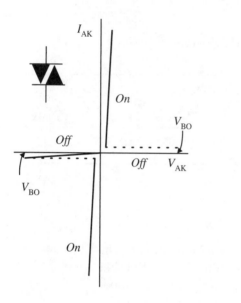

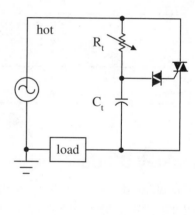

Figure 8-19 Diac symbol and characteristic curve

Figure 8-20 Phase angle firing circuit

The circuit shown in Figure 8-20 is a simple application of the diac. The load is connected to neutral and has a much smaller resistance than the resistor in the *RC* network. The voltage across the capacitor is smaller than the line voltage and lags in phase. The waveforms are shown in Figure 8-21. Initially, as the line voltage rises, the diac is *off*.

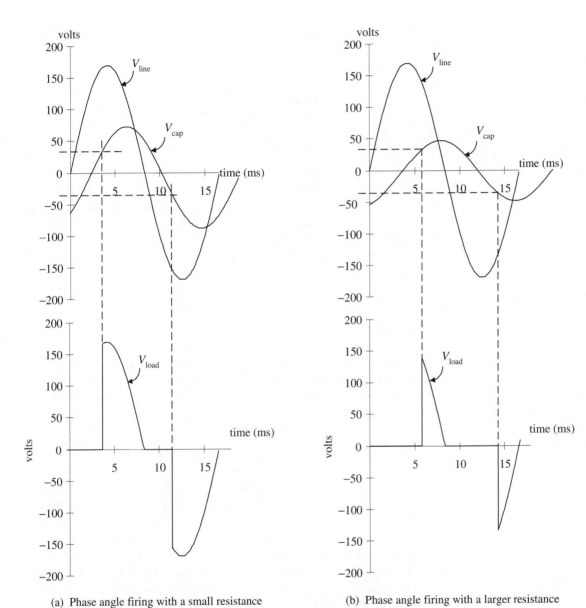

(a) Phase angle firing with a small resistance

(b) Phase angle firing with a larger resistance

Figure 8-21 Phase angle firing wave forms

This leaves the gate open, and the triac *off*. No voltage is applied to the load.

For Figure 8-21 (a), at about 3.5 ms, the voltage across the capacitor exceeds the diac's breakover voltage. The diac shorts. The capacitor discharges rapidly through the gate of the triac, turning the triac *on*. The load voltage, shown in the lower graph, jumps up to the value of the line voltage, since the triac now only drops a volt or two. Since the capacitor is discharging through a forward biased gate-to-MT1 junction, it discharges rapidly. Once the discharge current from the capacitor falls below the holding current of the diac, the diac turns *off*. This may only take 100 μs. The diac is reset, ready for the next cycle. The triac, however, stays *on* for the rest of that line half cycle. It continues to pass the line voltage to the load until the next zero crossing. At that time, the line current falls below the triac's holding current and the triac too goes *off*.

During the negative line cycle the procedure repeats, in the opposite direction. Sine the triac is a switch it is fired *on* at the same point in the negative half-cycle. If the triac is replaced with an SCR, then no negative pulses are passed to the load.

Increasing the resistance of the timing resistor, R_t, lowers the voltage across the capacitor, and increases the lag angle. This is shown in Figure 8-21 (b). It takes longer for the voltage across the capacitor to reach the diac's breakover voltage. So, the triac is fired later each half-cycle, providing less of the line voltage to the load. This is the technique used by most manually adjusted light dimmers.

Optically Coupled Triac Triggers

The diac senses the control signal, produces a high current pulse when the control reaches the breakover voltage, and turns itself *off* as soon as the thyristor is *on*. There is only one feature missing. The control signal is referenced to the top of the load. When the load is *off* this is zero volts. But when the thyristor turns *on*, this reference point jumps to the line voltage. Creating the control signal from any low level analog or digital electronics may destroy those ICs the first time the thyristor is fired. There is also the major safety concern of having the common of a circuit over one hundred volts above earth.

The solution is to isolate the thyristor from the control circuit, and couple the gating pulse from the control electronics optically. There is a series of optically coupled triac triggers that contains an LED on one side and a light activated diac on the other. The schematic symbol is shown in Figure 8-22.

MOC3010

Figure 8-22 Opto-coupler schematic symbol

A sample of the available triggers is given in Table 8-2, along with key specifications.

Table 8-2 Optically coupled thyristor triggers

	LED			Output		Zero Crossing
	I_{FT}	I_{fmax}	V_F @ 30 mA	V_R	V_{DRM}	V_{ih}
	mA	mA	V	V	V_p	V
MOC3010	15	60	1.5	3	250	
MOC3011	10	60	1.5	3	250	
MOC3012	5	60	1.5	3	250	
MOC3031	15	60	1.5	3	250	20
MOC3032	10	60	1.5	3	250	20
MOC3033	5	60	1.5	3	250	20

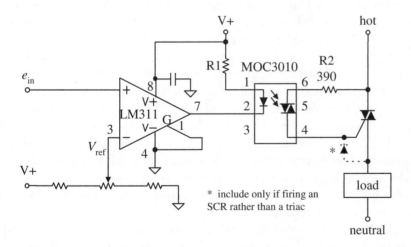

Figure 8-23 Typical use of an optically coupled thyristor trigger

Figure 8-23 is the schematic of a typical application. As long as the input voltage, e_{in}, is above the reference, v_{ref}, the output of the LM311 is open. No current flows through the MOC3010's LED. It does not shine on its internal diac, so the diac is *off*. With the diac *off*, the triac is not triggered. No power is sent to the load.

When e_{in} falls below the reference, the output of the LM311 is shorted to common. Current flows through the MOC3010's LED. It shines on the internal diac, causing the diac to fire. The voltage across the diac falls to typically 1.8 V_p. The rest of the line voltage is across the 390 Ω resistor. Current rises through the diac, and into the gate. This fires the triac. As soon as the triac is *on*, its MT2 to MT1 voltage falls to about 1.6 V_p. This is below the voltage across the diac, leaving no voltage across the 390 Ω resistor. Current through it drops to zero, and the diac turns *off*. Even though the LED continues to shine on the diac, the diac stays *off* once the triac goes *on*, because the voltage across the triac is too small to keep the trigger *on*. Even though the diac and the 390 Ω resistor are exposed to the line voltage and pass over 100 mA$_p$, they do not overheat because they are *on* for less than 100 μs every half-cycle of the line voltage.

There are two key points in driving the LED. The circuit that turns the LED *on* sinks current. Between 5 mA and 15 mA of LED current are needed to assure that a trigger pulse is generated. Many ICs cannot be sure of sourcing that much current. But most logic gates and comparators can sink at least this current. Secondly, the reverse voltage rating of the LED is only 3 V_p. So, be sure to drive this part with an open collector IC. The positive output from an IC with an active high side drive may be high enough above the LED's anode voltage to damage it.

The triggers in Table 8-2 are divided into two groups. Those in the top section apply a trigger pulse to the thyristor as soon as they are lit by the LED. However, the second set of triggers are called **zero crossing** triggers. The diac in these ICs only turns *on* when lit by the LED *and* when the voltage across it is less than 20 V. They only fire the triac when the line voltage is very near the zero crossing. Figure 8-24 shows the effect of each type trigger.

Zero crossing firing has both advantages and limitations. Since the triac or SCR is fired at the zero crossing of the line voltage, and stays *on* until the line voltage falls to zero at the end of that half-cycle, power must be applied to the load in full half-cycle packets. You cannot turn the load *on* a little more each cycle. It is either *on* or *off* for the entire half-cycle.

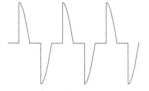

(a) Phase angle firing

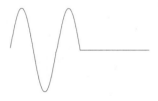

(b) Zero crossing firing

Figure 8-24 50% load power

But, near the zero crossing, line voltage and current are both very small. Turning the main switch *on* then means that there are no transients generated in the line. No electromagnetic interference is created. Current rises slowly as the low frequency line voltage climbs. There are no *di/dt* problems. This also means that the load is treated very gently. Transients of hundreds of volts are *not* applied to the load in 100 μs. All in all, if you just want to turn the load full *on*, zero crossing firing is better. To assure that the pulse to the thyristor occurs at the zero crossing, just select a MOC3031 trigger rather than the MOC 3011. No special zero crossing detection and pulse shaping circuitry are needed.

8.4 Proportional Power Circuits

The thyristor is a switch. When it is *off*, no power is applied to the load. When it is *on*, the full line power is connected. But to dim the lights, warm the room a little, or speed up the drill, more sophisticated control is needed.

For loads with large inertia, power can be controlled by applying half-cycle "packets" of energy. If you want fifty percent power to the load, turn the triac *on* for three half-cycles, and then *off* for three. If that's not quite enough, then apply power four half-cycles of the line and then remove it for six half-cycles. This technique is called **time proportioning**. Energy is passed in entire half-cycles, by firing at the zero crossing. This was shown in Figure 8-24 (b). It is quiet, and easy on both the load and the thyristor.

But time proportioning is a very poor way to control the intensity of a light bulb. Turning a light bulb *on* for three half-cycles, and then *off* for seven half-cycles produces an easily seen flicker. The bulb does not just glow softly. Your eye can see it as it turns *on* and *off*. For low inertial loads, like the light bulb, some power must be applied every half-cycle of the line. **Phase angle firing** waits a set time after the zero crossing and then fires the thyristor. The switch stays *on* for the rest of that half-cycle, then goes *off*, just to be fired *on* again after another delay. Look back at Figure 8-24 (a). The load appears to operate much more smoothly. The light does not seem to flicker, because some power is applied every 8–10 ms.

Phase angle firing may cause the thyristor to turn *on* at the line's peak, though. This would step the current through the switch and voltage to the load from zero to maximum in less than 100 μs. Without proper snubbing, the thyristor may fail and there will be massive electromagnetic interference spread through the power distribution system.

Time Proportioning

The schematic of a time proportioning circuit is shown in Figure 8-25. Only a few changes have been made from the typical use schematic of Figure 8-23. The most important is that a zero crossing trigger (MOC3031) is used. This assures that the switch is triggered *on* only at the zero crossing. To assure that enough gating current is available from the low line voltage near the zero crossing, R2 is lowered to 180 Ω. It may be appropriate to lower it further.

Snubbing for *dv/dt* has been added. Time proportioning uses zero crossing triggering, so it does not have a problem with the critical rate of rise of current. The triac may still be falsely fired *on* by noise at MT2. The resistor and capacitor across the triac short out any such noise.

The key to this circuit's operation is the V_{ref} ramp generator. It begins at zero volts, and rises linearly to the largest voltage that is to be applied at E_{in}. The period of the ramp is *slow*, covering an integer number of line half-cycles. The input voltage, E_{in}, is dc. It indicates how much power is to be applied to the load. When E_{in} is at half of the ramp's peak amplitude, the triac is *on* half of the time. Lowering E_{in} means that the switch is turned *on* for a shorted time, sending less power to the load. A higher E_{in} causes the triac to be turned *on* longer.

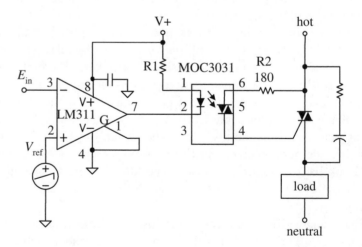

Figure 8-25 Time proportioning circuit

The timing diagram in Figure 8-26 illustrates the circuit's operation. The input, E_{in}, is a slowly varying dc. The reference, V_{ref}, is a ramp with a period equal to 10 line half-cycles. For 60 Hz, that is 83.3 ms. The voltage delivered to the load is shown on the bottom trace.

At time **0**, the ramp begins upward. But, initially V_{ref} is below E_{in}. So, the output of the comparator is low, the LED is lit, at each zero crossing a trigger pulse is delivered to the triac, it fires, and power is delivered to the load for the following half-cycle.

At time **1** the ramp exceeds the input. The comparator's output goes high. The LED turns *off*. This has no immediate effect on the triac, since once it is fired, it remains *on* for the rest of that half-cycle. So, power continues to be sent to the load until the first zero crossing after time **1**. At that point the triac turns *off* and is not triggered back *on*. Six half-cycles have been delivered to the load.

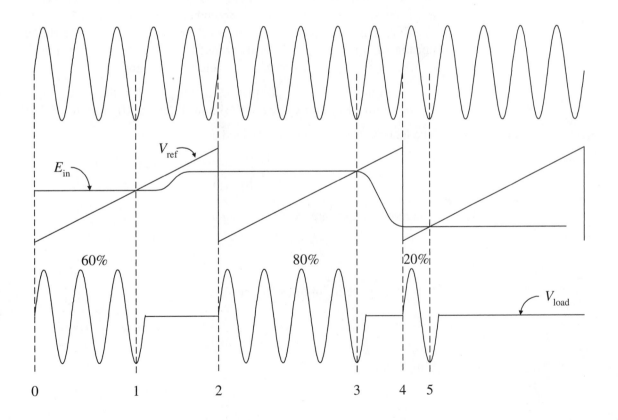

Figure 8-26 Timing diagram for the time proportioning circuit

For the remainder of the time before time **2**, the ramp stays above the input, the comparator output stays high, the LED stays *off*, and the triac stays *off*. Six of the ten half-cycles have been delivered to the load, providing a duty cycle of 0.6, and 60% power.

At time **2** the ramp falls below E_{in}, and the cycle begins over again. To illustrate the effect of changing the control voltage, E_{in} has been raised. From time **2**, at every zero crossing the triac is fired and power is applied to the load for an entire half-cycle. This continues until time **3**. Then the ramp exceeds the input and the LED goes *off*. That half-cycle continues to be passed to the load, but after the triac turns *off* at the next zero crossing, it is not triggered *on* again. So eight half-cycles (80%) are passed to the load. Increasing the input control voltage has increased the power provided to the load.

Lowering the input control voltage means that the ramp takes less time to exceed that input. The triac is fired *on* for less time, so less voltage is passed to the load. This is shown between times **4** and **5**.

Changing the input voltage has a linear and direct effect on the proportion of time that power is delivered to the load. To determine the numerical effect this has on $V_{rms\ load}$, begin by defining

a = number of half-cycles that the line is passed to the load
b = number of half-cycles of the control period (10 in Figure 8-26)

The load voltage can then be defined as

$$v_{load} = \begin{cases} V_p \sin\theta & 0 \le \theta < a\pi \\ 0 & a\pi \le \theta < b\pi \end{cases}$$

The root mean squared value of a signal, $v(t)$, is

$$V_{rms} = \sqrt{\frac{1}{T}\int_0^T \left[v(t)\right]^2 dt}$$

Substituting the definitions for the signal,

$$V_{rms\ load} = \sqrt{\frac{1}{b\pi}\left[\int_0^{a\pi} V_p^2 \sin^2\theta d\theta + \int_{a\pi}^{b\pi} 0 d\theta\right]}$$

Simplifying

$$V_{rms\ load} = \sqrt{\frac{V_p^2}{b\pi}\int_0^{a\pi} \sin^2\theta d\theta}$$

The standard integral of $\sin^2 x$ is

$$\int \sin^2 x\, dx = \frac{1}{2}x - \frac{1}{4}\sin 2x$$

Applying this definition

$$V_{\text{rms load}} = \sqrt{\frac{V_\text{p}^{\,2}}{b\pi}\left(\frac{1}{2}\theta - \frac{1}{4}\sin 2\theta\right)\Big|_0^{a\pi}}$$

Next, the limits must be applied.

$$V_{\text{rms load}} = \sqrt{\frac{V_\text{p}^{\,2}}{b\pi}\left[\left(\frac{a\pi}{2} - \frac{1}{4}\sin 2a\pi\right) - \left(\frac{0}{2} - \frac{1}{4}\sin 0\right)\right]}$$

Since a is an integer, the sine of 2π, 4π, 6π, ..., are all zero. So the equation simplifies considerably.

$$V_{\text{rms load}} = \sqrt{\frac{V_\text{p}^{\,2}}{b\pi}\frac{a\pi}{2}} = \frac{V_\text{p}}{\sqrt{2}}\sqrt{\frac{a}{b}}$$

It is common to define the duty cycle, D, as the ratio of the time *on* to the total time. This is precisely what was done for all of the switching power supplies.

$$D = \frac{a}{b}$$

Making this substitution gives

$$V_{\text{rms load}} = V_{\text{rms line}}\sqrt{D}$$

Time proportioning rms voltage

Since the power delivered to a *resistive* load is

$$P_{\text{load}} = \frac{V_{\text{rms}}^{\,2}}{R_{\text{load}}}$$

Time proportioning power

$$P_{\text{load}} = \frac{V_{\text{rms line}}^{\,2}}{R_{\text{load}}}D$$

Applying voltage to the load 60% of the time lowers the load rms voltage to 77% $V_{\text{rms line}}$ and drops the power to 60%. There is a direct, linear relationship between the control voltage, the percent of time power is applied to the load, and the level of that power.

Example 8-6

a. Calculate the rms voltage and the power applied by a triac to a 10 Ω load from a 120 V_{rms}, 60 Hz line with a control period of 83.3 ms, first using an *on* time of 15 ms, then using an *on* time of 55ms.

b. Verify your answers by simulation.

Solution

a. The 60 Hz signal has a period of

$$T_{line} = \frac{1}{60\,Hz} = 16.7\,ms$$

Since the triac can pass both the positive and the negative signal to the load, you must account for each half-cycle.

$$T_{half-cycle} = \frac{16.7\,ms}{2} = 8.3\,ms$$

A control period of 83.3 ms means that there are

$$\#\,of\,half\text{-}cycles = \frac{83.3\,ms}{8.3\,\dfrac{ms}{cycle}} = 10\,cycles$$

The zero crossings occur at
   0 ms,   8.3 ms, 16.7 ms, 25.0 ms, 33.3 ms, 41.7 ms, 50.0 ms, 58.3 ms, 66.7 ms, 75.0 ms, 83.3 ms.

When the triac is fired for 15 ms, it is triggered at 0 ms, and 8.3 ms. The first two half-cycles are passed to the load. The third zero crossing, at 16.7 ms, comes too late.

$$D = \frac{2\,half\text{-}cycles}{10\,half\text{-}cycles} = 0.2$$

$$V_{rms\,load} = 120\,V_{rms}\,\sqrt{0.2} = 54\,V_{rms}$$

$$P_{load} = \frac{\left(V_{rms\,load}\right)^2}{R_{load}} = \frac{\left(54\,V_{rms}\right)^2}{10\,\Omega} = 292\,W$$

When the triac is fired for 55 ms, it is triggered at 0 ms, 8.3 ms, 16.7 ms, 25.0 ms, 33.3 ms, 41.7 ms, and 50.0 ms. The first seven half-cycles are passed to the load. The eighth zero crossing, at 66.7 ms, comes too late.

$$D = \frac{7 \text{ half-cycles}}{10 \text{ half-cycles}} = 0.7$$

$$V_{\text{rms load}} = 120\,V_{\text{rms}}\sqrt{0.7} = 100\,V_{\text{rms}}$$

$$P_{\text{load}} = \frac{\left(V_{\text{rms load}}\right)^2}{R_{\text{load}}} = \frac{\left(100\,V_{\text{rms}}\right)^2}{10\,\Omega} = 1000\,W$$

b. Neither simulation software supports the MOC series of optical couplers. So a rectangular pulse is applied directly between the triac's gate and MT1. The schematic is given in Figure 8-27.

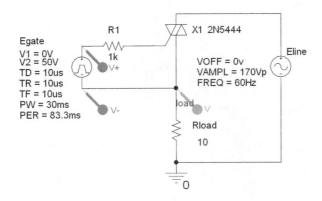

Figure 8-27 Schematic for simulation of Example 8-6

The simulation results (Figure 8-28) match the calculations. well. Even though the gate is *on* for 15 ms and 55 ms of the 83.3 ms control period (18% and 66%), you have to count the number of zero crossing triggers that occur (2 of 10 and 7 of 10).

Practice: Calculate the load voltage and power for a 30 ms gating pulse.

Answer: $D = 0.4$, $V = 76$ V, $P_{\text{load}} = 576$ W (twice the power for $D = 0.2$)

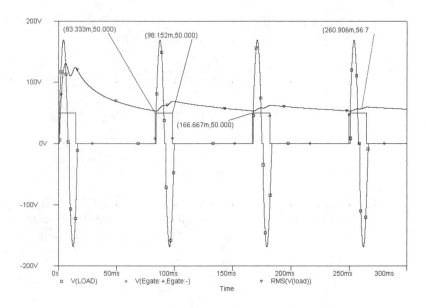

(a) Probe output for 15 ms *on* time

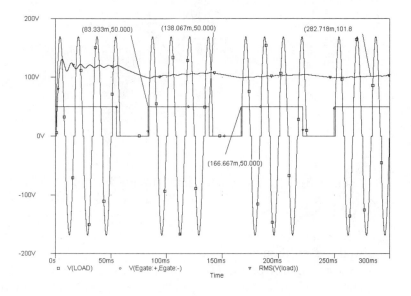

(b) Probe output for 55 ms *on* time

Figure 8-28 Simulation results for Example 8-6

Phase Angle Firing

Time proportioning does allow you to provide different amounts of power to the load. It is electrically quiet, and gentle on the load and thyristor. But the power is sent in discrete packets. Often this means that the load jerks or flickeres every control period. This is fine for high inertia loads such as heaters. But for lighting or motor speed or position control, some power must be applied every cycle of the line voltage. That can be done by phase angle firing.

A phase angle fired signal is shown in Figure 8-29. The first zero crossing happens at **a**. The thyristor is *not* fired then. Instead the firing pulse is delayed. At time **b**, the thyristor *is* triggered *on*. Power is applied for the rest of that cycle. At the next line current zero crossing, time **c**, the switch goes *off*. It is not fired *on* until a later time, **d**. Each cycle, some power is sent to the load. In Figure 8-29, the firing angle is delayed more and more each cycle, to illustrate progressively lower amounts of power.

Phase angle firing allows you to apply some power every line cycle, so the jerkiness associated with time proportioning is overcome. You can adjust the amount of power applied to the load every half-cycle, making phase angle firing more responsive.

But there are two major problems with phase angle firing. First, since the thyristor may be turned *on* at any instant, snubbing is critical, and electromagnetic interference is a real possibility. Secondly, most of

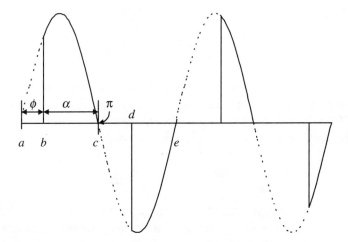

Figure 8-29 Phase angle firing waveforms for a resistive load

power is applied as part of a sine wave. The relationship between the time delayed, the firing angle, ϕ, and the power delivered to the load is *very* nonlinear. The vast majority of the power occurs in the center of the sine, between 50° and 130°. This is only a small part of the half-cycle.

An analog circuit that produces phase angle firing is shown in Figure 8-30. The associated waveforms are shown in Figure 8-31. To begin, you must know when to begin, that is, when the zero-crossing occurs. So the transformer couples the line voltage to the control circuit and the full-wave bridge rectifies that sine wave. The signal at node A is some positive voltage at all times *except* at the zero-crossing. Only at that time does the voltage drop to zero.

At all times, except the zero crossing, the positive voltage from the bridge rectifier turns Q1 *on*, holding the base of Q2 just above ground. Q2 is *off*. At each zero crossing, for a few tens of microseconds, the voltage from the bridge falls to zero, and Q1 turns *off*. At that time, the current through the 1 kΩ resistor flows into the base of Q2, saturating it. When Q2 turns *on*, it shorts out the 0.33 µF capacitor, discharging it. The purpose of the transformer, the bridge rectifier, and the two transistors is to discharge the capacitor quickly at every zero crossing.

Just after the zero crossing, the transistor across the capacitor turns *off*. The capacitor's charge then begins to ramp up. The inverting pin of U1a is held at virtual ground. So current flows from the output of that op amp, right to left through the capacitor, then right to left through the rate potentiometer, to −15 V_{dc}. With the potentiometer set to 37.7 kΩ, that current is

$$I = \frac{15\,V_{dc}}{37.7\,k\Omega} = 398\,\mu A_{dc}$$

This is a constant current and charges the capacitor − to +. Since the left end of that capacitor is held at virtual ground, the right end (the output of U1a) ramps up as the capacitor charges.

The performance of a capacitor is described by

$$i = C\frac{dv}{dt}$$

Since both the capacitor and the current are constants, the rate of change of the voltage across the capacitor, and at the output of U1a must also be a constant. A constant rate of change describes a ramp.

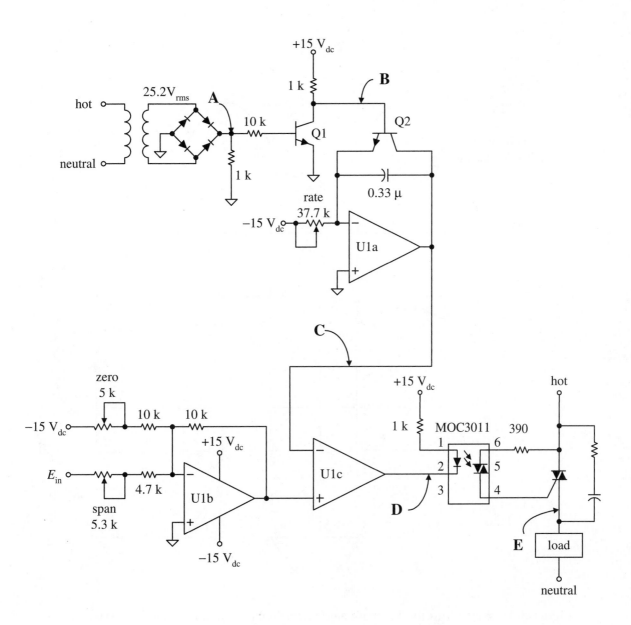

Figure 8-30 Analog controlled phase angle firing circuit

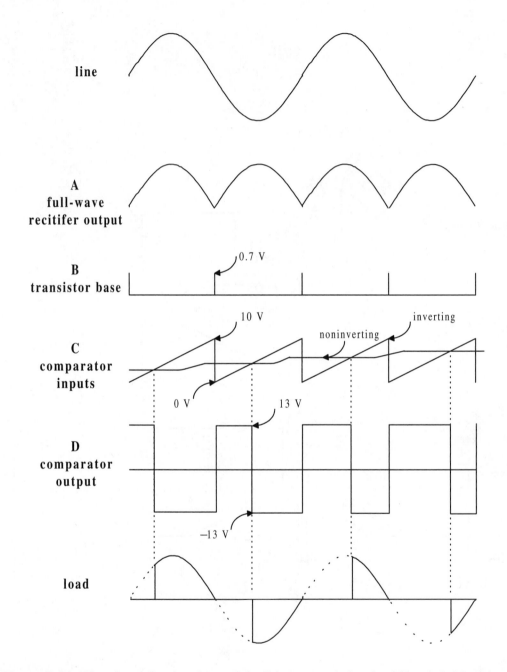

Figure 8-31 Waveforms for the phase angle firing control circuit of Figure 8-30 with a resistive load

$$ramp\ rate = \frac{dv}{dt} = \frac{I}{C}$$

$$= \frac{398\,\mu A}{0.33\,\mu F} = 1206\,\frac{V}{s}$$

This ramp begins at 0 V, at the zero crossing, and charges for one half-cycle, until the next zero crossing. It that time the capacitor's voltage reaches

$$dv = 1206\,\frac{V}{s} \times dt = 1206\,\frac{V}{s} \times 8.3\,ms$$

$$dv = 10.0\,V_p$$

This is the ramp at the inverting input of U1c. The dc level at the noninverting input comes from the control input and is inverted and offset by U1a. The reason for that will be discussed shortly.

Just following the zero crossing, the ramp falls to zero, below the dc input to U1c. This drives the output of U1c to positive saturation. The LED in the MOC3011 does *not* shine, and the triac is *not* fired. This condition continues until the ramp climbs above the dc input to U1c. When the ramp on the inverting input exceeds the voltage on the noninverting input, the output of U1c steps to negative saturation. The LED in the MOC 3011 is lit, and its diac is fired. That fires the triac. Power is applied to the load for the remainder of the cycle.

Several half-cycles are shown in Figure 8-31. The dc level at the noninverting input of U1c is increased each cycle to illustrate the effect on the voltage passed to the load. The triac goes *off* at the zero crossing and remains *off* until the ramp climbs above the dc input to U1c. The larger that dc input, the longer it takes for the ramp to reach it. This means a longer delay until the triac is fired and less voltage to the load.

This circuit allows a 0 V_{dc} to 10 V_{dc} to inversely proportionally adjust the power to the load. Op amp U1a inverts the control input and offsets it by 10 V_{dc}. When the control input is 0 V_{dc}, U1a outputs 10 V_{dc} to the comparator, and no power is sent to the load. When the control input rises to 10 V_{dc}, U1a send 0 V_{dc} to the comparator. This causes the triac to be fired at the very beginning of the cycle, providing full power.

multi**SIM**

Example 8-7

Confirm the performance of the control circuit.

Solution

Simulation schematics of the control circuit through the LED are shown in Figures 8-32 (a) and (b). Neither simulation software provides the MOC3011.

In Figure 8-32 (a), the control voltage is set to 2.5 V_{dc}. The voltage out of the bridge rectifier is a full wave rectified signal. The base of Q2 is an 0.8 V_p positive pulse occurring around that zero crossing. At each zero crossing, the ramp out of U1a drops to zero and then starts again, rising to 10 V in 8.3 ms.

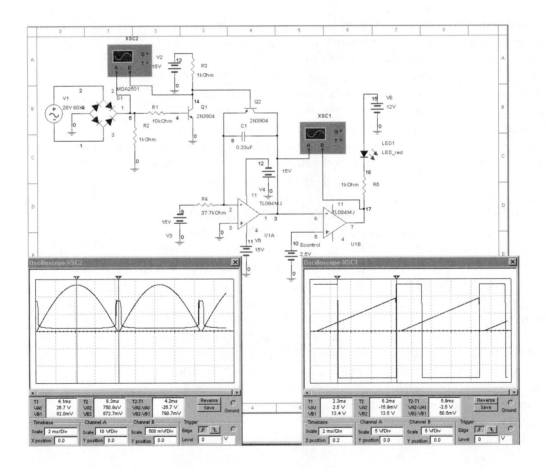

Figure 8-32a Simulation of a phase angle firing control circuit at 2.5 V_{dc} control input

With the control input (the voltage at the noninverting input of the comparator, U1b) set to 2.5 V_{dc}, the output of U1b drops from +13 V_p to −13 V_p at 2.3 ms. That is when the ramp crosses 2.5 V. When U1b's output goes negative, the LED is lit. In the actual circuit this fires the triac early in the cycle, sending lots of power to the load.

Figure 8-32 (b) shows the effect of changing the control voltage to 7.5 V_{dc}. The output of the comparator stays high for 6.4 ms. This delays the firing pulse to the triac to near the end of the cycle, providing only a small amount of power to the load.

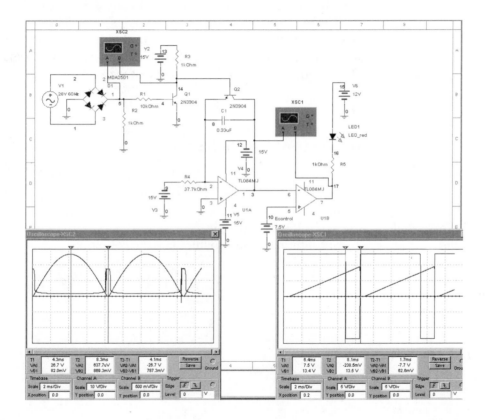

Figure 8-32b Simulation of a phase angle firing control circuit at 7.5 V_{dc} control input

Practice: What must be done to the circuit if the control voltage only goes from 0 V$_{dc}$ to 5 V$_{dc}$? Verify your answer by simulation with V$_{control}$ = 2.5 V$_{dc}$.

Answer: Set the ramp potentiometer to 75.4 kΩ, and the zero potentiometer to 20 kΩ. The comparator goes low at about 4.2 ms.

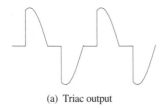

(a) Triac output

(b) SCR output

(c) SCR full wave rectified output

Figure 8-33 Phase angle fired waveforms

Phase angle firing a triac sends energy to the load on both half-cycles, in both directions. Phase angle firing a single SCR provides a signal to the load only on the positive half-cycle. If that SCR is combined with a full-wave rectifier, positive voltage is applied to the load every half-cycle. These are shown in Figure 8-33.

The average value of the triac output is zero. The signal above the x axis matches the signal below. So for dc loads such as power supplies and dc motors, the SCR signals of Figure 8-33(b) or (c) must be used. These loads respond to average value of the signal.

$$V_{dc} = \frac{1}{T}\int_0^T v(t)dt$$

The phase angle fired SCR signal, $v(t)$ is defined as

$$v(t) = \begin{cases} 0 & 0 \le \theta < \phi \\ V_p \sin\theta & \phi \le \theta < \pi \end{cases}$$

where ϕ is the **firing angle**.

Substituting these variables into the definition

$$V_{dc} = \frac{1}{\pi}\left(\int_0^\phi 0 dt + \int_\phi^\pi V_p \sin\theta d\theta\right)$$

$$V_{dc} = \frac{V_p}{\pi}\left(-\cos\theta\Big|_\phi^\pi\right)$$

$$V_{dc} = \frac{V_p}{\pi}\left(-\cos\pi + \cos\phi\right)$$

V_{ave} for a full-wave rectified phase angle fired SCR output

$$V_{dc} = \frac{V_p}{\pi}\left[\cos(\phi) + 1\right]$$

This is the average value for the first half-cycle. For a full-wave rectified signal, this signal occurs *every* half-cycle. So the equation is correct for the full-wave rectified output.

The half-wave rectified signal is this value for one half-cycle, and zero for the other. So across the full 2π, the average value is halved.

$$V_{dc} = \frac{V_p}{2\pi}[\cos(\phi)+1]$$

V_{ave} **for a half-wave rectified phase angle fired SCR output**

For resistive loads, such as heaters or lights, the root mean squared voltage is used to determine the power delivered to the load.

$$V_{rms} = \sqrt{\frac{1}{T}\int_0^T [v(t)]^2 dt}$$

A half-cycle of the phase angle fired signal is defined as

$$v(t) = \begin{cases} 0 & 0 \le \theta < \phi \\ V_p \sin\theta & \phi \le \theta < \pi \end{cases}$$

Making this substitution gives

$$V_{rms} = \sqrt{\frac{1}{\pi}\left[\int_0^\phi 0^2 d\theta + \int_\phi^\pi (V_p \sin\theta)^2 d\theta\right]}$$

$$V_{rms} = \sqrt{\frac{V_p^2}{\pi}\int_\phi^\pi \sin^2\theta\, d\theta}$$

But

$$\int \sin^2 x\, dx = \frac{1}{2}x - \frac{1}{4}\sin 2x$$

Applying this,

$$V_{rms} = \frac{V_p}{\sqrt{\pi}}\sqrt{\left(\frac{1}{2}\theta - \frac{1}{4}\sin 2\theta\right)\Big|_\phi^\pi}$$

$$V_{rms} = \frac{V_p}{\sqrt{\pi}}\sqrt{\left(\frac{1}{2}\pi - \frac{1}{4}\sin 2\pi\right) - \left(\frac{1}{2}\phi - \frac{1}{4}\sin 2\phi\right)}$$

$$V_{rms} = \frac{V_p}{\sqrt{\pi}}\sqrt{\frac{1}{2}(\pi - \phi) + \frac{1}{4}\sin 2\phi}$$

V_{rms} **for a triac and a full-wave rectified phase angle fired SCR output**

This is the rms value for both the triac or the full-wave rectified SCR. The half-wave rectified signal has a period of 2π. So its rms value is

$$V_{rms} = \frac{V_p}{\sqrt{2\pi}}\sqrt{\frac{1}{2}(\pi - \phi) + \frac{1}{4}\sin 2\phi}$$

V_{rms} **for a half-wave rectified phase angle fired SCR output**

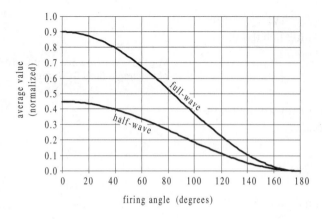
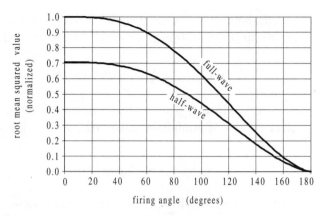

Figure 8-34 Normalized average and rms values versus phase angle firing angle

Notice that the 2 is inside the radical. These equations require that you complete the computation in *radians*, not degrees.

Given the firing angle, you can complete these calculations to determine the average value or the root mean squared value of the load voltage. It may be a bit tedious, but it can be done accurately. However, if you know the average value or the rms load voltage you want, determining the firing angle is much more involved. Figure 8-34 is a plot of the normalized average and rms voltages versus firing angle. The vertical axes have been normalized (divided by $V_{\text{rms line}}$). So, to denormalize the result, multiply the value on the vertical axis by the root mean squared value of the line voltage.

The larger the firing angle, the lower the output voltage. That is, the longer you delay before firing the thyristor, the lower the voltage that remains to be passed to the load. All four curves are nonlinear, very S shaped. Between 50° and 140°, though, the curves fall steadily, covering the vast majority of the load signal's amplitude. Changes in firing angle near 90° have the most effect. Little is accomplished at the extremes.

Example 8-8

Determine the average value of the voltage delivered to the load by a half-wave (single SCR) with a firing angle of 60° and the rms value of a full wave (triac) with a firing angle of 60°.

$V_{\text{line}} = 120 \text{ V}_{\text{rms}}$

a. by calculation.   **b.** from the curves.   **c.** with a simulation.

Solution

Average value of a half-wave 60° firing angle:

From the left graph, the lower curve in Figure 8-34, at 60° the normalized factor is 0.35.

$$V_{\text{ave half-wave}} = 0.35 \times 120\,V_{\text{rms}} = 42\,V_{\text{dc}}$$

From the equation for the average of a half-wave,

$$V_{\text{dc}} = \frac{V_{\text{p}}}{2\pi}\left[\cos(\phi)+1\right]$$

$$V_{\text{dc}} = \frac{120\,V_{\text{rms}} \times \sqrt{2}}{2\pi}\left[\cos(60°)+1\right] = 41\,V_{\text{dc}}$$

The OrCAD schematic is shown in Figure 8-35. Notice that it is necessary to edit the model of the SCR to set its forward and reverse breakover voltages above 170 V_{p}.

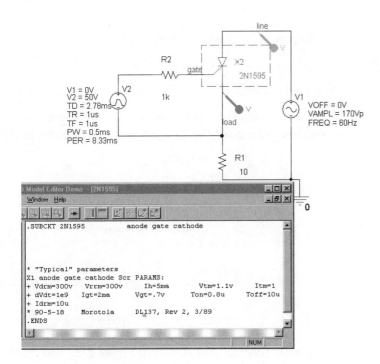

Figure 8-35 Simulation schematic and SCR model

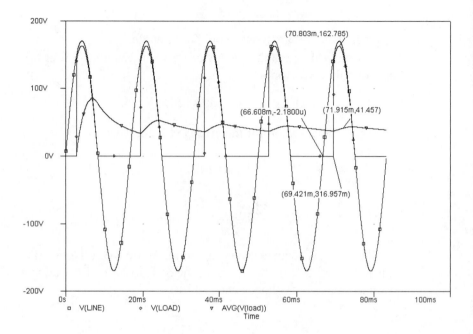

Figure 8-36 Simulation V_{dc} for the SCR

The markers indicate $V_p = 163$ V$_p$ $V_{ave} = 41$ V$_{dc}$

For the rms value of a full-wave signal, the upper line of the graph on the right side of Figure 8-34 indicates a factor of 0.9.

$$V_{\text{rms full-wave}} = 0.9 \times 120\,V_{\text{rms}} = 108\,V_{\text{rms}}$$

From the equation for the rms of a full-wave,

$$V_{\text{rms}} = \frac{V_p}{\sqrt{\pi}}\sqrt{\frac{1}{2}(\pi - \phi) + \frac{1}{4}\sin 2\phi}$$

$$V_{\text{rms}} = \frac{120V_{\text{rms}} \times \sqrt{2}}{\sqrt{\pi}}\sqrt{\frac{1}{2}(\pi - 1.047\,\text{rad}) + \frac{1}{4}\sin 120°}$$

$$V_{\text{rms}} = 95.91\,\text{V}\sqrt{1.047 + 0.217} = 108\,V_{\text{rms}}$$

The simulation schematic is shown in Figure 8-37 and the result in Figure 8-38. The markers indicate

$$V_p = 169 \text{ V}_p \qquad V_{ave} = 106 \text{ V}_{\text{rms}}$$

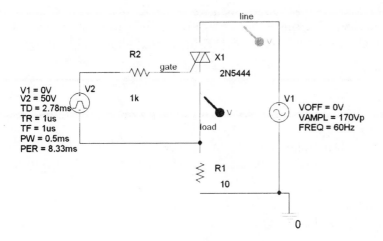

Figure 8-37 Simulation schematic for full-wave triac phase
angle firing

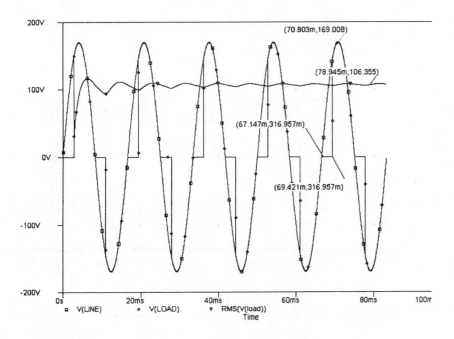

Figure 8-38 Simulation results for 60° phase triac firing

Practice: Determine the average value and rms for a full-wave rectified signal with a 120° phase angle firing angle, with $V_{line} = 120\ V_{rms}$.

Answer: $V_{ave} = 27\ V_{dc}$, $V_{rms} = 53\ V_{rms}$

Example 8-9

Determine the firing angle needed to deliver 1.3 kW to a 15 Ω resistive load using a full-wave (triac) circuit operating from a line voltage of 220 V_{rms}.

Solution

$$P = \frac{V_{rms}^{2}}{R_{load}}$$

Solving for the rms voltage needed to deliver the power,

$$V_{rms} = \sqrt{P \times R_{load}}$$

$$V_{rms} = \sqrt{1.3\,kW \times 15\Omega} = 140\,V_{rms}$$

This load voltage represents a normalized factor of

$$\frac{140\,V_{rms}}{220\,V_{rms}} = 0.64$$

Look at the upper curve on the right graph in Figure 8-34. To obtain a normalized rms factor of 0.64 requires a firing angle of about 98°.

You can double check this by substituting 98° into the rms equation.

$$V_{rms} = \frac{220\,V_{rms}\,\sqrt{2}}{\sqrt{\pi}}\sqrt{\frac{1}{2}(\pi - 1.710\,rad) + \frac{1}{4}\sin 196°} = 141\,V_{rms}$$

Practice: What firing angle is required to cut the power to the load in half?

Answer: $P = 650\ W$, $V_{rms} = 99\ V_{rms}$, normalized factor $= 0.45$, $\phi = 119°$

Summary

Thyristors allow you to control power directly from the commercial power line. The silicon controlled rectifier is a four layer semiconductor with an anode, a cathode, and a gate. It is modeled as a *pnp* transistor driving an *npn* transistor. Normally the SCR is *off*, blocking current flow from anode to cathode. However, a small pulse between the gate and the cathode turns the switch *on*. After it has been turned *on*, the SCR holds itself *on* by internal regenerative feedback. The gating pulse may be removed. Once *on*, though, the SCR stays *on* as long as current continues to be supplied by the external circuit. It only goes *off* when the anode-to-cathode current falls below a small level, called the holding current. When used on an ac line, this happens at every zero crossing. So, you trigger the SCR *on*, and it stays *on* for the rest of that half-cycle and then turns itself *off*. You must then trigger it back *on* again. Key SCR parameters were defined in the first section of the chapter.

The triac may be viewed as a bidirectional SCR. It blocks current flow in both directions when *off*. It may be triggered *on* with either a positive or a negative signal between the gate and MT1. Once *on*, the triac conducts in either direction until its main path current falls below the holding current.

Thyristors should be connected on the high side of the load and the load should be connected to neutral. When the switch is *off*, no voltage is passed to the load. This places neutral on both sides of the load, making it safe for the operator.

There are several precautions that must be observed when operating a high voltage, high current semiconductor switch. If the current is allowed to increase too quickly through the crystal, there can be large, localized overheating, burning the thyristor open. To assure that the critical rate of rise of current specification is not exceeded, you may have to add an inductor in series with the thyristor.

High frequency noise coupled into the power grid may be coupled through the parasitic capacitance from anode to gate. A signal at the gate could erroneously fire the thyristor, applying power to the load when it should be *off*. The manufacturer specifies a critical rate of rise of voltage at the anode. Placing a capacitor between anode and cathode assures that any noise at the anode does not exceed this dv/dt rating and falsely fire the SCR. A resistor must be placed in series with the capacitor to assure that the capacitive discharge when the SCR is properly triggered does not damage the switch. But this resistor must be by-passed with a diode. This allows the capacitor to charge rapidly to prevent *dv/dt* false triggering, but to discharge slowly through the resistor. These snubbing circuits also reduce the electromagnetic interference generated by the switch and sent out onto the power line.

The best way to turn a thryristor *on* is with a trigger. The trigger must provide a *pulse* of current at enough voltage to forward bias the gate junction. It must also sense when the SCR or triac is *on*, however, and stop sending gating current. The Shockley diode is an SCR without a gate. When the voltage is large enough, it breaks down and shorts, passing current to the gate. When that current falls below its holding current it turns *off*. The diac is a bidirectional Shockley diode. Combined with an *RC* phase lag circuit it can trigger a thyristor every cycle, depending on the setting of the resistor.

The MOC series of opto-couplers combine a light sensitive diac with an LED. When the LED is lit, the diac fires and triggers the SCR or ti-rac. This has the advantage of allowing isolation between the control circuit and the power switch and load. There are couplers that fire whenever the LED is *on*, and others that provide a trigger pulse only at the zero crossing of the line voltage.

Zero crossing firing has the advantage of turning on the power to the load when the voltage and current are very low. This means that there are no *di/dt* problems, the load and thyristor do not receive repeated shocks as hundreds of volts are applied to the load in a few microseconds, and little electromagnetic interference is generated.

Time proportioning uses zero crossing to apply a proportionally variable voltage to the load. The control interval must be long compared with the period of the line voltage. For the first part of this interval, the switch is *on*, sending power to the load. At some point, the thyristor is no longer fired, so the load is not powered during the last part of the control interval. Adjusting how much of the control interval applies power and how much does not proportionally adjusts the power sent to the load. Though gentle to the load, these discrete packets of energy cause the load to appear to flicker.

Phase angle firing provides some power every cycle. The zero crossing is detected, and a timing circuit started. At some time after the zero crossing the thyristor is fired. The longer the delay after the zero crossing, the smaller the part of each half-cycle is applied to the load, and the lower the output. This is a *very* nonlinear relationship. Most of the power is contained between 50° and 130°. For dc loads, such as dc motors and power supplies the average value is needed. For heating and lighting, you must calculate the rms. Both trigonometric formulas and a set of curves were developed.

Problems

Thyristor Device Characteristics

Locate a full set of specifications for the Teccor S4006LS SCR and the Teccor Q4006LH triac. Use them to answer the following questions.

8-1   What is required to turn the Teccor S4006LS SCR *on*?

8-2   What is required to turn the Teccor Q4006LH triac *on*?

8-3   What is required to turn the Teccor S4006LS SCR *off*?

8-4   What is required to turn the Teccor Q4006LH triac *off*?

8-5   Explain four characteristics that the SCR and the triac have in common.

8-6   What are two major differences between the SCR and the triac?

8-7   A 200 W light bulb is powered from a 120 V_{rms} line voltage, and is switched *on* and *off* with a S4006LS SCR.

   a.   Prove that the light bulb's nominal resistance is 72 Ω.

   b.   When the SCR is *on* continuously, calculate the rms current through the light bulb.

   c.   Calculate the maximum instantaneous voltage across the SCR when it is *off*.

   d.   What is the maximum power that the light dissipates?

   e.   When the SCR is *on* how much voltage is dropped across it?

   f.   How much power does the SCR dissipate when it is continuously *on*?

   g.   Given an ambient temperature of 50°C, calculate the thermal resistance of the SCR's heat sink.

   h.   Is that heat sink electrically connected to one of the terminals of the SCR or is it floating? Explain why this is important to know.

8-8   Repeat Problem 8-7 using a Q4006LH triac.

Snubbing

8-9 How do you prevent positive spikes at the anode of an SCR from falsely triggering it *on*?

8-10 Describe three sources of noise spikes on a commercial power line.

8-11 What is the purpose of the inductor in a triac snubber circuit?

8-12 Explain why critical rate of rise of current snubbing is *not* usually needed when driving a solenoid.

8-13 What is the purpose of the resistor in the snubbing circuit?

8-14 What would be the effect if the resistor in the snubbing circuit opened?

8-15 What is the purpose of the diode in the snubbing circuit?

8-16 What would be the effect if that diode was installed around the capacitor instead of the resistor?

8-17 Design a snubbing network for the S4006LS SCR driving a 200 W light bulb from a 120 V_{rms} line voltage. Determine the:
  a. inductor's inductance.

  b. capacitor's capacitance and peak voltage.

  c. minimum resistor.

  d. diode's peak current.

8-18 Repeat Problem 8-17 for the Q4006LH triac.

Triggers

8-19 Locate the specifications for a diac with a breakover voltage of 20 V. Determine its maximum *on* state current and its holding current.

8-20 For the diac in Problem 8-19, used in the circuit shown in Figure 8-39, calculate the minimum size of R_t for a 120 V_{rms} line.

8-21 Determine the setting of R_t using a 20 V diac and a 0.22 µF capacitor to fire the SCR at 60° after the zero crossing.

8-22 Determine the setting of R_t using a 20 V diac and a 0.22 µF capacitor to fire the SCR at 120° after the zero crossing.

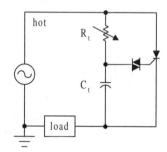

Figure 8-39 Diac trigger circuit (also Figure 8-20)

8-23   Calculate the current limiting resistor needed for a MOC3011 operated from a +5 V supply.

8-24   Calculate the current limiting resistor needed for a MOC3033 operated from a comparator that is powered from ±12 V.

8-25   Explain the difference between a MOC3011 and a MOC3031.

8-26   The resistor used in series with the diac of a MOC 3011 is much larger than the one recommended for the MOC3031. Explain why.

Proportional Power: Time Proportioning

8-27   A zero crossing, time proportioning circuit, in Figures 8-25 and 8-26, has a control interval of 167 ms. Explain why 85% power *cannot* be applied to the load.

8-28   What changes must be made to the circuit of Figures 8-25 and 8-26 to allow 85% power to be applied to the load?

8-29   Using an SCR, a ramp of 5 V_p, with a period of 167 ms, and a control voltage of 3 V_{dc}:

  a. Accurately draw a timing diagram similar to Figure 8-26.

  b. Calculate the V_{rms} delivered to the load from a 120 V_{rms} line.

  c. Calculate the power delivered if the load is 72 Ω.

  d. Calculate the V_{rms} if the control voltage drops to 1.5 V_{dc}.

  e. Calculate the power delivered to the 72 Ω load for a 1.5 V_{dc} control voltage.

8-30   Using a triac, a ramp voltage of 12 V_p with a period of 83.3 ms, and a control voltage of 8 V_{dc}:

  a. Accurately draw a timing diagram similar to Figure 8-26.

  b. Calculate the V_{rms} delivered to the load from a 120 V_{rms} line.

  c. Calculate the power delivered if the load is 72 Ω.

  d. Calculate the V_{rms} if the control voltage drops to 4 V_{dc}.

  e. Calculate the power delivered to the 72 Ω load for a 4 V_{dc} control voltage.

Proportional Power: Phase Angle Firing

8-31   Modify the phase angle firing circuit in Figure 8-30 to use a 5 V_p ramp, ± 5 V_{dc} supplies, and a maximum control input of +5 V_{dc} to produce maximum power to the load.

8-32   Using as little of Figure 8-30 (and your answer to Problem 8-31), replace as much of the analog circuitry as possible with a microprocessor to provide the phase angle firing. Assume that the value to set the firing angle is contained within a register inside the microprocessor.

8-33   Draw a diagram similar to Figure 8-31 for your circuit from Problem 8-31, given an input control voltage of 2 V_{dc}.

8-34   Draw a diagram similar to Figure 8-31 for your circuit from Problem 8-31, given an input control voltage of 4 V_{dc}. Replace the triac with an SCR.

8-35   For a line voltage of 120 V_{rms}, and a firing angle of 30°, what is the:

   a. V_{dc} from a half-wave rectified SCR (using the graph)?

   b. V_{dc} from a half-wave rectified SCR (manual calculation)?

   c. V_{rms} from a triac (using the graph)?

   d. V_{rms} from a triac (manual calculation)?

8-36   For a line voltage of 220 V_{rms}, and a firing angle of 100°, what is the:

   a. V_{dc} from a half-wave rectified SCR (using the graph)?

   b. V_{dc} from a half-wave rectified SCR (manual calculation)?

   c. V_{rms} from a triac (using the graph)?

   d. V_{rms} from a triac (manual calculation)?

8-37   For a line voltage of 120 V_{rms}, what firing angle must be used to deliver 30% of the maximum power to a 72 Ω load from a single SCR?

8-38   For a line voltage of 220 V_{rms}, what firing angle must be used to deliver 75% of the maximum power to a 15 Ω load from a triac?

Triac Lab Exercise

> **Warning!**   Operation at 120 V_{rms} may be lethal.   You *must* assure that accidental contact with the circuit cannot be made when power is applied.   An enclosure such as shown in Figure 8-40 is highly recommended.

(a) Safety enclosure photograph

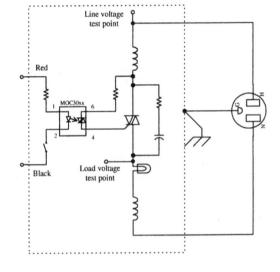

(b) Safety enclosure schematic

Figure 8-40 Safety enclosure for the triac circuit

A. Simple *On/Off* Control
   1. Obtain a triac-light control box

   2. With the box's power cord *unplugged*, remove the safety cover. With an ohmmeter, verify the proper configuration, as shown in Figure 8-40.

3. Install your triac and your MOC3030 (or 3031 or 3032). Double check to verify that it is properly installed in the socket.

4. Determine the value of the three resistors, the inductors, and the capacitor. Record these values.

5. Prove with a calculation that each component is an appropriate value.

6. Remove (or tape back) the ground clip from your oscilloscope. Connect the probe to the hot end of the light. There is a test point available there.

Warning! Assure that the probe is set to attenuate the voltage to the oscilloscope **x** 10. This lowers the signal to the oscilloscope to 17 V_p. Without the **x** 10 attenuator, the input voltage to the oscilloscope is 170 V_p. That may damage the input channel of the oscilloscope.

7. Set the switch on the box to *off*.

8. Replace the plexiglass cover.

9. Connect the +5 V_{dc} power supply and common to the LED side of the opto-coupler.

10. Plug the box into the line voltage.

11. Turn the +5 V_{dc} supply *on*.

12. With the oscilloscope measure the rms and dc values of the signal across the light when it is *off*.

13. With an *insulated* tool, flip the switch to *on*.

14. With the oscilloscope measure the rms and dc values of the signal across the light when it is *off*.

15. Turn the +5 V_{dc} supply *off*. Verify that the light goes *off*.

16. Unplug the box. Assure that the line plug is in clear view on the bench top, *not* connected to the line voltage.

B. Time Proportional Zero Crossing Firing
1. Build the circuit in Figure 8-41.

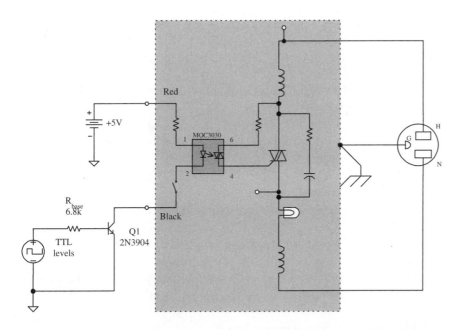

Figure 8-41 Time proportional zero crossing firing schematic

2. Leave channel one of the oscilloscope, with its ground clip *removed*, set to **×** 10. Connect it to the hot side of the light (the lower test point)

3. Connect channel two of the oscilloscope to display the voltage from the function generator, with its 0 V level across the center of the screen. Tie the oscilloscope common to the common of the function generator. Trigger on this channel.

4. Set the function generator to provide a positive going TTL compatible signal at 6 Hz, with a 20% duty cycle.

5. Adjust the oscilloscope controls to display about two cycles of the 6 Hz TTL signal from the function generator.

6. Move channel two of the oscilloscope to the transistor's collector. Verify that the signal is an inverted version of the TTL input.

7. Plug the box's power cord into the line voltage.

8. With an *insulated* tool assure that the box's switch is *on*.

9. Adjust the oscilloscope to display the rms value of the voltage across the light.

10. For each of the following input duty cycles, measure and record the following:
    duty cycles: 20%, 40%, 60%, 80%
    measure and record:
       number of line half-cycles applied to the light
       $V_{rms\ measured}$ of the voltage applied to the light
       $V_{rms\ theory}$ of the voltage applied to the light.

11. Accurately record the oscilloscope display of the transistor voltage and the voltage across the light at the 60% duty cycle.

12. Turn the function generator *off*. Verify that that light goes out.

13. Turn the +5 V_{dc} supply *off*.

14. Unplug the box. Assure that the line plug is in clear view on the bench top, *not* connected to the line voltage.

C. Phase Angle Firing Proportional Power
   1. Look at the circuit in Figure 8-42. This is a simplified version of Figure 8-30. Begin by building the transformer, the bridge rectifier, and R1 only.

   2. Connect channel one of the oscilloscope across R1. The oscilloscope's ground clip must be connected to circuit common.

   3. Plug in the transformer. Verify that the signal across R1 is a full-wave rectified sine wave at the proper amplitude and dc level. Compare this to Figure 8-31.

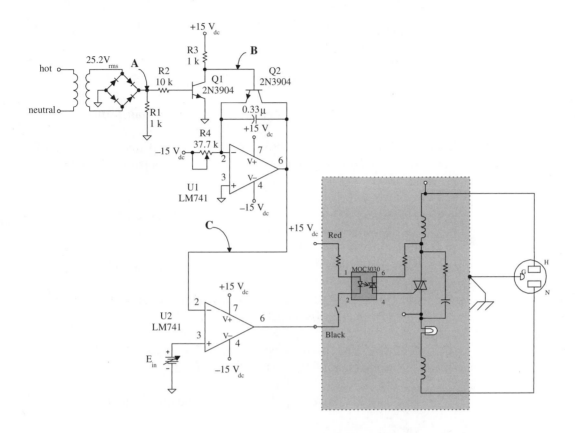

Figure 8-42 Phase angle firing schematic

4. Unplug the transformer.

5. Add R2, Q1, R3, and the +15 V_{dc} power supply.

6. Place the oscilloscope's channel two at the collector of Q1.

7. Turn the +15 V_{dc} supply *on*. Plug in the transformer.

8. Verify that the collector signal is a positive pulse at every zero crossing of the line voltage. Compare this with Figure 8-31.

9. Unplug the transformer and turn the +15 V_{dc} supply *off*.

10. Add U1, R4 (set to 37.7 kΩ), the feedback capacitor, Q2, and the −15 V$_{dc}$ supply.

11. Move channel two of the oscilloscope to the output of U1.

12. Turn the ±15 V$_{dc}$ supplies *on*. Plug in the transformer.

13. Confirm that the output of U1 is a ramp that starts at the line zero crossing and reaches 10 V$_p$. Compare this with Figure 8-31.

14. Adjust R4 to determine its effect. Be sure its last setting provides the appropriate ramp rate.

15. Unplug the transformer and turn the power supplies *off*.

16. Connect U2 and a second dc supply (or signal generator with only the dc offset enabled).

17. Move the oscilloscope's channel one to the collector of Q2. Move channel two to the output of U2.

18. Turn the power supplies *on*. Plug the transformer *in*.

19. Set E$_{in}$ = 5 V$_{dc}$.

20. Verify that the output of U2 goes to positive saturation at each zero crossing (at the positive pulse from Q2's collector) and falls to negative saturation about half-way through the cycle. Compare this with Figure 8-31.

21. Alter E$_{in}$ to assure that a lower input voltage produces a smaller duty cycle and that a larger input voltage produces a longer duty cycle. Check the duty cycle at E$_{in}$ = 2 V$_{dc}$, 4 V$_{dc}$, 6 V$_{dc}$, and 8 V$_{dc}$.

22. Set E$_{in}$ = 0 V$_{dc}$. Unplug the transformer. Turn the power supplies *off*.

23. Leave the ground clip for the oscilloscope's channel one connected to circuit common. Remove or tape back channel two's ground clip.

24. Connect channel two to the test point immediately above the light in the triac box.

25. Connect the +15 V_{dc} and the output of U2 to the triac box's opto-coupler.

26. Turn the power supplies *on*.

27. Plug in the transformer, then the triac box.

28. Set $E_{in} = 5$ V_{dc}.

29. With an *insulated* tool, assure that the switch in the triac box is *on*.

30. Verify that the light shines at about half brightness, and that the waveform across the load is a phase angle fired signal, firing at about 90°. Compare this to Figure 8-31.

31. Sweep E_{in} from 0.5 V_{dc} to 9.5 V_{dc} in steps of 0.5 V_{dc}. At each input voltage level, complete the following steps:
    a. measure the time from the zero crossing pulse on channel one to when the triac turns *on*.

    b. calculate the firing angle, ϕ, using this delay time.

    c. measure and record the rms voltage delivered to the load.

    d. compare this voltage to that obtained from calculation and from the graph in Figure 8-34.

32. Complete two graphs from your data; E_{in} (*x* axis) versus ϕ (*y* axis), and ϕ (*x* axis) versus $V_{rms\ load}$.

33. Once you are convinced that your measurements are correct, unplug the triac box and the transformer. Set $E_{in} = 0$ V_{dc}. Turn the power supplies *off*.

9

Power Conversion and Motor Drive Applications

Introduction

Converting power from one form to another or using it to control the motion of a motor are two major applications of power electronics. Operation must be as efficient as possible because thousands of watts are being controlled, and only a small percentage loss means that the circuit may overheat and fail, or precious resources might be squandered. Accurate control is required to assure that the load is properly serviced and the source is not overloaded. Large currents and voltages, rapidly switched, generate considerable electromagnetic interference. Careful circuit implementation is required to restrain this noise.

This chapter begins with a review of three switching power supplies. Electrical energy, however, is usually provided as a sinusoid. Although you already understand diodes, when dealing with considerable power more attention to details is needed. Phase angle firing of single-phase and three-phase rectifiers provides direct control of the output dc level.

We are increasingly becoming disconnected from the sinusoidal power grid. Much of the equipment that supports business and industry still requires sine wave power. The dc-to-ac inverter allows a battery to provide that ac, either in a vehicle or as utility company backup.

Precise and accurate control of position is central to the operations of much in our daily lives. From the disk drive of your personal computer, to the printer head, to the performance of the anti-lock brakes on your car, you assume that each reliably moves to its assigned position at your command. Power electronics switches assure that this happens.

Objectives

By the end of this chapter, you will be able to:
- Apply buck, boost, and flyback converters.
- By manual calculation and simulation, determine the performance of single-phase and three-phase, phase controlled rectifiers.
- Draw the block diagram and explain the operation of the elements of a dc-to-ac inverter.
- Discuss the characteristics of, and apply an IC to control the motion of, a dc brush and a dc brushless motor.

9.1 DC-to-DC Converters

Electrical power is commonly available as either dc or sinusoidal ac. Its shape and magnitude are usually *not* what you need. So you must convert what is readily available to what your circuit or your load requires. Batteries, automobile generators, and solar cells all provide widely varying energy. The details of the design and analysis of buck, boost and flyback dc-to-dc converters are covered in Chapter 7, Switching Power Supplies. They are the most efficient way to convert a widely varying dc voltage into a steady, defined, reliable dc supply. Their key points are reviewed in the following three sections.

Buck

The buck regulator outputs a dc voltage that is *below* its input. A MOSFET driven buck regulator is shown in Figure 9-1. This is also Figure 7-2.

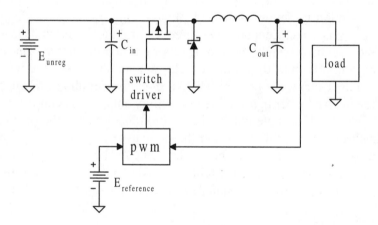

Figure 9-1 Basic buck regulator

The input dc voltage is chopped into a rectangle by alternately turning the transistor switch *on* and *off*. The duty cycle of the switch is D.

$$D = \frac{t_{on}}{T_{period}}$$

It is always a fraction, less than one.

The load voltage, then, is

$$V_{\text{load}} = DE_{\text{unreg}}$$

Since $D < 1$, the load voltage must be less than the input. At the most, with the switch *on* all of the time, $D = 1$, and $V_{\text{load}} = E_{\text{unreg}}$.

The pulse width modulator (pwm) compares a sample of the load voltage to a fixed, accurate reference. If the load voltage is too low, the pwm increases the duty cycle. If it is too large, the duty cycle is reduced. This part of the circuit is usually an IC. But its role may be fulfilled by a microprocessor.

The basic frequency is usually set well above the audio range. This removes any annoying hum you might hear. The higher the frequency, the smaller the filtering inductor and capacitor can be. Therefore, it is not unusual to find switching frequencies of 100 kHz to 1 MHz. The higher the switching frequency, the more rapidly the transistor must be driven *on* and *off*. Although a MOSFET gate draws no current, there are several hundred picofarads of capacitance between the gate and the source. Charge must be moved onto and off of this capacitance to turn the transistor *on* or *off*. Moving charge rapidly requires lots of current, since $i = dQ/dt$.

Switch drivers are presented in detail in Chapter 6. A high side driver, the LT1160, is capable of driving a MOSFET with 3000 pF of gate capacitance *on* and *off* at over 100 kHz. Its schematic is shown in Figure 9-2. This is also Figure 6-27.

> **The pulse width modulator controls the operation.**

> **A switch driver is needed for high speed operation.**

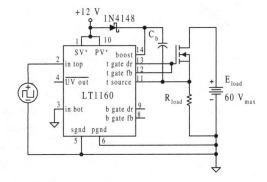

Figure 9-2 LT1160 driving an *n*-channel high side switch (*courtesy of Linear Technology*)

A *p*-channel MOSFET is simple, but the *n*-channel is faster and cheaper.

The transistor shown in Figure 9-1 is a *p*-channel. It is easy to drive. Pull its gate several volts below its source (to common) and it goes hard *on*. Set the gate at the same voltage as the source (E_{unreg}) and the transistor is *off*. But *p*-channel transistors are more expensive, slower, and less efficient than *n*-channel devices. The driver in Figure 9-2 allows you to use an *n*-channel high side switch.

The diode at the transistor's output (Figure 9-1) is *off* when the transistor is *on*, since the entire positive input voltage is passed through the transistor and onto the diode's cathode. So be sure to select a diode that can tolerate at least this much reverse voltage.

Select a Schottky diode if its PIV is adequate.

When the transistor turns *off*, the diode goes *on*, connecting that node to common. This must happen very quickly, in tens to hundreds of nanoseconds. So a high speed diode is needed. When it is *on* it furnishes the common return path for the entire load current. It must be able to handle that much repetitive peak forward current. Power dissipated by the diode during this time is a major factor in the converter's losses. So the forward voltage drop of the diode is important. Select a Schottky diode. They are fast, handle lots of current, and have a low voltage drop.

The inductor charges, storing energy in its magnetic field, when the switch is *on*. When the switch goes *off*, the field collapses, flipping the polarity of the inductor's voltage (this turns the diode *on*), and generating current to the load. So current ripples through the inductor in a triangle, ramping up as the inductor charges and down as it discharges. It is typical to set this peak-to-peak ripple current to about 25% of the load current, but that is not critical and may vary considerably. The *type* of inductor you choose has a large effect on the size, cost, and the amount of magnetic interference the buck converter produces.

$$L = \frac{\left(E_{unreg} - V_{load}\right)DT_{period}}{\Delta i_{pp}}$$

The output capacitor smoothes out any ripple voltage, discharging into the load as the inductor discharges to try to hold the output voltage constant. Since it is this output voltage that the pulse width modulator senses to control the drive to the transistor, the capacitor plays a key role in establishing the stability of the converter. The lower the capacitance, the smaller the capacitor, the less expensive it is, and the more quickly it can respond to changes. But a small capacitor also means more ripple voltage on the output, and the possibility that the converter will oscillate. Conversely, more capacitance means a steadier load voltage and a more stable operation. But it also requires a larger capacitor that is more expensive. The system may become sluggish. Some experimentation is needed.

$$C_{out} > 500\frac{\mu F}{A} \times I_{dc\ load}$$

Electrolytic capacitors often have poor high frequency performance. Therefore, it is a good idea to place a 0.1 µF film capacitor in parallel with C_{out}.

If your load requires less than 1 A_{dc}, then most of the electronics is provided within the LM2595 IC. Its use as a buck converter is shown in Figure 9-3. This is also Figure 7-5.

The switch, pulse width modulator, support electronics, and the reference are all provided within the IC. There are several versions. The **adj** version has a reference of 1.23 V. The negative feedback is provided through resistors R_f and R_i. You can view the entire circuit as a sophisticated noninverting amplifier, with 1.23 V_{dc} as the input. These components set the load voltage.

The inductor and the diode in Figure 9-3 are selected just as they are for the discrete version of the buck converter. The LM2595 sets the frequency internally.

Both C_{out} and C_f contribute to the stability of the system. Their value depends on the output voltage, type, and manufacturer of the capacitors. The data sheets for the LM2595 provide a table of recommended values. That table is repeated in Table 7-1.

$$V_{load} = 1.23\,V_{dc}\left(1+\frac{R_f}{R_i}\right)$$

f = 150 kHz

85 µF < C_{out} < 330 µF

220 pF < C_f < 4.7 nF

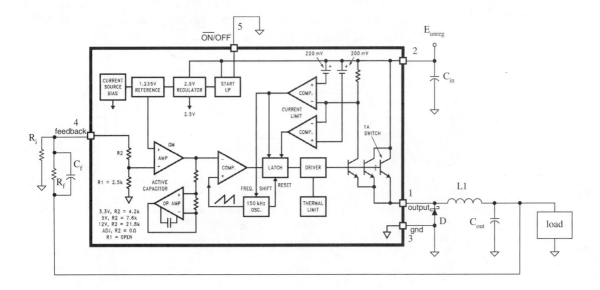

Figure 9-3 LM2595 in a buck converter (*courtesy of National Semiconductor*)

Boost

The buck converter steps the dc voltage *down*. The **boost** converter provides a load voltage that is *greater* than the input. Its schematic is shown in Figure 9-4. This is also Figure 7-7.

It contains the same elements you saw in the buck converter. But they have been rearranged. The switch now is *low* side. When the switch is *on*, the inductor charges, with the entire E_{unreg} across it + to −. The diode is *off* since its anode is pulled to common by the switch. The output voltage is held up by C_{out}, as it discharges through the load.

When the switch turns *off* the inductor's magnetic field begins to collapse. The field cuts the inductor, generating current, and reversing the polarity of the inductor's voltage (− to +). This voltage is now in series aiding with E_{unreg}. These two together produce a voltage that is higher than the input source alone. The output has been boosted above the input by the inductor's flyback voltage. This voltage forward biases the diode, allowing current to flow onto the capacitor, to recharge it.

$$V_{load} = E_{unreg} \frac{D}{1-D}$$

Adjusting the duty cycle alters the time (and therefore the amount) that the inductor can charge. This, then, sets the output voltage.

As with the buck converter, the inductor sets the peak-to-peak ripple current that passes through the circuit. When the switch goes *on*, the full input voltage is across the inductor. So the value of the inductor is a little different than that for the buck.

$$L = \frac{E_{unreg} D T_{period}}{\Delta i_{pp}}$$

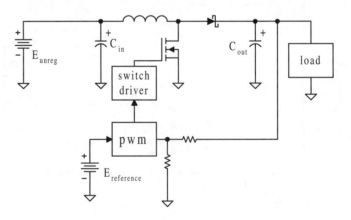

Figure 9-4 Boost converter

During the time the switch is *on* and the inductor is charging, the diode is *off*. The output is being provided entirely by C_{out}. Accordingly, it discharges, and its voltage drops. How much variation in the output voltage you can tolerate determines the value of that output capacitor.

Since the transistor is a low side switch, an *n*-channel MOSFET is a good choice. A switch driver is required to provide the current needed to charge and discharge the gate capacitance quickly. An IC, the MAX4420, is discussed in detail in Chapter 6. Its schematic is shown in Figure 9-5 (also Figure 6-26). The IC can source and sink up to $6 A_p$ while driving a 2500 pF capacitance with rise and fall times of less than 60 ns. When you take into account the duty cycle from the pulse width modulator, and the MAX4420's delay times, this means it can drive a MOSFET switch at over 300 kHz.

At such speeds, decoupling, grounding, and layout are critical for the entire converter. A ground plane is important. Rigorously decouple the power pins of the switch, the switch driver, the pulse width modulator, and the reference. These decoupling capacitors must be located as close to each component as possible, since lead inductance counters the benefit of the capacitance. Use both high quality tantalum or electrolytic capacitors to provide the main charge, paralleled with film capacitors for high speed transient response. Keep returns in the main current path as short as possible. Return the load to E_{unreg} in a lead separate from the control voltage common. Keep all of the high current paths as short as you can and away from the sensitive feedback and reference signals. Locate the inductor away from these small signals too. This reduces the effect that stray magnetic interference may have on these critical signals.

$$C_{out} = \frac{DT_{period}\, I_{dc\ load}}{\Delta v_{pp\ load}}$$

Decoupling, grounding, and layout are particularly important in a boost converter.

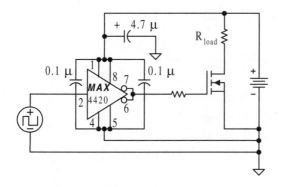

Figure 9-5 Low side switch driver

$$I_{dc\ in} = \frac{V_{dc\ load}}{0.9 E_{inunreg}} I_{dc\ load}$$

$I_{input} < 3$ A
4 V $< E_{in\ unreg} < 40$ V$_{dc}$
$V_{load} < 65$ V

$$V_{load} = 1.23\,V_{dc}\left(1 + \frac{R_f}{R_i}\right)$$

f = 100 kHz

The boost circuit steps the voltage passed to the load *up*. This is done at the expense of requiring a proportionally higher *input* current. The 0.9 in the equation assumes a 90% efficient conversion.

If your load requires less than 3 A$_{dc}$ from the *input,* then most of the electronics is provided within the LM2585 IC. Its use as a boost converter is shown in Figure 9-6. This is also Figure 7-12.

The switch, pulse width modulator, support electronics, and the reference are all provided within the IC. There are several versions. The **adj** version has a reference of 1.23 V. The negative feedback is provided through resistors R$_f$ and R$_i$. You can view the entire circuit as a sophisticated noninverting amplifier, with 1.23 V$_{dc}$ as the input. These components set the load voltage.

The inductor, capacitors, and the diode in Figure 9-6 are selected just as they are for the discrete version of the boost converter. The LM2585 sets the frequency internally.

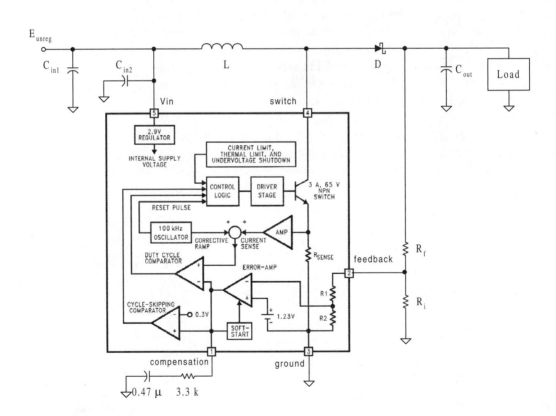

Figure 9-6 LM2585 in a boost converter (*courtesy of National Semiconductor*)

Flyback Converter

The buck converter only steps the voltage down. The boost only steps the voltage up. If the input voltage may vary, starting above the required output but dropping below it, neither converter works alone. One solution is to cascade a boost and a buck. First boost the input up to a point that it is always above the required output, then buck that intermediate voltage down to the final level needed.

This approach is both cumbersome and inefficient. A simpler circuit is shown in Figure 9-7 (also Figure 7-13). The inductor of the boost converter has been replaced by the primary of a transformer. A key feature is that the secondaries are all out-of-phase with the primary. When the switch within the LM2585 turns *on*, the dots are all induced positive, turning all of the Schottky diodes *off*. Then the primary functions only as an inductor.

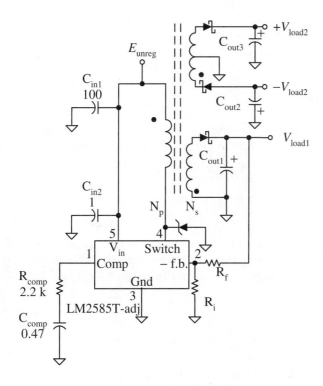

Figure 9-7 Simple flyback converter

When the switch goes *off*, the field collapses, reversing the polarities of the secondary voltages, and turning the diodes *on*. Energy is then coupled to the output capacitors, recharging them.

A turns ratio greater than 1 allows the output voltage to be either larger or smaller than the input.

Load voltage

$$V_{load} = E_{unreg} \frac{D}{\frac{N_{pri}}{N_{sec}}(1-D)}$$

This equation can be manipulated to determine the duty cycle needed to produce a particular load voltage from a given input.

Duty cycle

$$D = \frac{\frac{N_{pri}}{N_{sec}} V_{load}}{E_{unreg} + \frac{N_{pri}}{N_{sec}} V_{load}}$$

The average current required from the unregulated input supply, assuming an efficiency of 90%, is

Average input current

$$I_{dc\ in} = \frac{I_{dc\ load} V_{dc\ load}}{0.9 E_{dc\ unreg}}$$

As with the boost converter, it is typical to restrict the current variation through the primary to about 25% of that average value. This is set by the inductance of the primary.

Pick $\Delta I_{pp\ pri} \approx 0.25\ I_{dc\ in}$

$$L_{pri} = \frac{E_{unreg} D T_{period}}{\Delta I_{pp\ pri}}$$

The output capacitor(s) hold the load voltage up while the primary inductance is charging and the diodes are *off*. This too is just like the boost converter. For stable control from the IC hold the load ripple voltage to

$\Delta V_{load} < 20\ mV_{pp}$

$$C_{out} = \frac{D T_{period} I_{dc\ load}}{\Delta v_{pp\ load}}$$

There are several companies that manufacture transformers designed specifically to match the LM2585. Consult the National Semiconductor specifications for a current list.

9.2 AC-to-DC Converters

The most common source of power is the sinusoidal, ac single-phase, commercial line voltage from the utility company. Practically all of the electronics that you build require a steady source of dc. To convert tens of watts from single-phase ac to dc is a straightforward application of the simple transformer-bridge rectifier-capacitive filter you learned in your earlier studies. However, several details are too often overlooked. Transformers are far from ideal. What is an appropriate model? On the first cycle, when the capacitor is fully discharged, how much current must the diodes pass? During nominal operation, the diodes only turn *on* for a short time each cycle. What is their *average* forward current? How do you pick an appropriate capacitor? What it the impact of the pulsing current on the utility line, and on neighboring equipment that shares the utility?

The larger the dc load current, the larger the filter capacitor you must provide to smooth out the rectified single phase ac. For hundreds of watts, or more, these become impractical. Industrial dc loads are often powered from *three*-phase ac. Simple rectification, without any capacitive filters, provides a surprisingly low ripple dc. Replace the simple rectifiers with silicon controlled rectifiers. Controlling the phase angle at which each SCR is fired changes the magnitude of the output dc, without a step-down transformer.

Single-Phase

The traditional solution of providing dc from single-phase ac is the transformer-bridge rectifier-capacitive filter circuit shown in Figure 9-8. The transformer drops the voltage from the level provided by the utility to an

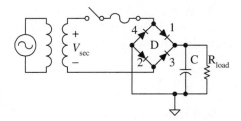

Figure 9-8 Simple single-phase ac-to-dc conversion

rms value about equal to the dc output required. On the positive half-cycle, current flows through D1 to the capacitor and the load, then through D2 back to the secondary. Diodes D3 and 4 are *off*. On the other half-cycle, diode D3 turns *on*. Current flows from the bottom of the transformer secondary, through D3, onto the capacitor and through the load, then through D4 and back to the secondary. During this half-cycle, D1 and 2 are *off*.

The bridge diodes convert the ac into a pulsating dc, shown as the dotted lines in Figure 9-9. The capacitor charges for a small part of each cycle, then slowly discharges through the load resistance for most of the cycle. This holds the load voltage more or less steady.

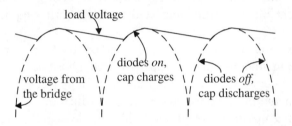

Figure 9-9 Bridge and filter waveforms

$$V_p = V_{rms\ sec} \times \sqrt{2} - 2V_{diode\ on}$$

$$V_{diode\ on} \approx 1V$$

The capacitor charges to the peak of the secondary voltage, less the drop across the diodes. At rated current, most power rectifiers drop about 1.0 V, not the 0.6 V assumed at their low current turn *on* point. Shortly after the peak, the voltage from the transformer (on the anode of the diodes) falls sinusoidally, while the voltage on the load (on the cathode of the diodes) is held up by the charge on the capacitor. So the diodes turn *off*. They stay *off* for the rest of that cycle and into the next cycle.

Finally the capacitor discharges below the rising voltage from the transformer, and the diodes can turn *on* again.

$$v_C = V_p e^{-\frac{t}{RC}}$$

While *on*, the diodes must carry enough current to provide the load and to recharge the capacitor. The capacitor is discharging exponentially from a value near V_p. The voltage from the secondary is rising sinusoidally from its negative peak.

$$v_{sec} = V_{rms\ sec} \sqrt{2} \sin(\omega t - 90°)$$

The diodes go *on* when the capacitor voltage is 2 V (two diode drops) below the primary voltage.

$$V_{rms\ sec} \sqrt{2} \sin(\omega t - 90°) = V_p e^{-\frac{t}{RC}} + 2\,V$$

To determine when the diodes turn *on*, the peak-to-peak ripple voltage, and the current that the diodes must carry, this equation must be solved for *t*. But it is transcendental. It cannot be solved in closed form.

A simpler approach is to assume that the capacitor charges and discharges *linearly*. For strongly filtered circuits in which the peak-to-peak ripple voltage is less than 10% of the average load voltage, this approximation produces reasonably accurate results.

$$\Delta V_{pp\ load} \approx \frac{I_{dc\ load}}{fC}$$

Light load approximations
$\Delta V_{pp} < 0.1\ V_{dc\ load}$

Remember that for a full wave rectifier, $f = 2\,f_{line}$, 120 Hz for a 60 Hz utility signal.

The average value of a triangle wave is halfway between its top and its bottom. So the dc voltage to the load can be calculated as

$$V_p = V_{rms\ sec} \sqrt{2} - 2V$$

$$V_{dc\ load} = V_p - \frac{\Delta V_{pp\ load}}{2}$$

Example 9-1

For a bridge rectifier, with a 28 V_{rms} secondary, a 4700 μF capacitor, and a 16 Ω load, calculate the V_p, $\Delta V_{pp\ load}$, and $V_{dc\ load}$ using the light load approximations.

Solution

The peak voltage to the load is

$$V_p = V_{rms\ sec} \sqrt{2} - 2V$$

$$V_p = 28\,V_{rms} \sqrt{2} - 2\,V = 37.6\,V_p$$

To calculate the load voltages, you must know the load current. But to calculate the load current, you need the load voltage. To resolve this issue, realize that the dc load voltage is a little below the load peak voltage. So

$$V_{dc\ load} \approx 0.95\,V_p = 0.95 \times 37.6\,V_p = 35.7\,V_{dc}$$

This now gives a load current of

$$I_{dc\ load} = \frac{V_{dc\ load}}{R_{load}} = \frac{35.7\,V_{dc}}{16\,\Omega} = 2.2\,A_{dc}$$

The ripple voltage is

$$\Delta V_{pp\ load} \approx \frac{I_{dc\ load}}{fC}$$

$$\Delta V_{pp\ load} \approx \frac{2.2\,A_{dc}}{120\,Hz \times 4700\,\mu F} = 3.9\,V_{pp}$$

A closer approximation of the dc load voltage is

$$V_{dc\ load} = V_p - \frac{\Delta V_{pp\ load}}{2}$$

$$V_{dc\ load} = 37.6\,V_p - \frac{3.9\,V_{pp}}{2} = 35.7\,V_{dc}$$

The 3.9 V_{pp} ripple is a little more than 10% of the 35.7 V_{dc}. So it is reasonable to expect some error in these numbers.

Practice: A 4700 μF capacitor is rather large and expensive. What results do the light load approximations predict for a 1000 μF capacitor?

Answer: ΔV_{pp} = 18 V_{pp} and V_{dc} = 28 V_{dc}. But at 28 V_{dc} the load current is only 1.8 A_{dc}, which means a ripple of 15 V_{pp} and V_{dc} = 30 V_{dc}. These approximations do not work well at lower capacitance and higher ripple.

The step-down transformer is a significant part of this type of ac-to-dc converter. It is tempting to just consider them to be an ideal voltage source. However, like relying only on the light load approximations, this simplification may introduce considerable error. A better approach is to determine the Thevenin equivalent model for the transformer. It is shown in Figure 9-10.

Begin by measuring the open circuit secondary voltage. It is usually a little above the transformer's rated voltage. The open circuit voltage is the Thevenin equivalent voltage, because with no load there is no current flow, and the impedances drop no voltage.

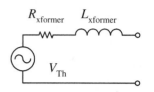

$$V_{Th} = V_{open}$$

Next, connect the primary to an auto transformer, or variac. This allows you to control the input voltage. With the primary voltage set to zero, short the secondary. Connect one channel of the oscilloscope to the primary voltage, and the other to a current probe measuring the shorted secondary current.

Figure 9-10 Transformer model

Increase the primary voltage until I_{short} is at the rated secondary current. Accurately measure the magnitude and phase shift of the primary voltage and the secondary current. Remember to account for the rated transformer turns ratio.

$$\overline{Z_{Th}} = \frac{\dfrac{V_{sec-rated}}{V_{pri-rated}} \times \overline{V_{pri-shorted}}}{\overline{I_{short}}}$$

Example 9-2

A step-down control transformer is rated at:

$V_{pri} = 120\ V_{rms},\ V_{sec} = 28\ V_{rms},\ I_{pri} = 2\ A_{rms},\ f = 60\ Hz$

Tests indicate:

$V_{open} = 28.8\ V_{rms},\quad V_{pri\text{-}short} = 12.1\ V_{rms},\quad \overline{I_{short}} = 2\ A_{rms}\angle -45°$

Determine the appropriate model.

Solution

$$V_{Th} = V_{open} = 28.8\ V_{rms}$$

$$\overline{Z_{Th}} = \frac{\dfrac{V_{sec-rated}}{V_{pri-rated}} \times \overline{V_{pri-shorted}}}{\overline{I_{short}}}$$

$$\overline{Z_{Th}} = \frac{\dfrac{28\ V_{rms}}{120\ V_{rms}} \times 12.1\ V_{rms}}{2\ A_{rms}\angle -45°} = 1\Omega + j1\Omega$$

This is a 1 Ω resistor in series with an inductor with a 1 Ω reactance at 60 Hz.

$$L = \frac{1\,\Omega}{2\pi \times 60\,\text{Hz}} = 2.7\,\text{mH}$$

Practice: Find the model for a 120 V_{rms}, 60 Hz transformer with a secondary rating of 12.6 V_{rms}, 1 A. Tests show V_{open} = 13.2 V_{rms}, I_{short} = 1 A_{rms} ∠ −58°, with V_{pri} = 15 V_{rms}

Answer: V_{Th} = 13.2 V_{rms}, $R_{xformer}$ = 0.84 Ω, $L_{xformer}$ = 3.6 mH

The simplest and most accurate way to predict the performance of a single-phase ac-to-dc converter is by simulation. You can accurately determine the average voltage delivered to the load, the load ripple voltage, the diodes' first cycle peak current, repetitive peak and average currents, and the current drawn from the utility. Alternate components can then be easily evaluated.

Example 9-3

Simulate a full-wave bridge rectified, capacitive-filtered ac-to-dc converter. Use the step-down transformer from Example 9-2, a 4700 µF capacitor, and a 16 Ω load.

Solution

The schematic is shown in Figure 9-11. The value of the transformer model source is set to the *peak* of its 28.8 V_{rms} value. Both the model's resistance and inductance have been included. The 16 Ω load draws close to the transformer's rated current.

The results of the transient analysis are shown in Figure 9-12. The simulation has been run for 83 ms. This gives the circuit time to stabilize. An even longer sample may be appropriate. The lower plot indicates a load voltage that ripples between 30.0 V and 28.3 V, or

Simulation $V_{p\,load}$ = 30.0 V_p $V_{pp\,ripple}$ = 1.7 V_{pp}

Equations (Ex. 9-1) $V_{p\,load}$ = 37.6 V_p $V_{pp\,ripple}$ = 3.9 V_{pp}

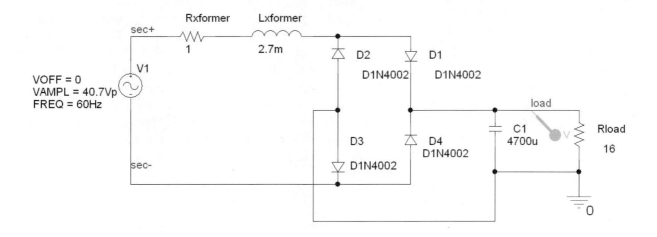

Figure 9-11 Simulation schematic for Example 9-3

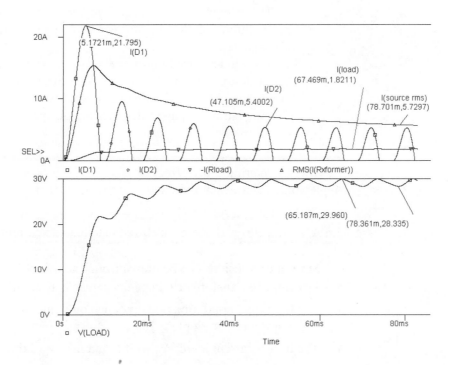

Figure 9-12 Simulation results for Example 9-3

Ignoring the transformer's impedance produces a 25% error in the peak voltage, causing you to expect more load voltage than the converter could actually produce. The light load approximations also predicted a ripple voltage that is 100% larger than that from the simulation. This error may lead you to select a larger and more expensive capacitor than actually required.

Of major concern is the initial inrush current. When the capacitor is completely discharged, it appears as a short across the transformer. Only the transformer's impedance limits that current. According to the simulation the first cycle current peaks at

$$I_{inrush} = 21.8 \text{ A}_p$$

The diodes and transformer must be able to handle this non-repetitive peak current. Fortunately, a rectifier diode's non-repetitive peak current rating is 20 to 40 times its continuous forward current. The massive capacitance, though providing very small ripple, pushes this initial inrush to very high levels and may damage the transformer or diodes.

The average load voltage and load current are within the specifications of the transformer and diodes, and are close to that expected.

The diodes only turn *on* for a part of each cycle. The larger the capacitor, the smaller that time. So the current from the source and through the diodes is a nonstandard shaped pulse. The simulation package calculates the rms value of that current.

$$I_{rms \text{ source}} = 5.7 \text{ A}_{rms}$$

Practice: Determine the effect of $C = 1000 \text{ μF}$ on peak load voltage, ripple voltage, inrush current, and source rms current.

Answers: $V_{p \text{ load}} = 34.3 \text{ V}_p$, $V_{ripple} = 8.2 \text{ V}_p$, $I_{inrush} = 13.1 \text{ A}_p$, $I_{source} = 3.6 \text{ A}_{rms}$. Repetitive diode peak current and load dc current change little.

Several conclusions can be drawn from Example 9-3.

- Ignoring the transformer's impedance results in significant error.

- The light load approximations produce only marginal results at best, and major errors at ripple $\geq 10\%$.

- The inrush current increases with C and may > 10 times $I_{dc \text{ load}}$.

- Simulation is the simplest way to predict the circuit's performance.

There is one further consideration. Power utilities measure the impact of a load on the power grid with the **power factor**. Although this is normally a measure of the lag of the sinusoidal current produced by an inductive load, it is also applicable to nonsinusoidal currents, such as that drawn by the ac-to-dc converter of this section.

$$F_p = \frac{P}{S}$$

Where $\qquad P =$ average power dissipated by the load

$$= I_{dc\,load} \times V_{dc\,load} = \left(I_{dc\,load}\right)^2 R_{load}$$

and $\qquad S =$ apparent power delivered by the source

$$= I_{rms\,source} \times V_{rms\,source}$$

The ideal value is $F_p = 1$. This results in minimum losses. The smaller the power factor, the larger the losses in producing and distributing the energy. In an industrial or commercial setting, power factors of < 0.8 may lead to line distortion, interference with other loads sharing the line, and even financial penalties levied by the power utility.

Example 9-4

Calculate the power factor for the converters from Example 9-3.

Solution

$$P = \left(1.8A_{dc}\right)^2 \times 16\,\Omega = 51.8\,W$$

$$S = 5.7\,A_{rms} \times 28.8\,V_{rms} = 164\,VA$$

$$F_p = \frac{51.8\,W}{164\,VA} = 0.32$$

Practice: Calculate the power factor if $C = 1000\,\mu F$.

Answer: $P = 51.8$ W, $S = 103.7$ VA, $F_p = 0.5$

These are very poor power factors. Although a single such circuit would not have a major impact on even a 15 A_{rms} residential line, multiple circuits, all on the same line, most certainly can create problems.

It is becoming common to replace the simple ac-to-dc converter of this section with an off-line switching power supply. These were covered in Chapter 7, Section 7.3. The power factor corrected version is built around an IC that adjusts the pulse width modulation to assure both a regulated dc output voltage and a power factor that is close to 1.

Three-Phase

Most industrial facilities have electrical power provided in three phases. These are shown in Figure 9-13. For the familiar voltage

$$\overline{E_{AN}} = 120\,\text{V}_{\text{rms}}\angle 0°$$

$$\overline{E_{BN}} = 120\,\text{V}_{\text{rms}}\angle -120°$$

$$\overline{E_{CN}} = 120\,\text{V}_{\text{rms}}\angle 120°$$

There are circumstances in which the neutral is not used, or even not available. The line-to-line voltages are defined as

$$\overline{E_{AB}} = \overline{E_{AN}} - \overline{E_{BN}} = 120\,\text{V}_{\text{rms}}\angle 0° - 120\,\text{V}_{\text{rms}}\angle -120°$$

$$\overline{E_{AB}} = 208\,\text{V}_{\text{rms}}\angle 30°$$

Similarly

$$\overline{E_{BC}} = 208\,\text{V}_{\text{rms}}\angle -90°$$

$$\overline{E_{CA}} = 208\,\text{V}_{\text{rms}}\angle 150°$$

$$\boldsymbol{E}_{\text{line}} = \boldsymbol{E}_{\phi} \times \sqrt{3}$$

In general, the line voltage is larger than the phase by $\sqrt{3}$, and advanced by 30°.

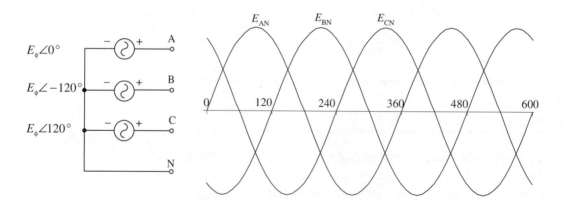

Figure 9-13 Three-phase power

The circuit in Figure 9-14 shows how these three phase ac signals can be converted to drive dc into a single, grounded load. Assuming typical 120 V$_{rms}$ phase voltages, the resulting signals from an OrCAD simulation are also included. The load voltage is a fairly smooth dc, even without *any* capacitance.

But that voltage is quite a bit larger than the generators', and not in phase with any of the sources. Look at the schematic closely. The load is connected to common, but neutral (N) is not. There is no direct connection between the load and the common of the sources. So, comparing the load voltage to the sources' phase signals is not appropriate. Instead, look at Figure 9-15. It is the same analysis, but plots the *line* voltages, instead of the phase.

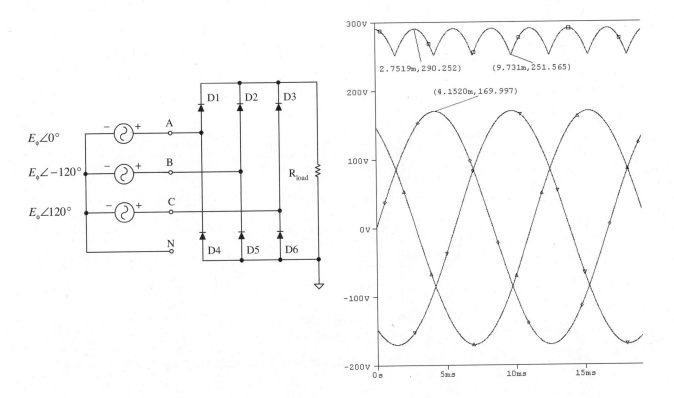

Figure 9-14 Three-phase ac-to-dc converter

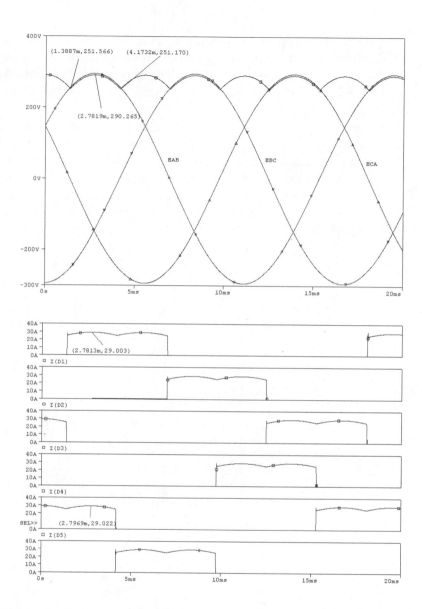

Figure 9-15 Three-phase ac-to-dc converter line voltages and currents

$$V_{\text{p load}} = E_{\text{p line}} - 2\,\text{V}$$

$$= \left(E_{\text{rms }\phi}\sqrt{3}\right)\sqrt{2} - 2\,\text{V}$$

$$PIV_{\text{diode}} = \left(E_{\text{rms }\phi}\sqrt{3}\right)\sqrt{2}$$

The load voltage follows the peaks of the *line* signals. The peaks are two diode drops below the line peak voltage. This also is the peak *inverse* voltage rating of each diode. Three phases of 120 V_{rms} produce a dc voltage of 291 V_{p}.

The currents for the six diodes are also shown in Figure 9-15. At the 2.8 ms peak, current flows from source A → D1 → R_{load} → D5 and to source B. At that instant, E_{BN} is negative. This forward biases D5 and allows the current to pass through source B and back to the negative side of source A. A similar analysis can be done for each of the six peaks in the load voltage.

The load voltage is the peak of a sine wave that starts 30° before the peak and ends 30° after the peak. So the minimum load voltage is

$$V_{min\ load} = E_{p\ line} \sin(60°)$$

$V_{min\ load} = 0.87 E_{p\ line}$

The three-phase ac-to-dc converter has a peak-to-peak ripple that is only 13% of the peak, without *any* filtering.

The average value of the load voltage is

$$V_{dc\ load} = \frac{1}{120° - 60°} \int_{60°}^{120°} V_{p\ line} \sin\theta\, d\theta$$

The diode turns *on* and passes the signal to the load at 60°, and goes *off* when the next phase comes *on*, at 120°. For the equation to work correctly, the angles must be expressed in radians.

$$V_{dc\ load} = \frac{1}{\dfrac{\pi}{3}} \int_{60°}^{120°} V_{p\ line} \sin\theta\, d\theta$$

$$V_{dc\ load} = \frac{3V_{p\ line}}{\pi} \int_{60°}^{120°} \sin\theta\, d\theta$$

$$V_{dc\ load} = \frac{3V_{p\ line}}{\pi} \left(-\cos\theta \Big|_{60°}^{120°} \right)$$

$V_{dc\ load} = \dfrac{3}{\pi} V_{p\ load}$

For the 120 V_{rms} phase voltage used in Figures 9-14 and 9-15 this corresponds well with the simulation.

$$V_{dc\ load} = 279\,V_{dc}$$

The peak current through each diode is the peak current through the load. From Ohm's law,

$$I_{p\ diode} = I_{p\ load} = \frac{291\ V_p}{10\Omega} = 29.1 A_p$$

$I_{p\ diode} = \dfrac{E_{p\ line} - 2V}{R_{load}}$

This is *much* less than the diode's peak current in the single-phase, capacitive-filtered ac-to-dc converter.

The other key specification for the diode is its continuous forward current; that is, its dc or average current. Look at Figure 9-15. Each diode is *on* for one third of the time. During that time, the load current passes through the diode. So, while the diode is *on*, its average current is the load average current. But the diode is only *on* for one third of the time.

$$I_{\text{dc diode}} = \frac{1}{3} \times \frac{V_{\text{dc load}}}{R_{\text{load}}}$$

$$I_{\text{dc diode}} = \frac{1}{3} \times \frac{279\,\text{V}_{\text{dc}}}{10\,\Omega} = 9.3\,\text{A}_{\text{dc}}$$

Finally, the power factor for this three-phase circuit is

$$F_{\text{p}} = \frac{P_{\text{load}}}{S_{\text{total}}}$$

Since the load is dc,

$$P_{\text{load}} = I_{\text{dc load}} V_{\text{dc load}}$$

The total apparent power comes from all three sources

$$S_{\text{total}} = 3 I_{\text{rms }\phi} E_{\text{rms }\phi}$$

The phase current is a pedestal, double hump shape. The simplest solution is to have the simulation software compute its value.

OrCAD▸

Example 9-5

Calculate the power factor for the three-phase ac-to-dc converter of Figures 9-14 and 9-15.

Solution

The cropped simulation result for the phase rms current is shown in Figure 9-16.

$$I_{\text{rms }\phi} = 22.6\,\text{A}_{\text{rms}}$$

This gives a total apparent power of

$$S_{\text{total}} = 3 I_{\text{rms }\phi} E_{\text{rms }\phi}$$

$$S_{\text{total}} = 3 \times 22.6\,\text{A}_{\text{rms}} \times 120\,\text{V}_{\text{rms}} = 8.14\,\text{kVA}$$

The load dissipates

$$P_{\text{load}} = I_{\text{dc load}} V_{\text{dc load}}$$

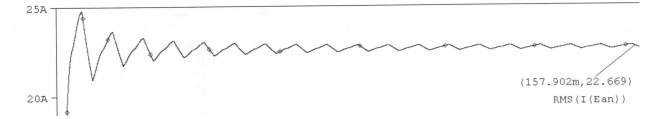

Figure 9-16 RMS phase current simulation results

$$P_{\text{load}} = 28\,\text{A}_{\text{dc}} \times 280\,\text{V}_{\text{dc}} = 7.84\,\text{kW}$$

The power factor is

$$F_{\text{p}} = \frac{P_{\text{load}}}{S_{\text{total}}}$$

$$F_{\text{p}} = \frac{7.84\,\text{kW}}{8.14\,\text{kVA}} = 0.96$$

Since the perfect power factor is 1.0, this circuit is nearly ideal. It introduces no problems onto the utility line.

In Chapter 8, you saw that the diode in a simple rectifier could be replaced with a silicon controlled rectifier (SCR). This allows much more control over the load voltage. By adjusting when the SCR is fired *on*, only a portion of the line voltage is passed to the load. This is called **phase angle firing**, and provides good control over the load's dc voltage.

Phase angle firing can also be used with the three-phase rectifier. The six diodes are replaced with six SCRs. Look at Figure 9-17. Be sure to include a snubbing inductor, capacitor, resistor, and diode for each SCR. These are discussed in Section 8.2. They have been omitted from Figure 9-17 for simplicity, but are critical to the practical operation of the converter.

It is the responsibility of the controller to provide a firing pulse to each of the SCRs at the proper time. That involves communication with

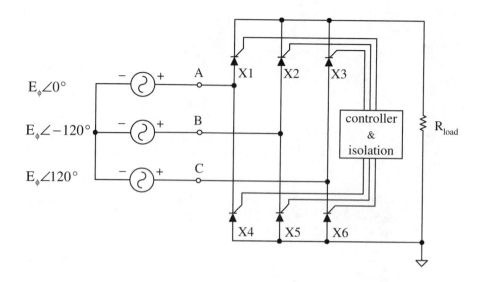

Figure 9-17 Three-phase SCR controlled ac-to-dc converter

Table 9-1 Three-phase diode and SCR nominal start times

Diode	Time ms	Angle degrees
1	1.39	30
6	4.18	90
2	6.96	150
4	9.74	210
3	12.52	270
5	15.30	330

the operator (or operating computer), calculations, and the generation of six properly timed pulses. Considerable electronics, perhaps even a microprocessor, is needed. The signals must be isolated from the large line voltages. Therefore, do not forget to include a coupler, such as the MOC3011, for each gating signal.

To determine the proper sequence for firing each of the SCRs, look back at Figure 9-15. This simulation shows the currents through each of the diodes in the simple three-phase rectifier. Diode 1 fires first, at 1.39 ms, or 30° with respect to E_{AN}. It then conducts until 6.96 ms, or for 120° before it turns *off*. However, halfway through its conduction, at 4.17 ms after D1 turns *on*, diode 6 also goes *on*. It also conducts for 120° before going off. Table 9-1 lists each of the diodes and their *nominal* starting time with respect to E_{AN}.

The diodes delivers the maximum dc voltage to the load. This is also the result from firing each of the SCRs at the angles (with respect to E_{AN}) indicated in Table 9-1.

Delaying the firing angle beyond these nominal angles, however, delays when each SCR goes *on*. Such a further delay lowers the average voltage passed to the load. The signals for an additional 20° delay are shown in Figure 9-18.

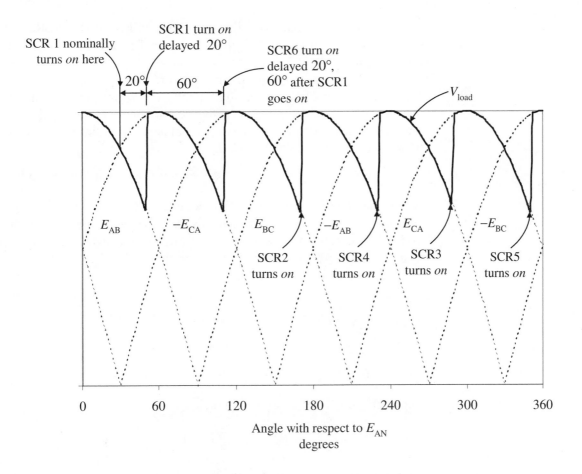

Figure 9-18 Three-phase, phase angle firing

Diode D1 in Figure 9-17 turns *on* when E_{AB} exceeds $-E_{BC}$, at 30°. That is also where SCR1 nominally turns *on*. A further delay of 20° allows $-E_{BC}$ (which is also the load voltage) to continue to fall. After the additional 20° delay, SCR1 is turned *on*. The load voltage jumps to E_{AB}, and tracks it for 60°.

The second SCR to be turned *on* is SCR6. It is also given an additional 20° lag. This causes SCR6 to be fired 60° after SCR1 is fired. Again, the load voltage has sagged below the point it would have if no delay had been introduced.

The same procedure is followed to fire SCR2, SCR4, SCR3, and SCR5. Each is turned *on* 20° after its nominal point. That is also 60° after the preceding SCR is fired.

Since V_{load} now drops much further before the next SCR is turned *on*, the average value of V_{load} has also dropped. The waveform is symmetric, so calculating the average value of the load voltage across the time when one SCR is *on* gives the same average value as the entire wave. When SCR1 goes *on*, the load voltage is E_{AB}.

$$v_{load}(t) = E_{p\,line}\sin(\omega t + 30°) \qquad 30° + \phi < \omega t \le 90° + \phi \quad \text{with respect to } E_{AN}$$

Notice that the *line* peak voltage is used, not the E_{AN} phase peak voltage. The line voltage is shifted 30° ahead of the E_{AN} phase voltage. The integral is a little easier to complete if you rewrite the equation with respect to E_{AB}. This shifts the zero phase point to the origin of E_{AB}. The SCR is turned *on* 60° after the E_{AB} origin.

$$v_{load}(t) = E_{p\,line}\sin(\omega t) \qquad 60° + \phi < \omega t \le 120° + \phi \quad \text{with respect to } E_{AB}$$

The integration is also simpler in terms of angle instead of time.

$$v_{load}(\theta) = E_{p\,line}\sin(\theta) \qquad 60° + \phi < \theta \le 120° + \phi \quad \text{with respect to } E_{AB}$$

The angles should be expressed in radians.

$$v_{load}(\theta) = E_{p\,line}\sin(\theta) \qquad \frac{\pi}{3} + \phi < \theta \le \frac{2\pi}{3} + \phi \quad \text{with respect to } E_{AB}$$

Now, the average value of a signal is

$$V_{dc} = \frac{1}{T}\int v(\theta)\,d\theta$$

The integration is evaluated over the period of the signal.

For the load voltage

$$V_{dc} = \frac{1}{\dfrac{2\pi}{3} - \dfrac{\pi}{3}}\int_{\frac{\pi}{3}+\phi}^{\frac{2\pi}{3}+\phi} E_{p\,line}\sin\theta\,d\theta$$

$$V_{dc} = \frac{3}{\pi}E_{p\,line}\int_{\frac{\pi}{3}+\phi}^{\frac{2\pi}{3}+\phi}\sin\theta\,d\theta$$

Simplifying

Taking the integral of $\sin \theta$ gives

$$V_{dc} = \frac{3}{\pi} E_{p \text{ line}} \left(-\cos\theta \Big|_{\frac{\pi}{3}+\phi}^{\frac{2\pi}{3}+\phi} \right)$$

$$V_{dc} = \frac{3}{\pi} E_{p \text{ line}} \left[\cos\left(\frac{\pi}{3}+\phi\right) - \cos\left(\frac{2\pi}{3}+\phi\right) \right]$$

$$V_{dc} = \frac{3\sqrt{2}}{\pi} E_{rms \text{ line}} \left[\cos\left(\frac{\pi}{3}+\phi\right) - \cos\left(\frac{2\pi}{3}+\phi\right) \right]$$

Although this may appear somewhat involved, Figure 9-19 is a plot of this equation versus delay angle. This is similar to Figure 8-34 for the average value of a single-phase, phase angle fired voltage. The vertical axis has been normalized (divided by $E_{rms \text{ line}}$). So to denormalize the result, multiply the value from the vertical axis by the root mean squared value of the *line* voltage.

$$V_{dc} = Y_{normalized} \times E_{rms \text{ line}}$$

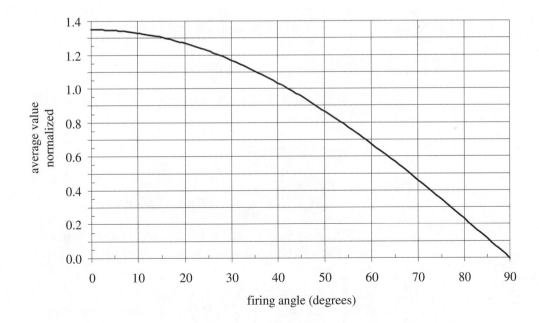

Figure 9-19 Three-phase, phase angle fired average load voltage

Example 9-6

Determine each of the following, for a 208 V_{rms} line voltage, three-phase system driving a 100 Ω load:

1. Find the firing angle necessary to deliver 150 V_{dc}.

2. Confirm that phase angle with a calculation.

3. Determine the firing time for *each* of the six SCRs.

4. Verify the answers with a simulation.

Solution

1. The ratio of dc voltage to line voltage is

$$\frac{150\,V_{dc}}{208\,V_{rms}} = 0.72$$

From Figure 9-19, a firing angle of $\phi = 58°$ is needed.

2. A firing angle of $58° = 1.01$ radians, produces

$$V_{dc} = \frac{3\sqrt{2}}{\pi} E_{rms\ line} \left[\cos\left(\frac{\pi}{3}+\phi\right) - \cos\left(\frac{2\pi}{3}+\phi\right) \right]$$

$$V_{dc} = \frac{3\sqrt{2}}{\pi} 208\,V_{rms} \left[\cos\left(\frac{\pi}{3}+1.01\right) - \cos\left(\frac{2\pi}{3}+1.01\right) \right]$$

$$V_{dc} = 281\,V_{dc}\left[-0.469 + 0.999\right] = 149\,V_{dc}$$

3. The firing angle of 58° can be converted to delay time with a ratio.

$$\frac{t_{delay}}{T} = \frac{\phi}{360°}$$

Or, $$t_{delay} = \frac{58°}{360°} \times 16.7\,ms = 2.69\,ms$$

This time must be added to each of the diode start times from Table 9-1 in order to reference them to E_{AN}. The results are:

SCR1, 4.08 ms   SCR6, 6.87 ms   SCR2, 9.65 ms
SCR4, 12.43 ms   SCR3, 15.21 ms   SCR5, 17.99 ms

4. The evaluation version of OrCAD allows only three SCRs. The rest must be diodes. SCR6, SCR2, and SCR4 are used. The simulation is then viewed only during the time that these SCRs are conducting. The schematic and the simulation setup are shown in Figure 9-20

The PSpice models for the diodes and the SCRs have been edited to set BV, V_{drm}, and V_{rrm} to 300 V.

The sources are configured for each of the phase voltages. A line voltage of 208 V_{rms} is produced from phase voltages of 120 V_{rms}. The simulator expects the source voltages to be specified in V_p. A phase voltage of 120 V_{rms} has a peak value of 170 V_p

Each SCR is fired by a VPUL, with a 2 V_p amplitude, 10 ns rise and fall times, and a 60 Hz period. The delay time for each sets when its SCR fires with respect to E_{AN}.

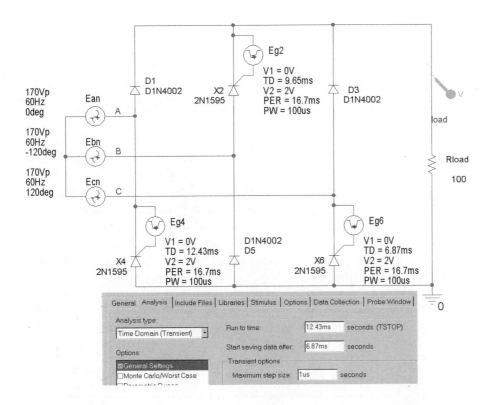

Figure 9-20 Simulation setup for Example 9-6

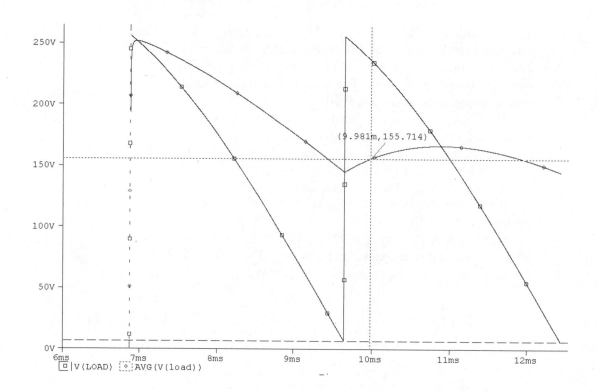

Figure 9-21 Simulation result for Example 9-6

The shape for this segment of time certainly looks correct, partial sinusoids phase angle fired *on* at a little before 7 ms (6.87 ms for SCR6) and at about 9.7 ms (9.65ms for SCR2).

The dc value is calculated using AVG(V(load)). It starts too high, but begins to settle out to 156 V during the second cycle. If all six SCRs could be used and the simulation run for 200 ms, the average should closely match the results from the manual calculations.

Practice: Calculate the average value delivered from a three-phase system with a line voltage of 480 V$_{rms}$ line, for a firing angle of 30°. Also determine the proper time to fire each SCR with respect to E_{AN}.

Answer: V_{dc} = 561 V$_{dc}$, SCR1, 2.78 ms; SCR6, 5.56 ms; SCR2, 8.34 ms; SCR4, 11.12 ms; SCR3, 13.90 ms; SCR5, 16.68 ms

---

9.3 DC-to-AC Converters

Once, uninterruptible power supplies and dc-to-ac conversion were reserved for a few critical facilities such as emergency services and key
elements in hospitals and industries. But the widespread use of the lap-
top computer, printer, fax, and other personal productivity electronics
has changed that. There is now a demand for battery backup and in-car
operation of equipment originally designed to be powered with a utility
supplied sinusoid. Every electronics mart, office supply depot, and department store now has racks of uninterruptible power supplies, deep
discharge batteries, and inverters.

There are a variety of ways to produce a high voltage ac from a battery level dc. If the quality of the wave shape and the accuracy of its
amplitude and frequency are not critical, the circuit is fairly simple.
This group of inverters delivers a modified square wave to the load.

Producing a clean sine wave at a precise voltage over a controlled
frequency range, regardless of the load, takes more effort. These inverters use the same basic architecture as the square wave output circuits.
But instead of delivering a 50 Hz to 60 Hz square wave, they drive the
load with a pulse width modulated rectangle wave at a frequency above
the audio range. Over the period of the output line voltage, the duty cycle of the output rectangle is varied so that its average value closely ap-
proximates a sinusoid. Severe filtering between the switching transistors
and the load recovers the sine wave by removing all of the high frequency harmonics, smoothing out the sharp edges.

The block diagram and partial schematic of a dc-to-ac converter is
shown in Figure 9-22. This technique can be used for either the square
wave output or the pulse width modulated inverter. Part selection and
the programming of the microcontroller are the key differences.

Either design actually uses two dc-to-ac inverters. The first converts
the input low voltage battery dc into a square wave with transistor Q1
driving the bottom of transformer T1. One set of secondary windings
steps the voltage up, delivering a high voltage square wave to the bridge,
D2, and filter, C1. The microcontroller, U3, controls the second inverter. Through the bridge driver, U4, and transistors Q2 through Q5,
the high voltage dc is chopped into a square wave and delivered to the
load. On the positive half-cycle, Q3 and Q4 power the load. Transistors
Q2 and Q5 turn *on* to reverse current flow to the load during the negative
half-cycle.

This schematic is intended for discussion of the *principle* of dc-to-ac
inversion. Precise component selection depends on the requirements of

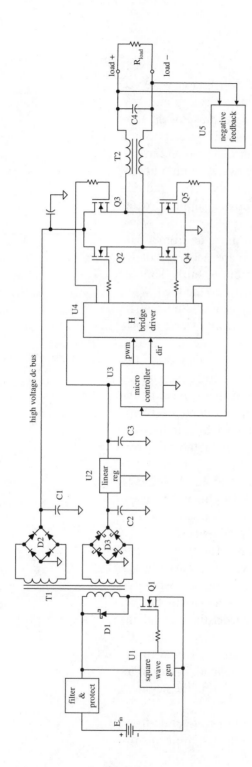

Figure 9-22 DC to ac converter

the design. You must consider the load demands, the supply available, the accuracy needed, the electrical and thermal environment, the degree of operator control, single point failure reliability, and the cost.

Input Inverter

The input filtering and protection is to protect the inverter from variations and weaknesses of the battery supply, *and* to protect any other external circuitry connected to the battery from the transient spikes caused by the inverter's primary switch. Two decoupling capacitors are needed. One provides the bulk of the stored charge and should be thousands of microfarads. The other must be able to respond to the high speed transients that the switch Q1 creates. A 1 to 10 μF *ceramic* capacitor is appropriate. Also, be sure to provide some form of series current limit protection, such as a fuse or semiconductor circuit breaker. Overvoltage protection in parallel with the decoupling capacitors is also a good idea. A MOV (metal oxide varistor) can absorb transient spikes above its rated voltage, clamping the input voltage to that level. They are faster and simpler to use than zener diodes.

The square wave generator drives the transistor switch *on* and *off*. Its output amplitude, frequency, and duty cycle are not critical. The signal must be large enough to drive the transistor hard *on*, but small enough to not exceed the transistor's maximum gate-to-source rating. The frequency should be above the audio range to avoid an annoying hum. The higher the frequency, the smaller the transformer and filter components can be. But high frequency also causes the transistor to dissipate more power because it spends a larger percentage of its time turning *on* and *off* (crossing the linear region). Higher frequencies also more easily generate electromagnetic interference. A frequency of 30 kHz to 100 kHz is a reasonable compromise.

The combination of battery voltage, transformer turns ratio, and duty cycle determine the secondary voltage. This is just like the flyback switching regulator of Section 9.1. Assure that at minimum battery input voltage and minimum duty cycle, the secondary voltage is above the peak needed for the output signal.

$$V_{dc\ sec} = \frac{E_{battery}D}{\frac{N_{pri}}{N_{sec}}(1-D)}$$

A 7555 timer IC is a simple choice to create the square wave. But it alone cannot provide the current quickly enough to properly drive the transistor switch. So, follow the timer with a switch driver IC such as the MAX 4420. This part is explained in Section 6.5.

The primary transistor switch, Q1, is a relatively low voltage, high current part. Assuming that the entire inverter is 90% efficient,

$$P_{\text{in}} = \frac{P_{\text{load}}}{0.9}$$

A dc voltage provides this input power. So,

$$E_{\text{battery}} I_{\text{dc in}} = \frac{P_{\text{load}}}{0.9}$$

$$I_{\text{dc max in}} = \frac{P_{\text{load}}}{0.9 \times E_{\text{battery min}}}$$

But the transistor is only *on* for a fraction of the time, defined as the duty cycle, *D*.

$$I_{\text{p max in}} = \frac{P_{\text{load}}}{0.9 \times D \times E_{\text{battery min}}}$$

The transistor dissipates power based on the rms value of this current and its *on* resistance. And this power dissipation determines the transistor's heat sink requirements. These are the same relationships developed for the transistor switch, in general, in Chapter 6.

$$I_{\text{rms}} = I_{\text{p max in}} \sqrt{D}$$

$$P_Q = I_{\text{rms}}^2 R_{Q\,\text{on}}$$

$$\Theta_{\text{SA}} = \frac{T_{J\,\text{max}} - T_A}{P_Q} - \Theta_{\text{JC}} - \Theta_{\text{CS}}$$

Example 9-7

Given a load power of 150 W, a minimum battery voltage of 8 V$_{\text{dc}}$, and an *on* resistance of 0.01 Ω, determine Q1's peak current and power dissipation.

Solution

Assuming a 50% duty cycle,

$$I_{\text{p max in}} = \frac{P_{\text{load}}}{0.9 \times D \times E_{\text{battery min}}}$$

$$I_{\text{p max in}} = \frac{150\,\text{W}}{0.9 \times 0.5 \times 8\,\text{V}_{\text{dc}}} = 41.7\,\text{A}_{\text{p}}$$

$$I_{rms} = 41.7\,\text{A}_p\sqrt{0.5} = 29.5\,\text{A}_{rms}$$

$$P_Q = \left(29.5\,\text{A}_{rms}\right)^2 \times 0.01\,\Omega = 8.7\,\text{W}$$

Practice: What is the effect when the battery voltage rises to 14 V_{dc}?

Answer: $I_{p\,max\,in}$ = 23.8 A_p, P_Q = 2.8 W

The values in Example 9-7 are typical of an inverter you might buy from a department store to power a laptop computer in your car. It is a fairly low level, consumer application. But the current in the primary circuit is considerable. This is the value that you must use to size the input wiring, connectors, printed circuit board traces, input protection, and fuse. The transistor switch must be carefully chosen to have very low *on* resistance (0.01 Ω in the example). Even at that level, the transistor in Example 9-7 has to dissipate a significant amount of power. A simple solution is to parallel several MOSFET switches. Paralleling ten transistors means that each carries only 4.2 A_p and dissipates less than a watt. Using surface mount components, it may be less expensive and take up less space to buy 10 transistors than one transistor with a much higher current rating and a monster heat sink.

When the transistor(s) goes *on*, it grounds the bottom of transformer T1's primary. With the battery voltage on the other side, current ramps up through the primary inductance, and the magnetic field builds. Then the transistor turns *off*. The magnetic field begins to fall. As it cuts the turns of the primary, current is generated. The voltage reverses polarity, placing a large positive on the drain of the transistor. But the transistor is *off* and there is no place for the induced current to flow.

That is the purpose of the Schottky diode, D1. The big positive turns the diode *on*. As the field falls, the current it induces returns to the other side of the primary coil through D1. The diode also clamps the voltage level at the transistor's drain to just a little above the battery. Select a high speed Schottky diode with a forward current rating comparable to the current through the transistor(s) Q1.

Flyback diode:
  On when Q1 is *off*.
  Path for *I* from collapsing field.
  Pick a Schottky.
  $I_{D1} \approx I_{Q1}$

Although the selection of switching transformers available as off-the-shelf stock items is increasing, it is still *very* limited. So the transformer manufacturer may design T1 to fit your needs. Provide the nominal frequency and its range of variation, the duty cycle, and the maximum applied primary voltage. Secondary current, maximum, over-current, and rms level are needed. The secondary filter capacitor value and the circuit's operating temperature may be needed as well.

The transformer may be custom.

Secondaries

The upper secondary is the source of high voltage. So, it is step-up and must provide the current delivered to the load. The diodes used in the bridge rectifier D2 must be fast and carry large currents. But they also must be able to withstand the high voltage of this secondary. Schottky diodes at these levels are unusual. Generally, fast recovery diodes, such as the UF5405 from General Instruments, are required.

D2 should be fast recovery (e.g., UF5405).

Filtering the secondary waves is much easier since they are square waves, rather than the traditional rectified sinusoid. Even providing load power of over 100 W, C1 need only be a few hundred microfarads. However, do not forget that the voltage rating must exceed the highest peak from the primary, at the largest value of E_{in}.

C1 ~ 220 μF at 400 V

Bridge D3, filter capacitor C2, and the linear regulator provide the low voltage needed to run the microcontroller, H bridge driver, and negative feedback. Depending on the components you choose, a combination of 5 V_{dc}, 12 V_{dc}, or 15 V_{dc} may be appropriate. You may even need two regulated low voltages, although only one is shown in Figure 9-22. Pick the lower secondary turns ratio to provide an output voltage several volts above the largest regulated low voltage you need, at the lowest value of E_{in}. This secondary voltage is small, and its current level may only be a few hundred milliamperes. So, Schottky diodes can be used for D3. Because of this lower current level, the square wave shape, and its relatively high frequency, filter capacitor C2 need only be a few microfarads.

Select a linear regulator(s) with a short circuit current limit just a little above your highest requirement. This allows the IC to protect the circuit should a short occur. A small, plastic packaged TO 92 or surface mount part may be fine. Be sure to do the heat sink calculation for these linear regulators at the maximum E_{in}, though.

In the secondary part of the inverter, large voltages are being switched every few microseconds. So transients are everywhere. It is very important to rigorously decouple *every* supply and every IC pin.

Microcontroller

The microcontroller provides the "brains" of the operation. Its program determines the type of signal eventually sent to the load. It adjusts the drive signals to provide the required frequency and amplitude as the load changes. It interacts with the operator, and supervises the overall performance of the circuit.

Its primary job is to produce two signals that are sent to the transistor switches. The **pulse width modulation (pwm) signal** determines how long each cycle the transistors in the H bridge are *on* sending energy to the load. The **direction (dir) signal** sets the direction that current flows through the load. This allows you to provide both positive and negative load voltages with only a positive high voltage supply.

The simple technique of creating a pseudo-sine wave is shown in Figure 9-23. The load voltage is a rectangular signal. The square wave's period matches that of the load voltage (16.7 ms for a 60 Hz line signal). There are two flat spots in the output. The width of these zero voltage *off* times is adjusted by the microcontroller to set the load rms value. Should the load voltage increase, as noted by the negative feedback circuit, the *off* intervals are widened. Lowering these *off* times sends more energy to the load, perhaps in response to a droop in the input battery's voltage.

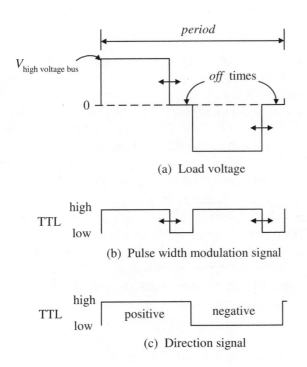

(a) Load voltage

(b) Pulse width modulation signal

(c) Direction signal

Figure 9-23 Pseudo-sine wave signals

The two signals that the microcontroller must generate are also shown in Figure 9-23. The pulse width modulation signal is at twice the load frequency (120 Hz for a 60 Hz line). It falls to a logic low near the end of its period. This assures that all of the transistors are *off*, creating the flat spots in the load voltage. Adjusting the width of this pulse, then, adjusts the rms voltage sent to the load. Many microcontrollers have a pwm register. You load two values, indicating the period and the *on* time, and then let it run. It creates the pwm output signal of Figure 9-23 (b) without any further attention from the controller. To adjust the load rms, just change the value loaded into the register that sets the *on* time.

The direction signal [Figure 9-23(c)] is synchronized with the desired load voltage frequency, alternating between a logic high and a logic low. The only critical issue is that this signal's transitions must be made during the *off* time. That way, when the pwm signal sends the transistors *on* again, the opposite pair is activated. This assures that both transistors on either side (Q2 and Q4 or Q3 and Q5) cannot turn *on*, or even be changing states, at the same time. The sequence is

Q3 and Q4 *on*,   all transistors *off*,   Q2 and Q5 *on*,   all transistors *off*

This technique can be extended to produce a much higher quality sine wave. Instead of turning the transistors *on* and *off* only once each cycle to produce a square wave, the pwm frequency is set hundreds of times higher than the line frequency. Then the pulse width is constantly adjusted to synthesize the low frequency sine wave. During that part of the cycle when the sine should be a low voltage, the duty cycle is set small, producing a low average voltage. As the sine angle increases, the pulse width is broadened, increasing the load's average voltage. After 180° the duty cycle decreases to zero. Then the negative half-cycle is created the same way, using the other pair of transistors in the H bridge.

This is illustrated in Figure 9-24(a). The switching frequency of the pulse width modulation signal has been set to just 20 times the sine it is trying to synthesize to show the pwm cycles. At a much higher, more practical pwm frequency, the switching wave would appear as a blur, high almost constantly near the peaks, wildly switching near the 45° and 135° points, thinning to nothing as the sine angle approaches 0° and 180°.

These signals are not difficult for the microcontroller to produce. Pick the pwm frequency at some multiple above the line frequency. This should be above the audio range to avoid an annoying hum. However, be sure that the microcontroller can actually execute the following code in the allotted time. Load the value that will create this period into the register controlling the pwm frequency. Store a look-up table in memory that contains the successive values of each pulse width.

The core program waits for the pwm period to end. It then accesses the next value for the pulse width from the look-up table, and loads that into the pwm register. It waits for that pwm period to end again.

This continues until the end of the table is reached. There the direction signal is complemented to drive the other pair of transistors in the H bridge. The pulse widths are looked up again and sent out to the pwm register, creating the negative half-cycle. These signals are shown in Figure 9-24(b) and (c).

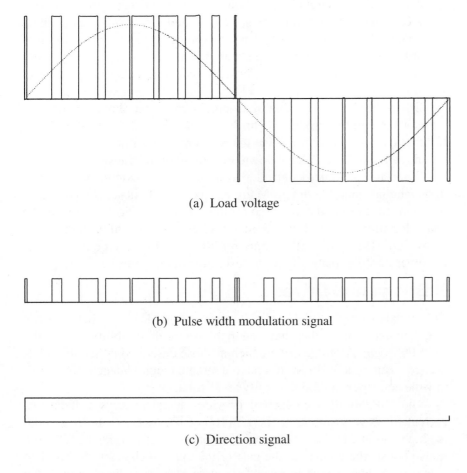

(a) Load voltage

(b) Pulse width modulation signal

(c) Direction signal

Figure 9-24 PWM synthesized sine wave

The microcontroller must be able to adjust the amplitude of the signal sent to the load. A sagging input battery or an increase in load current requires that each pulse must be held *on* a little longer, supplying more energy to keep the load rms voltage constant. So, enter values in the look-up table that create a sine *smaller* than the largest value the pwm register can accept. (e.g., if the pwm register is 8 bits and the value 255_{10} sets the output high for almost the entire pwm period, then adjust the data in the look-up table so that its maximum value is 200.) In every iteration, read the next value from the table. Then add or subtract an offset before sending the value to the pwm register.

This offset is determined by measuring the load voltage with the negative feedback circuit and an analog-to-digital converter. A few times each second, that input must be read and a new offset calculated. The load voltage is a low frequency and must be rectified and filtered. Accordingly, there is lag from the change in offset, until the a-d actually sees a different load voltage. Consider that time delay in the control scheme.

Other channels of the analog-to-digital converter should be used to monitor the current through the bridge transistors and their temperature. If either exceeds a safe level, the microcontroller should warn the operator and ramp the load voltage down, or turn all of the transistors *off*.

When using a microcontroller to drive power devices, it is wise to implement a watchdog timer. Many ICs have this built-in. As part of its program execution, the microcontroller must periodically service the watchdog timer. Should the microcontroller become confused, the timer times out. This drives all outputs *off*, the operator may be notified, or the timer may automatically try to restart the microcontroller.

H Bridge and Driver

The signals from the microcontroller are at logic levels. These are not large enough, nor do they have enough current drive ability to turn the MOSFETs *on*. A bridge driver, such as those described in Section 6.5 is needed. But none of those ICs have a large enough voltage rating to allow the converter to produce utility level voltages.

The HIP2500IP from Intersil provides adequate drive voltage and current. It is very similar to the LT1160 of Figure 6-27. It drives both the high and low transistors for one side of the bridge. Like the LT1160, it also boosts the voltage to the gate of the upper *n*-channel MOSFET to about 10 V above that transistor's source voltage (whatever the high voltage dc bus makes it). This allows you to use less expensive, faster *n*-

channel MOSFETs for all four transistors in the bridge. The capacitor and diodes connected to pins 5, 6, and 7 create this boosted supply.

The microcontroller creates a pwm signal and a dir signal. Each of the four transistors, though, must have its own signal. When the dir signal is high, Q3 and Q4 should receive the pwm signal and Q2 and Q5 should be driven with a logic low. When dir goes low, Q2 and Q5 must get the pwm signal, and Q3 and Q4 must be *off* from a logic low at their gate driver inputs. Under no circumstances can Q2 and Q4, or Q3 and Q5 be *on* at the same time. That would short out the high voltage dc bus.

The inverter U1 and the AND gates U2a and U2b accomplish this. It may also be possible to do this within the microcontroller, and drive the inputs to U3 and U4 directly from an output port. Program execution speed determines whether you need the inverter and gate.

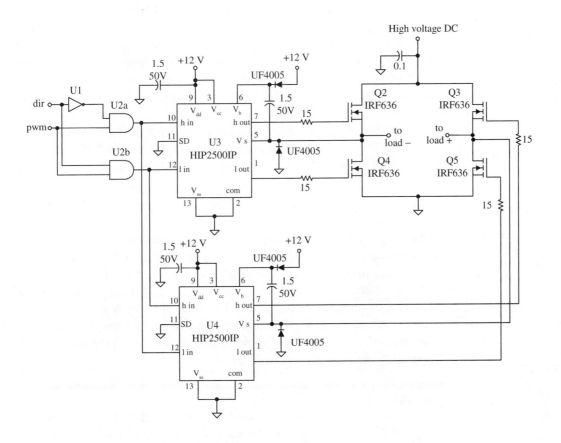

Figure 9-25 H bridge and drivers (*courtesy of Intersil*)

Output Filter

The output filter consists of T2 and C4. It smoothes the sharp edges of the square wave from the transistors, attenuating upper frequency harmonics and lowering the electromagnetic interference sent to the load.

The puesdo-sine wave of Figure 9-23 is produced for those applications where signal quality is *not* critical and circuit simplicity is important. For these applications, the filter attenuates the sharp edges created by the sub-microsecond high voltage and current switching by the transistors. The bifilar-wound choke (T2) is commonly available to protect the line input of computer equipment, passing the utility frequency, but looking like an open to any high frequency noise common to both leads. Be sure to select a choke that can pass the maximum load current.

Capacitor C4, at the output of the filter, shorts any high frequency transients that escape the choke. Its reactance must be very large as compared to the load resistance at 50–60 Hz (the fundamental output frequency). That reactance, though, must be small compared to the load resistance at the significant upper frequency harmonics. Be sure to select a ceramic or metal film capacitor with a voltage rating twice the maximum expected output peak voltage. A nominal value of 0.1 μF is common.

The pulse width modulation technique shown in Figure 9-24 is chosen when the quality of the output sine wave is important. For this scheme, the *LC* circuit forms a second order low pass filter, with

$$f_{-3dB} = \frac{1}{2\pi\sqrt{LC}}$$

Look at Figure 9-26. The frequency to be passed is that of the load voltage sine wave, 10 Hz to 200 Hz. The fundamental of the pwm signal is above the audio range, tens to hundreds of kilohertz. All of the upper frequency harmonics are even higher. So, set f_{-3dB} a few octaves above the highest frequency of the sinusoid to be sent to the load. This places it six to seven octaves below the pwm fundamental. The roll-off rate for a second order filter is −12 dB/octave, passing the sine wave but severely attenuating the pwm upper harmonics.

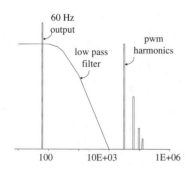

Figure 9-26 Effect of filtering

Example 9-8

The sinusoid to be delivered to the load has a frequency of 60 Hz. The pwm fundamental frequency is 31 kHz, and has an amplitude of 150 V$_p$. Calculate the pwm amplitude at the load.

Solution

Set $f_{\text{-3dB}}$ two octaves above 60 Hz.

$$f_{-3dB} = 60\,\text{Hz} \times 2 \times 2 = 240\,\text{Hz}$$

The pwm fundamental is then seven octaves above this cutoff.

$$240\,\text{Hz} \times 2^7 = 30.7\,\text{kHz}$$

This second order filter rolls off at –12 dB/octave. Seven octaves provides –84 dB of attenuation.

The *dB* is defined as

$$dB = 20\log\frac{V_{\text{out}}}{E_{\text{in}}}$$

Manipulate this to find V_{out}.

$$V_{\text{out}} = E_{\text{in}} \times 10^{\frac{dB}{20}}$$

$$V_{\text{out}} = 150\,\text{V}_{\text{p}} \times 10^{\frac{-84}{20}} = 9.5\,\text{mV}_{\text{p}}$$

Compared to the 170 V_{p}, 60 Hz sinusoid output, the filter has removed virtually *all* of the noise. This is overkill. Even a few volts of noise are acceptable on a 170 V_{p} signal.

Exploration: Select $f_{\text{-3dB}}$ to pass < 2 V_{p} of the 150 V_{p}, 30 kHz noise.

Answer: Move the filter to 4 octaves below 30 kHz, $f_{\text{-3dB}}$ = 1.88 kHz

---

The quality of the inductors and capacitor has a big impact on the filter's performance. Since the load current passes through the inductors, their resistance must be very low. Any resistance drops voltage across the inductors, dissipates power in the inductors, heats them up, and lowers the voltage passed on to the load. The inductors must also be able to handle the peak load current without having their core saturate. Once the core saturates, the magnetic field cannot grow, and the inductor looses its ability to oppose a change in current. Filtering stops and the noise is passed to the load. Torroidal cores help keep the rf magnetic fields contained and do not overheat as much as other type cores.

The filter capacitors should be film or ceramic with X7R characteristics. These perform better at high frequencies and are more stable across a wider voltage and temperature range than other type capacitors. It is wise to select a voltage rating that is twice that expected.

Negative Feedback

The negative feedback signal indicates the load voltage's amplitude. It must be a smooth dc so that the analog-to-digital converter can provide a correct conversion. The diode bridge and the *RC* low pass filter accomplish this. Look at Figure 9-27(c).

At the largest possible load voltage, the negative feedback signal should be below the a-d's maximum input. The two resistors divide the load voltage to that level. Values below 100 kΩ provide good attenuation without being too susceptible to noise. Zener diode D2 breaks down and clamps the input just *below* the a-d's maximum input voltage. Should some failure occur, driving the voltage into the converter too high, the microcontroller is not damaged. Instead, it senses an invalid input, and takes corrective action.

In addition to dividing the voltage to a level that the a-d converter can handle, the resistors combine with the capacitor to form a low pass filter. It must eliminate the variation in the rectified signal from the load and diode bridge. A steady level allows the a-d to make a correct conversion. However, the more severe the filtering, the larger the capacitor, and the slower it can charge and discharge in response to changes in the load's amplitude. Seen from the top of the capacitor, looking back toward the load, the filter's equivalent resistance is R1 in *parallel* with R2. So, pick

$$f_{-3\text{dB neg feedback}} = \frac{1}{2\pi(\text{R}1 // \text{R}2)\text{C}} << f_{\text{load}}$$

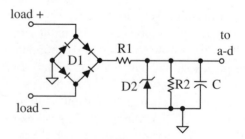

Figure 9-27 Signal conditioning for the negative feedback

Example 9-9

The load voltage should be a 120 V_{rms}, 60 Hz sine wave, but it may become as large as 200 V_p. The a-d converter has a maximum input of 5.00 V_{dc}.

1. Select values for R1, R2, D2, and C.
2. Verify the circuit's performance by simulation.
3. Determine how long it takes the output voltage to settle to within 5% of its final value.

Solution

1. The average value of a full-wave rectified sine wave is

$$V_{dc} = \frac{2V_p}{\pi} = \frac{2 \times 200\,V_p}{\pi} = 127\,V_{dc}$$

Choose the maximum voltage to the a-d to be

$$V_{\text{a-d @ 200 V in}} \approx 4.2\,V$$

This gives the zener room to work should a failure occur.

Pick   R1 = 47 kΩ

At the maximum input, the current through the dividers is

$$I = \frac{127\,V - 4.2\,V}{47\,k\Omega} = 2.61\,mA$$

This calls for

$$R2 = \frac{4.2\,V}{2.61\,mA} = 1.61\,k\Omega$$

Pick   R2 = 1.6 kΩ

The 1N4732A zener diode has a voltage of 4.7 V.

The filter is a first order low pass. Its roll-off rate is −6 dB/octave. Moving $f_{\text{-3dB}}$ five octaves below the filter's output frequency of 120 Hz produces −30 dB of attenuation. This leaves about 3% variation. Set $f_{-3dB} \approx 3.8\,Hz$.

$$C = \frac{1}{2\pi\,f_{-3dB}\,(R1//R2)}$$

$$C = \frac{1}{2\pi \times 3.8\,\text{Hz} \times 1.55\,\text{k}\Omega} = 27\,\mu\text{F}$$

2. The setup and the results of the simulation are in Figure 9-28. The diode bridge's model has been edited to change the breakdown voltage, BV = 300 V.

   The average value out of the bridge is 126 V_{dc}. Ignoring the diode drops, it was calculated to be 127 V_{dc}.

   The oscilloscope shows a full-wave rectified 200 V_p signal at the output of the diode bridge.

   The resistors drop this voltage to 4.2 V. This is confirmed by cursor 2. There is a 0.1 V_{pp} ripple. This may require further filtering or special programming.

3. Cursor 1 indicates that it takes the filter about 0.11 seconds to rise to 4.0 V, 95% of the final 4.2 V.

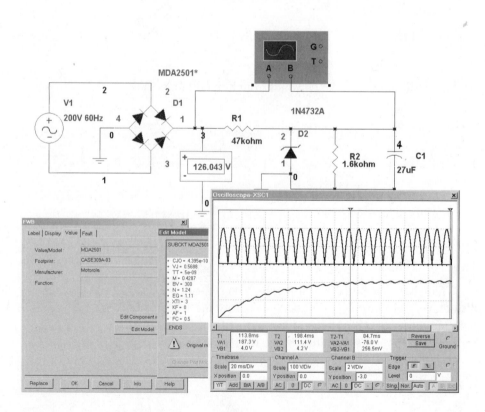

Figure 9-28 Simulation for Example 9-9

Practice: How would you change the circuit, if the maximum load voltage is 370 V_p, and the a-d has a maximum in of 10.24 V? R1 = 82 kΩ.

Answer: $V_{\text{in max}}$ = 9.2 V, R2 = 3.3 kΩ, C = 16 µF, D2= 1N4740, t = 0.11 s

---

9.4 Permanent Magnet Motor Drivers

Small, rapidly responding motors driven from dc voltages are used extensively. You will find them in the vibrators in pagers, computer disk drives and faxes, in a host of places in cars, and even in ceiling fans. Two types are common; the traditional dc motor with brushes, and the brushless dc motor. Each has distinct advantages and limitations. Each also has unique drive requirements.

Common Concerns

There are several characteristics common to driving either dc motor. They are all driven with a series of pulses. The switching speed of that signal affects the performance of each type of permanent magnet motor. Each of these motors turns by controlling the energy applied to an electromagnet. But an electromagnet is just a large coil of wire, an inductor. Trying to rapidly *change* the current through an inductor creates significant problems for the drive electronics. The worst case current through a motor far exceeds the current needed to run it at full speed. Safety and reliability demand that you plan for this maximum, worst case current. Finally, power is applied to the motor by MOSFET switches. Their power dissipation, temperature, heat sinking and possible parallel operation must be considered.

   Pulses are applied to the dc motor. Their frequency has a major effect on the drive design and the interference generated. For years it was common to set the pulse frequency in the kHz range. This low frequency has several advantages. Look at Figure 9-29. Most of the power that a transistor switch has to dissipate occurs on the edges, while the switch is changing states. During those transitions there is considerable voltage across the switch, *and* lots of current through it. Since power is the product of voltage and current, that is when power is dissipated. At a low pulse frequency, the transistor spends only a very small percentage of its time making these transitions. So the *average* power is small. Driving the same transistor at a higher frequency means that it spends the same time on the edges, but that time is a larger percentage of the

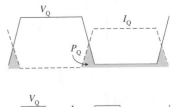

Figure 9-29 Effect of transistor switching speed on power dissipated

period. The *average* power that the transistor must dissipate goes up when you drive it at a higher frequency.

Electromagnetic interference is also generated by these transitions. Even with the same rise and fall times, a transistor switched at a lower frequency creates less emi than one switched more often, because there are few edges. So a lower switching frequency is electrically quieter.

But any frequency in the audio range produces an annoying hum. In consumer electronics this is a serious problem. Also, if it is necessary to filter the switched signal, or pass it through a transformer, the higher frequency allows physically smaller, lighter, and less expensive capacitors, inductors, and transformers.

The current compromise is for switching frequencies just above the audio range, around 50 kHz. But the explosion of portable consumer electronics has pushed switching power supply frequencies through 100 kHz to over 1 MHz. It is reasonable to expect these switching frequencies to be applied to dc motor control as well.

A motor spins because it contains an electromagnet. A major characteristic of an electromagnet is its self-inductance. Inductors respond to *changes* in the current through them.

$$v_{\mathrm{L}} = L\frac{di}{dt}$$

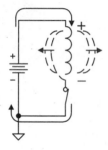

(a) switch closed

This presents a significant problem when the pulse width modulation signal tries to turn the motor *off*. Look at Figure 9-30.

While the motor is *on*, voltage is impressed across the inductance, and the magnetic field builds as the current ramps up. When the transistor switch suddenly interrupts the current, the magnetic field lines fall, cutting the windings of the motor's electromagnet (inductor), inducing a voltage of the *opposite* polarity. The inductor has turned into a generator as it opposes the change in current, trying to keep current flowing in the same direction. The more rapidly the current falls, the larger this induced flyback voltage. Current is forced through any parasitic paths, such as the switch's *off* state capacitance and the capacitance that exists between conductors.

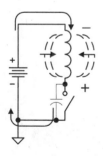

(b) switch opened

Figure 9-30 Motor inductance flyback

This flyback creates three problems. First, the induced voltage may be 100 V_p or more, even for small, logic driven motors, and it lasts for hundreds of nanoseconds. Such large voltages can easily damage the drive electronics, controlling op amps, microcontroller, or computer. Secondly, such large, rapid transients radiate through the circuitry as electromagnetic interference, contaminating equipment even several meters away. Finally, though the switch is *off*, current continues to flow.

For high speed circuits this current may still be flowing when the next cycle begins. The motor never really turns *off*, and control is lost. If you are using an H bridge to reverse direction, the entire power supply may be shorted by this **shoot-through** as one side turns *on* while the other continues to fly back.

The solution is to add a Schottky diode directly across the terminals of the motor. Connect it to be reverse biased during normal operation, cathode to the motor's positive terminal. When the switch is *on* and the motor is conducting, the diode is *off*. But when the switch turns *off* and the flyback voltage reverses polarity across the motor, the diode turns *on*, providing a path for the current generated by the motor's inductance as the field lines fall back. The diode must be fast enough to catch these spikes. Its peak forward current rating must match or exceed the current through the motor, since that is the current that the inductance tries to sustain the instant the transistor switch turns *off*. The diode's peak inverse voltage rating must be larger than the largest voltage you anticipate placing across the motor when it is *on*.

Example 9-10

Investigate the effect of flyback on a 5 V dc motor that draws 100 mA$_{dc}$ during steady state operation, driven by a 50 kHz signal with a 50% duty cycle. Look at the performance:

1. without the motor's inductance.

2. with a 100 μH internal inductance.

3. with a 100 μH internal inductance and a flyback diode.

multiSIM

Solution

1. The simplest model for a 5 V, 100 mA motor is a 50 Ω resistor. The resulting waveforms are shown in Figure 9-31. They indicate a well-behaved circuit. When the input is high, the transistor turns *on*. When the input goes low, the transistor turns *off* and the voltage across the transistor rises cleanly to 5 V.

2. Adding the 100 μH inductance in series with the motor's resistance reveals the problem, shown in Figure 9-32. While the input is high the switch is *on*. The voltage across the switch is low. But as soon as the switch turns *off*, the inductor flies back, producing a 40 V$_p$ kick. This is large enough to damage the logic circuitry driving the motor.

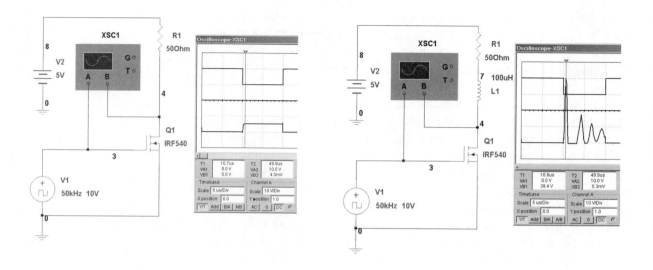

Figure 9-31 Desired motor response

Figure 9-32 Motor flyback

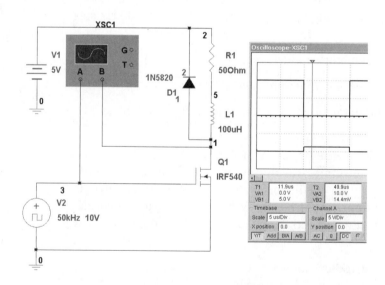

Figure 9-33 Effects of the flyback diode

3. Figure 9-33 shows the effect of adding a 1N5820 diode across the motor. The kickback is shorted out and the circuit performs correctly.

As long as the motor turns in only one direction, a single flyback diode works well. But, as soon as you reverse the direction the motor turns, perhaps with an H bridge, the flyback diode forward biases when the current direction reverses. The motor is shorted out.

The solution is shown in Figure 9-34. Protective diodes are placed across each of the transistors of the H bridge. When Q1 and Q4 turn *on*, voltage is impressed across the motor + to −, turning the motor in one direction. When these transistors turn *off*, at the end of that pulse, the voltage across the motor flies back, − to +. This turns D2 and D3 *on*. The flyback current flows through D2, the decoupling capacitor, and D3.

When Q2 and Q3 turn *on* to reverse the direction that the motor turns, the supply voltage is impressed across the motor − to +. At the end of that pulse, Q2 and Q3 go *off*. The collapsing magnetic field induces a fly back voltage + to − across the motor. Current then flows through D1, the decoupling capacitor, and D4.

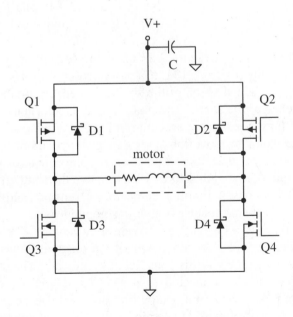

Figure 9-34 Flyback diodes for an H bridge

DC Brush Motor

The traditional, permanent magnetic dc motor consists of permanent magnets positioned on the frame (called the stator). DC current is coupled through brushes to an electromagnet attached to the shaft to be rotated (the armature). The current energizes the electromagnet. It interacts with the permanent magnets on the frame, causing the shaft to rotate in an attempt to align the magnets. As the armature rotates, the brushes move around a set of rings, continuing to feed current into the electromagnet on the shaft. Just as the armature coasts past the point where the electromagnet of the shaft is aligned with the permanent magnets on the frame, the brushes cross a split in the rings. On the other side of the split, the external voltage is connected in the opposite direction. The current reverses in the electromagnet. The magnetic field reverses, and the armature continues to rotate, seeking alignment in the opposite direction.

These motors are inexpensive to build and relatively easy to control. They are used in everything from toys to industrial manufacturing robots. If your goal is to just turn the shaft, more or less, at a given speed, then connect the motor to a voltage supply and adjust the energy to the armature by pulse width modulating the low side switch. This is illustrated in detail in Chapter 6 and also in Example 9-10 and Figure 9-33. Adjusting the duty cycle controls speed. If you want to control direction as well as speed, use the H bridge shown in Figure 9-34. The decoding circuitry and drive ICs in Figure 9-25 or Figure 6-29 allow you to command the motor directly from TTL ICs or a microprocessor.

However, the actual speed of the shaft also depends on the mechanical load attached to the shaft. The more drag the motor must overcome, the slower it turns for a given armature voltage and pulse width. If you are using the motor to move the shaft to a given position, how do you slow the armature as that desired position is approached, or provide an opposing torque if an external force tries to move the shaft? Do not forget the need for current limiting, temperature sensing, and shutdown. Additionally, a soft-start feature is always a good idea.

Clearly, there is a bit more to driving a dc motor. The details are outlined in Figure 9-35. At the center of the diagram is the dc motor, driven by the H bridge with its four Schottky diodes. The shaft of the motor is mechanically connected to a sensor. If you are controlling speed this may be a generator, outputting a dc voltage proportional to how fast the shaft is rotating. Alternately, it may be an optical encoder,

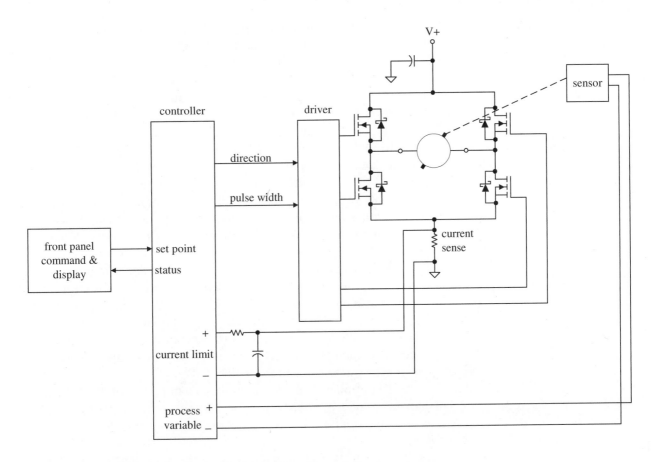

Figure 9-35 Servo motor control block diagram

providing a set of pulses. The faster the shaft rotates, the higher the pulses' frequency. A pair of pulses offset by 90° and an index pulse that happens once a rotation can be counted to determine the position of the shaft. Or an absolute optical encoder may be used for the sensor. It outputs a parallel code that directly specifies position. A potentiometer outputs a dc voltage that also indicates the position of the shaft. Whether analog voltage or digital code, specifying the speed or the position of the shaft, this information is sent to the **process variable** input of the controller.

The other signal sent to the controller is a small analog voltage that indicates the current through the motor. Regardless of which transistors are *on*, current always flows down through the current sense resistor to circuit common. This resistor should be small enough to develop only a fraction of a volt at full current, leaving almost all of the supply voltage for the motor. The small current sense voltage is sent to a severe low pass filter. Be sure to twist these leads to improve noise rejection. The filter removes the high frequency variation caused by the pulse width modulation and any line frequency interference picked up along the way. A differential amplifier may also be needed to help recover the small signal from the common mode noise.

The controller creates the pulse width modulated signal needed to command the bridge driver. It adjusts the width, as necessary, to bring the process variable (the measure of shaft speed or position) into alignment with the **set point**, the data sent from the front panel indicating the desired motor performance. This algorithm may be quite complicated, and is best left to control systems or digital signal processing texts. The controller must also watch for excessive current and shut down the driver when appropriate. Temperature sensors on the transistors' heat sinks and on the motor itself may also be fed to the controller's analog-to-digital converter. A soft-start feature may be included in the controller code. This allows the pulse width to increase gradually if the shaft is stopped. Although soft-start slows response, it provides a much gentler start for the transistors and the motor.

The controller also may provide **dynamic braking**. To bring the shaft to a sudden stop, both upper transistors are turned *off* and both lower transistors are turned *on*. This places a short directly across the motor.

The controller must interact with the operator through the front panel, receiving its set point and providing status information about the motor's performance. A watch-dog timer is also needed to allow the microcontroller to recover if it becomes confused.

The driver converts the low level logic direction and pulse width commands from the controller into the high voltage, fast, high current pulses needed to drive the transistors. It also provides a boosted supply, allowing *n*-channel MOSFETs to be used for the high side switches. The Allegro 3971 is presented in Chapter 6, Figure 6-29. A higher voltage driver, the Intersil HIP2500 is discussed earlier in this chapter, and is shown in Figure 9-25.

If the microcontroller and all of its programming seems like digital overkill, the Texas Instruments (Unitrode) UC1637 offers a single IC, analog alternative. Built within the IC are a triangle wave generator, two comparators, two power switches with controlling logic, a current sense comparator, a shutdown comparator, and an op amp. Figure 9-36 shows how these can all be combined to provide motor speed control.

The 7.5 kΩ and 5 kΩ resistors define the upper and lower height of the triangle wave, while the 25 kΩ resistor and the 0.0022 μF capacitor determine how rapidly the wave rises and falls. Together, these components set the frequency of the pulse width modulator.

The string of 10 kΩ resistors creates the set point voltage. This is connected to the noninverting input of the **error amp, E/A** (an op amp). The negative feedback (the process variable) comes from the generator and is connected to the inverting input of the error op amp through the 47 kΩ resistor (R_i). That amplifier's R_f is the 4.7 MΩ resistor. It is paralleled with a 0.22 μF capacitor. This lowers the amplifier's high frequency response and provides better stability.

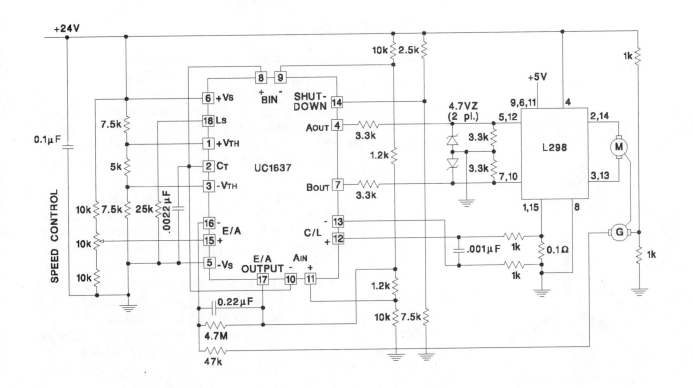

Figure 9-36 UC1637 configured as a motor speed controller (*courtesy of Texas Instruments*)

The output of the error amplifier (E/A output – pin 17) is divided by the 1.2 kΩ and 10 kΩ resistors and sent to the A and B comparators. The noninverting phase of A is driven while B is inverting. This assures that Aout goes high and Bout goes low to send the motor one direction, or Aout goes low and Bout goes high to send the motor in the opposite direction.

The shutdown and the current limit comparators provide protection. Pulling the shutdown pin more than 2.5 V below the supply voltage turns the output drivers *off*. An external comparator and temperature sensors, or an *RC* start-up delay circuit are common. The current limit comparator turns the output drivers *off* when there is more than a 200 mV difference between its input pins. This combines with the 0.001 µF capacitor and 1 kΩ resistor (filter), and 0.1 Ω current sense resistor to set the maximum peak current through the motor and bridge. The UC 1637 can output 100 mA. Additional drive is provided by the L298.

DC Brushless Motor

The traditional dc brush motor has three major problems. Placing the electromagnet on the rotating shaft causes them all. First, the power dissipated by the electromagnet produces heat that must be removed. It must flow from the shaft, through the air gap or bearing to the magnets on the frame, to the frame itself and finally to the external air. This is a long way, and very inefficient. So the motor heats up, and the power it can deliver to the shaft is limited. If this heat could be removed more efficiently, more power could be delivered to the shaft.

Current must be coupled to the rotating electromagnet through brushes bouncing along on a set of split rings. This mechanical connection requires maintenance. The brushes get dirty and they wear down. They must be routinely inspected, and replaced. Over the life of the motor, more may be spent on its repair than on its original purchase. The brushes are also constantly arcing as one connection is broken and the next made. Arcs generate a lot of electromagnetic interference. They also prevent the motor from being used in hazardous locations.

Turning the motor inside out can solve all these problems. Place the electromagnets on the frame, and attach the permanent magnets to the shaft. This is the essence of the dc brushless motor. Look at Figure 9-37. Three pairs of coils are evenly spaced around the frame. The coils are connected in a Y. The three outer leads are provided to the user. The center common point is not.

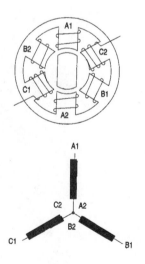

Figure 9-37 Brushless dc motor (*courtesy of Parker Motion and Control*)

To make the shaft spin, the magnetic field must be rotated. The permanent magnet then chases the spinning field. That rotation is shown in Figure 9-38. Initially current is sent from A1 to A2, then B2 to B1. This makes A1 a north pole, A2 south, B2 north, and B1 south. The composite magnetic stator field points southeast.

Current is then switched to enter the motor at A1 and exit at C1. Now A1 is a north pole, A2 a south, C2 north, and C1 south. The resulting field is now pointing south west. It has rotated 45° clockwise.

Switching the current so that it enters B1 and exits C1 causes B1 to be a north pole, B2 to be south, C2 north, and C1 south. The stator field now points due west, having rotated another 45° clockwise. To complete the full rotation, current is sent through the same sequence but in the opposite direction; B1 to A1, C1 to A1, then C1 to B1.

In order to keep the shaft spinning smoothly, the stator field must be rotated when the permanent magnets are in the correct position. So the position of the shaft must be sensed. This is done with three Hall-effect sensors placed on the frame. The Hall-effect sensor outputs a voltage proportional to the magnetic field impressed upon it. Some brushless motors place the sensors at 60° intervals, others at 120°. Once the outputs from the Hall-effect sensors have been conditioned into logic levels, these signals can be combined with a command indicating the desired direction of rotation to decide which electromagnets to energize. Table 9-2 is an example truth table, with 120° sensor placement.

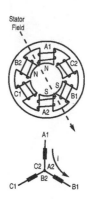

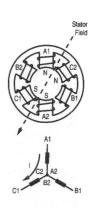

Table 9-2 Brushless motor commutation logic truth table *(courtesy of Allegro MicroSystems)*

	Hall Sensor Input			Outputs		
Direction	H1	H2	H3	Out A1	Out B1	Out C1
0	1	0	1	open	0	1
0	1	0	0	1	0	open
0	1	1	0	1	open	0
0	0	1	0	open	1	0
0	0	1	1	0	1	open
0	0	0	1	0	open	1
1	1	0	1	open	1	0
1	1	0	0	0	1	open
1	1	1	0	0	open	1
1	0	1	0	open	0	1
1	0	1	1	1	0	open
1	0	0	1	1	open	0

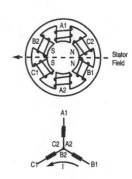

Figure 9-38 Brushless motor sequence *(courtesy of Parker Motion and Control)*

Example 9-11

Compare Figure 9-38 with Table 9-2. Determine which row in Table 9-2 corresponds to which stator field orientation from Figure 9-38.

Solution

In the top position of Figure 9-38, current is flowing *into* A1 and *out of* B1. So, A1 must have a logic high, to source current (1) and B1 must have a logic low, to sink current (0). Winding C is open. This corresponds to the second row of Table 9-2.

The second diagram of Figure 9-38 shows A1 sourcing current and C1 sinking it. B1 is open. This, indeed, is the next row (the third) of Table 9-2.

The last diagram of Figure 9-38 shows B1 sourcing current and C1 sinking it. A1 is open. This is row four in Table 9-2.

Practice: The logic is the same for both directions. Explain why the logic should *not* be reversed to reverse the direction of rotation.

Answer: In the upper half of the table, the shaft is turning clockwise, and you step *down* the table. In the lower half of the table the shaft is turning the opposite direction. The H inputs occur in the opposite sequence, so for counterclockwise rotation you are moving *up* the table.

The Hall-effect sensors and the logic circuit, along with three half H bridge transistors and drivers combine to make the shaft spin. The *speed* of the shaft depends on the characteristics of the motor, the load applied to the shaft, and the strength of the stator fields. These fields, in turn, depend directly on the amount of current through their coils. So, for a given motor, with a fixed shaft load, controlling the stator field current directly affects the speed of rotation.

The mechanics of the brushless motor are attractive. The power dissipating elements are on the frame where they are easily cooled. There are no brushes to arc, creating electromagnetic noise or hazardous sparks. There are no brushes to wear and to require maintenance.

But the electronics necessary to support all of this is significant. The shaft position must be determined with Hall-effect sensors. The small signals from the sensors must be conditioned into logic signals. These must be combined with a direction signal to command the correct stator fields at the correct time. A source and a sink transistor (half of an H bridge) are needed for each coil, and these transistors must be driven. Each output requires flyback diodes. The current must be sensed and its level controlled to provide the correct speed. If you want dynamic braking, or soft-start, additional logic is needed. Finally, be sure to measure the temperature of the transistors and turn them all *off* if any get too hot.

All of these circuits are provided within the Allegro MicroSystems 2936 three-phase brushless dc motor controller/driver. Its block diagram is given in Figure 9-39. The Hall–effect sensor inputs, H1–H3, are conditioned in the IC, so either low level linear or digital sensors will work. There are two versions of the IC, one for a 60° sensor placement and the other for 120°. Logic level inputs for direction and brake (active low) are provided.

The voltage provided to each stator coil is V_{BB}. It may be as high as 45 V_{dc}. The output transistors and flyback clamping diodes can nominally pass 2 A, and can withstand peaks of 3 A. The external resistor, R_s, is used to sense the current through the coils. The voltage developed across R_s by the coil current is then compared to $^1/_{10}$ of an external voltage applied to the THS input (pin 4).

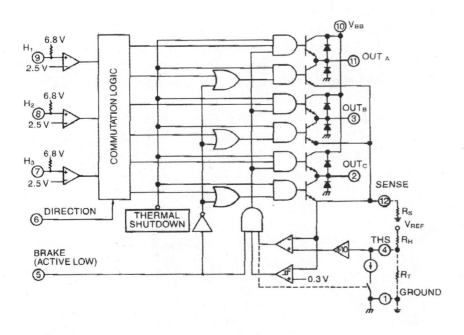

Figure 9-39 Brushless dc motor controller/driver IC *(courtesy of Allegro MicroSystems)*

Current is allowed to increase through the coils' inductance until $^1/_{10}$ of the voltage on the THS input is exceeded. The transistors are then turned *off*. Current decays through the flyback diodes as the magnetic field falls back. After a 7.5% drop in current, the transistors are turned *on* again, and the coil current ramps back up. This sawtooth ripple is repeated many times while each coil is *on*. Effectively, the coil current is limited to

$$I_{coil} = \frac{\frac{1}{10} V_{THS}}{R_s} \qquad \text{where} \quad V_{THS} < 3\,\text{V}$$

There is an internally set reference that overrides any level above 3 V.

The V_{THS} input allows several features. External resistors R_H and R_T divide an externally provided voltage (e.g., 0 V to 10 V is typical for control circuits) into the range needed to set each coil current (e.g., 0 V to 3 V). External op amps, or even a microcontroller, can sense the speed or position of the shaft and command the motor's speed.

If you need extra current to overcome start-up inertia, parallel R_H with a capacitor. When V_{REF} is first applied, the capacitor is a short. Full voltage is applied to V_{THS}. The IC limits the current to its maximum, with its internal 3 V override level. After the capacitor charges, it appears as an open. The voltage division between R_H and R_T then sets the run-current level (with $V_{THS} < 3$ V). Conversely, if you want a soft-start, parallel R_T with a capacitor. Initially, the capacitor holds V_{THS} at zero, then allows it (and the current it controls) to grow slowly. Figure 9-40 shows a typical application and IC pin connections.

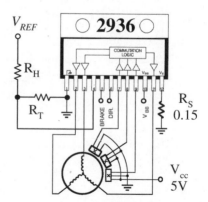

Figure 9-40 2936 application *(courtesy of Allegro Microsystems)*

Summary

The efficient conversion of one dc voltage to another is usually done by pulse width modulation of a power transistor switch, often a MOSFET. The large variations in current are then smoothed with a series inductor, and the ripple in the load voltage is reduced by a parallel capacitor. Variation in the duty cycle of the switch directly affects the load voltage.

The buck converter uses a high side switch to chop the input voltage into a lower load voltage. The inductor and capacitor follow. A Schottky diode is included to provide a discharge path for the inductor when the transistor turns *off*. The LM2595 IC from National Semiconductor provides the reference, control, protection, driver and switch all in a single IC. You simply add the Schottky diode, inductor, capacitor, and negative feedback divider.

A boost converter rearranges the same components to provide an output voltage that is larger than its input. This is accomplished by connecting the inductor directly to the input dc supply, and using the transistor as a low side switch. When the switch is *on* the inductor is grounded, storing energy. The Schottky diode is *off*, and the capacitor holds up the load. When the switch goes *off*, the collapsing magnetic field reverses the polarity of the voltage across the inductor, allowing it to *add* to the input supply. The diode turns *on* and the load and capacitor are charged. The LM2585 IC from National Semiconductor provides the reference, control, protection, driver, and switch. You connect an external inductor, diode, capacitor and feedback divider.

The flyback converter replaces the boost inductor with the primary of a switching transformer. There may be a variety of secondary windings. This allows both step-down and step-up. The LM2585 may also be used as the core of a flyback converter.

Conversion of sinusoidal ac-to-dc is simply done with a diode bridge rectifier followed by a capacitor in parallel with the load. Precise calculation of the load dc voltage is complicated by the series inductance of the step-down transformer. So simulation is the most expedient and reliable technique for analysis or design.

Three-phase voltages, when rectified by a six-diode bridge, provide a low ripple dc voltage to the load without a smoothing capacitor. At any instant two of the diodes are *on*, one is switching *off*, and another is switching *on* as the phase progresses. Again, simulation is the most accurate way to predict the circuit's performance. Replacing the diodes with SCRs and delaying when each turns *on* controls the value of the dc delivered to the load.

The synthesis of a high power sinusoid from dc is a more involved process. A flyback converter steps the low voltage input dc to the required peak value. A microcontroller then drives an H bridge to create the positive and negative half-cycles. The bridge is followed by an *LC* filter to remove the high frequency harmonics. Feedback is required. It senses the load sine wave amplitude, converting it back into a low level dc that is sent back to the microcontroller. The pulse width to the H bridge is then adjusted to control the load's sine wave amplitude.

If the shape of the load voltage is not critical, the microcontroller switches the H bridge at the frequency of the output signal, turning one set of transistors *on* to create the positive half-cycle, and the other set *on* for the negative output. A dead time is provided when all transistors are *off*. Adjusting the length of this time alters the output rms amplitude. A much higher quality sine wave can be synthesized by switching the transistors at a super-audio frequency. Increasing the pulse width causes the load voltage to climb, and reducing the width lowers the load voltage. Modulating this width creates a surprisingly good sine wave (after filtering).

Permanent magnet dc motors create intelligent motion in everything from pager vibrators to industrial robots. Every personal computer, copier, and car rely on them. Rapidly switching current to these inductive loads creates several concerns. Switching speed is a balance between the responsiveness and reduced size available at high speeds and the resulting power losses. The kick used in dc-to-dc converters when current is suddenly removed from an inductance is a major problem when controlling motors. The correct use of flyback diodes is critical.

DC brush motors put the magnets on the frame and couple current into the electromagnet on the shaft. Increasing the current increases the magnetic field and spins the motor more quickly. A controller commands an H bridge with a direction and a pulse width. Measurements of the current through the motor and the position or speed of the shaft are sent back to the controller to allow it to correctly govern the shaft's motion. The controller may be a microprocessor or a dedicated, analog IC such as the Texas Instruments UC1637.

DC *brushless* motors create much less interference, run cooler, and require little maintenance. The permanent magnets are placed on the frame and the electromagnetic field is rotated. Commutation requires Hall-effect sensors to determine the shaft's position. The 2936 from Allegro Microsystems handles all of the drive requirements.

Problems

DC-to-DC Converters

9-1   Calculate the duty cycle needed in a buck converter with an input of 12 V_{dc} and a 5 V_{dc} output.

9-2   What output is produced from a buck converter with a 67% duty cycle and an input of 18 V_{dc}?

9-3   Design a buck converter using an LM2595 that converts 12 V_{dc} to 5 V_{dc} at 0.75 A_{dc}. Specify all component values.

9-4   A buck converter with an LM2595 has the following components:

E_{unreg} = 18 V_{dc}, R_f = 8.8 kΩ, R_i = 1 kΩ, L = 220 μH, C = 330 μF

Calculate: V_{load}, D, $f_{switching}$, $\Delta i_{pp\ load}$, approximate I_{load}

9-5   Calculate the duty cycle needed in a boost converter with an input of 5 V_{dc} and a 12 V_{dc} output.

9-6   What output is produced from a boost converter with a 67% duty cycle and an input of 12 V_{dc}?

9-7   Design a boost converter using an LM2585 that converts 5 V_{dc} to 12 V_{dc} at 0.75 A_{dc}. Specify all component values.

9-8   A boost converter with an LM2595 has the following:

E_{unreg} = 12 V_{dc}, R_f = 18.5 kΩ, R_i = 1 kΩ, L = 560 μH, C = 100 μF

Calculate: V_{load}, D, $f_{switching}$, $\Delta i_{pp\ load}$, approximate I_{load}, $\Delta v_{pp\ load}$

9-9   A flyback converter (LM2585) has the following components:

E_{unreg} = 12 V_{dc}, N_p/N_s = 1.25, R_f = 3.1 kΩ, R_i = 1 kΩ, L_p = 85 μH, C_{out} = 220 μF

Calculate: V_{load}, D, $f_{switching}$, $\Delta i_{pp\ pri}$, approximate I_{load}, $\Delta v_{pp\ load}$

9-10   Design a flyback converter that converts 5 V_{dc} to 12 V_{dc} at a load of 0.75 A_{dc}. Specify all component values.

AC-to-DC Converters

9-11   For a bridge rectifier, with a 12.6 V_{rms} secondary, a 1000 μF capacitor, and a 36 Ω load, calculate the V_p, $\Delta V_{pp\ load}$, and $V_{dc\ load}$ using the light load approximations.

9-12 For a bridge rectifier, with a 25.2 V_{rms} secondary, a 2200 μF capacitor, and a 30 Ω load, calculate the V_p, $\Delta V_{pp\ load}$, and $V_{dc\ load}$ using the light load approximations.

9-13 A step-down transformer is rated at:
$V_{pri} = 120\ V_{rms}$, $V_{sec} = 12.6\ V_{rms}$, $I_{pri} = 0.5\ A_{rms}$, $f = 60\ Hz$

Tests indicate that:
$V_{open} = 13.4\ V_{rms}$, $V_{pri\text{-}short} = 5.3\ V_{rms}$, $\overline{I_{short}} = 0.5\ A_{rms}\ \angle{-42°}$

Determine the appropriate model.

9-14 A step-down transformer is rated at:
$V_{pri} = 120\ V_{rms}$, $V_{sec} = 25.2\ V_{rms}$, $I_{pri} = 1\ A_{rms}$, $f = 60\ Hz$

Tests indicate that:
$V_{open} = 27.4\ V_{rms}$, $V_{pri\text{-}short} = 6.8\ V_{rms}$, $\overline{I_{short}} = 1\ A_{rms}\ \angle{-51°}$

Determine the appropriate model.

9-15 Using the transformer model from Problem 9-13, a 1000 μF capacitor, and a 36 Ω load, determine the V_p, $\Delta V_{pp\ load}$, and $V_{dc\ load}$ by simulation. Compare the results to the light load approximations in Problem 9-11.

9-16 Using the transformer model from Problem 9-14, a 2200 μF capacitor, and a 30 Ω load, determine the V_p, $\Delta V_{pp\ load}$, and $V_{dc\ load}$ by simulation. Compare the results to the light load approximations in Problem 9-12.

9-17 Determine the power factor for Problem 9-15.

9-18 Determine the power factor for Problem 9-16.

9-19 A three-phase system with $E_{line} = 480\ V_{rms}$ is rectified and the rectified voltage is applied to a 33 Ω load. Calculate the following:

E_ϕ, $V_{p\ load}$, PIV_{diode}, $V_{min\ load}$, $V_{dc\ load}$, $I_{p\ load}$, $I_{dc\ load}$, P_{load}, S_{load}, F_p

9-20 A three-phase system with $E_{line} = 290\ V_{rms}$ is rectified and the rectified voltage is applied to a 14 Ω load. Calculate the following:

E_ϕ, $V_{p\ load}$, PIV_{diode}, $V_{min\ load}$, $V_{dc\ load}$, $I_{p\ load}$, $I_{dc\ load}$, P_{load}, S_{load}, F_p

9-21 Determine the following for a 480 V_{rms} line voltage, three-phase system driving a 33 Ω load:

    a. Find the firing angle necessary to deliver 270 V_{dc}.

    b. Confirm that phase angle with a calculation.

    c. Determine the firing times for *each* of the six SCRs.

    d. Verify the answers with a simulation.

9-22 Determine the following for a 290 V_{rms} line voltage, three-phase system driving a 14 Ω load:

    a. Find the firing angle necessary to deliver 130 V_{dc}.

    b. Confirm that phase angle with a calculation.

    c. Determine the firing times for *each* of the six SCRs.

    d. Verify the answers with a simulation.

DC-to-AC Converters

9-23 For the input converter part of a dc-to-ac converter, D = 0.5, given a load power of 230 W, a minimum input dc of 20 V_{dc}, and an *on* resistance of 0.005 Ω, determine Q1's peak current and power dissipation.

9-24 Calculate the root-mean-squared value of a pseudo-sine wave that is 220 V at $t = 0$ s, and stays at 220 V until $t = 6.5$ ms. It is 0 V from 6.5 ms to 8.3 ms, then goes to –220 V at 8.3 ms. It stays at –220 V until 14.8 ms. At 14.8 ms it goes back to 0 V and stays at 0 V for the rest of the cycle, until 16.7 ms.

9-25 Calculate the root-mean-squared value of the pseudo-sine wave from Problem 9-24, if the signal stays at 220 V until 7.2 ms and stays at –220 V until 15.5 ms.

9-26 Determine the number of steps per *half*-cycle in the look-up table of a pwm synthesized sine wave if the microprocessor can send a new value to its pwm register every 45 μs.

9-27 Create the look-up table from Problem 9-26. Set the maximum value to 200. Round the values to integers.

9-28 Explain why the maximum value in Problem 9-27 is chosen to be 200 for a 8-bit pwm register, and not 255.

9-29 The sinusoid to be delivered to the load is 60 Hz. The pwm fundamental frequency is 22 kHz, and has an amplitude of 220 V_p.

Calculate the pwm amplitude at the load if the filter's cutoff is set at 344 Hz.

9-30 For the circuit in Problem 9-29, design a filter that reduces the pwm amplitude to less than 4 V_p.

9-31 Design a feedback circuit, assuming that the load voltage is nominally 220 V_{rms} but may be as large as 350 V_p. Apply no signal larger than 4.8 V_{dc} to the a-d converter.

9-32 How long does it take for your design from Problem 9-31 to settle to within 5% of its final value?

Permanent Magnet Motor Drivers

9-33 Determine the effect of flyback on a 24 V_{dc} motor that draws 1.2 A_{dc} during steady-state operation, driven by a 30 kHz signal with a 50% duty cycle. Look at performance:

a. without the motor's inductance.

b. with a 300 µH internal inductance.

c. with a 300 µH internal inductance and a flyback diode.

9-34 Repeat Problem 9-33. Replace the single transistor driver with an H bridge.

9-35 Draw the flow diagram for the program needed to run the processor in the motor controller block diagram. Provide for initialization, front panel service, current limit interrupt, reading of the process variable, calculation of the pulse width, service of the pwm register, ramping to a new set point, dynamic braking, and a watch-dog timer.

9-36 Design the circuit to control the speed of the motor in Problem 9-34 using a UC1637. Determine appropriate component values for all items in the circuit. The generator outputs 10 V_{dc} at full speed. Replace the L296 IC H bridge with four discrete transistors and the appropriate drive circuitry.

9-37 For the brushless motor drive, 2936, select the correct resistors to allow a command of 0 V_{dc} to 10 V_{dc} to a motor with a maximum current of 1.2 A_{dc}.

Appendix
A

Answers to
Odd-Numbered Problems

Chapter 1

1-1 a. 5 mV **b.** 333 μV

1-3 20

1-5 a. 130 mV$_{rms}$ **b.** 495.4 mV$_{rms}$

1-7 $I1 = 0$ A, $v_a = 0.1$ v$_{rms}$, $I2 = 0$ A,
$I3 = 100$ μA$_{rms}$, $I4 = 100$ μA$_{rms}$,
$v_c = 1.10$ V$_{rms}$, $I5 = 2.2$ mA$_{rms}$,
$I6 = 3.3$ mA$_{rms}$

1-9 $I2 = 0$ A, $v_a = 0$ V, $v_b = 0$ V,
$I1 = 100$ μA$_{rms}$, $I3 = 0$A, $I4 = 100$ μA$_{rms}$,
$v_c = 1$ V$_{rms}\angle 180°$, $I6 = 2$ mA$_{rms}$,
$I5 = 2.1$ mA$_{rms}$

1-11 pick R$_f$ = 10 kΩ, R$_{1i}$ = 7.2 kΩ,
E$_{os}$ = −15 V$_{dc}$, R$_{i2}$ = 135 kΩ

1-13 $I1 = 0$ A, $V_a = 5$ V$_{dc}$, $V_b = 5$ V$_{dc}$,
$I2 = 425.5$ μA$_{dc}$, $I3 = 425.5$ μA$_{dc}$,
$V_c = 11.38$ V$_{dc}$

1-15 a. pick R$_i$ = 1 kΩ, R$_f$ = 4.86 kΩ

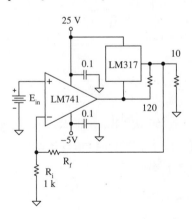

b. $V_{Ri} = 1.28$ V$_{dc}$, $I_{Ri} = 1.28$ mA$_{dc}$,
$V_{load} = 7.50$ V$_{dc}$, $I_{load} = 750$ mA$_{dc}$,
$V_{out\ op\ amp} = 6.3$ V$_{dc}$, $I_{120\Omega} = 10$ mA$_{dc}$,
$I_{out\ LM317} = 760$ mA$_{dc}$, $I_{out\ op\ amp} = 10$ mA$_{dc}$

1-17 $V_a = 0$ V$_{dc}$, $V_b = 0.15$ V$_{dc}$, $V_c = 0.15$ V$_{dc}$,
$I1 = 50$ μA$_{dc}$, $I2 = 50$ μA$_{dc}$,
$V_d = -350$ mV$_{dc}$, $I3 = 7.45$ μA$_{dc}$

1-19 $V_{U1\ NI} = 3.5$ V$_{dc}$, $V_{U1\ INV} = 3.5$ V$_{dc}$,
$I_{R1} = 3.18$ mA$_{dc}$, $I_{R2} = 3.18$ mA$_{dc}$,
$E_{out} = 35.3$ V$_{dc}$, $V_{Q1\ B} = 43.3$ V$_{dc}$,
$I_{R4} = 1.27$ mA$_{dc}$, $V_{Q2\ B} = 27.3$ V$_{dc}$,
$I_{R5} = 8.33$ mA$_{dc}$, $I_{R3} = 7.06$ mA$_{dc}$,
$V_{U2\ out} = 36$ V$_{dc}$, $V_{Q1\ E} = 42.6$ V$_{dc}$,
$V_{U2\ V+} = 42.6$ V$_{dc}$, $V_{Q2\ E} = 28$ V$_{dc}$,
$V_{U2\ V-} = 28$ V$_{dc}$, $V_{U2\ INV} = 31.4$ V$_{dc}$,
$V_{U2\ NI} = 31.4$ V, $I_{Rb} = 2.57$ mA$_{dc}$,
$I_{Rf} = 4.62$ mA$_{dc}$, $I_{U2\ out} = 2.57$ mA,
$I_{Ra} = 2.57$ mA$_{dc}$, $I_{Rb} = 2.57$ mA$_{dc}$,
$V_{U1\ out} = 5.69$ V, $I_{U1\ out} = 2.57$ mA$_{dc}$

Chapter 2
Printed circuit board layout is omitted.

Chapter 3
3-1 omitted

3-3 a. $V_{ave} = \dfrac{V_p}{2\pi}(1+\cos\phi)$ **b.** $V_{ave} = 40.6$ V$_{dc}$

3-5 a. $V_{rms} = \dfrac{V_p}{2\sqrt{\pi}}\sqrt{\pi - \phi + \tfrac{1}{2}\sin 2\phi}$

b. $V_{rms} = 76.29$ V$_{rms}$ **c.** omitted

3-7 a. $W = 97.62$ J **b.** $P = 48.81$ W
c. $P_{motor} = 0.163$ hp, $I = 1.02$ A

3-9 a. $P_{ave} = \dfrac{V_p I_p}{4}$ **b.** $P = 9$ W

Chapter 4
9-29 $V_A = 28$ V_{dc}, $V_B = 28$ V_{dc}, $V_C = 28$ V_{dc},
$V_D = 0$ V_{dc}, $v_A = 600$ mV_{rms}, $v_B = 600$ mV_{rms},
$v_C = 20.4$ V_{rms}, $v_D = 20.4$ V_{rms}

4-3 $V_A = 10$ mV_{dc}, $I_A = 1$ μA_{dc}, $I_B = 1.25$ μA_{dc},
$V_B = 422.5$ mV_{dc}, $I_C = 21.1$ mA_{dc}

9-29 $P_{load} = 28.8$ W, $P_{supply} = 33.6$ W,
$P_{IC} = 4.8$ W

4-7 $P_{load} = 14.4$ W, $P_{supply} = 21.4$ W,
$P_{IC} = 7.0$ W

4-9 $T_J = 120°C$

4-11 omitted

4-13 a. film musical score at 20 feet
b. $p_{ear} = 3.56$ P   **c.** 27 clicks

4-15 $P = 1175$ W

4-17 *gain* $= 25.3$

4-19 $C_{out} = 470$ μF, $f_{low} = 85$ Hz, $V_{out} = 6$ V_{dc},
$V_{out\ max} = 11$ V, $V_{out\ min} = 1$ V,
$V_{out\ rms} = 2.83$ V_{rms} (sine), $P_{load\ max} = 2.0$ W,
$P_{IC\ worst} = 1.8$ W, $\Theta_{SA} = 24.0$ °C/W

4-21 $V_{load\ max} = 16.3$ V_{rms}, $P_{load\ max} = 33.1$ W,
$P_{IC\ worst} = 19.7$ W, $\Theta_{SA} = 4.3$ °C/W,
A = 32.5,   pick $R_i = 150$ Ω, $R_f = 4.7$ $k\Omega$

Chapter 5
5-1   omitted

5-3   omitted

5-5   **a.** $P_{load} = 40.5$ W, $P_{supply} = 81$ W,
$P_Q = 40.5$ W,   **b.** $v_G = 3.45$ V_{rms},
$v_{load} = 3.22$ V_{rms}, $P_{load} = 5.18$ W,
c. $\eta = 6\%$

5-7 $V_{G1} = 3.6$ V_{dc}, $I_{Q1} = 0$ A_{dc}, $V_{G2} = -3.6$ V_{dc},
$I_{Q2} = 0$ A_{dc}, $V_{load} = 0$ V_{dc}, $P_{supply} = 0$ W,
$P_{Q1} = 0$ W, $P_{Q2} = 0$ W

5-9   Use Figure 5-19. $E_{supply} = \pm18$ V_{dc},
pick $R_i = 1$ $k\Omega$, $R_f = 47$ $k\Omega$,   pick R2 = R3
= 1.5 $k\Omega$,   R1 = R4 = 12 $k\Omega$,
$P_{Q\ worst} = 4.1$ W,   for $T_A = 120°C$,
$\Theta_{SA} = 18.2$ °C/W

5-11 $GBW = 960$ kHz, $V_p = 13$ V_p,
$SR = 1.64$ V/μs

5-13 a. $P_{load\ @\ Q\ wc} = 129$ W, $P_{Q\ @\ wc} = 65$ W,
$P_{Q\ each} = 5.8$ W,   11 n and 11 p transistors
b. $P_{load\ max} = 281$ W
c. $R_{each\ source} < 0.26$ Ω,   75 mW

5-15 a. $R_{sc} = 0.12$ Ω,   **b.** $P_{R\ current\ limit} = 1.5$ W

5-17 $T_{J\ with\ wafer} = 142°C$, $T_{J\ w/o\ wafer} = 124°C$

5-19 $V_R = 31.1$ $V_{rms}\angle-49°$, $P_R = 80.4$ W,
$P_{each\ supply} = 83.9$ W, $P_Q = 43.7$ W

Chapter 6
9-29 omitted

6-3 $I_{load} = 48$ mA_p, $I_B = 2.4$ mA_p, $R_B < 667$ Ω,
$P_{load} = 184$ mW, $P_Q = 7.7$ mW

6-5   omitted

6-7 $R_{Q\ on} = 0.14$ Ω, $i_{load\ p} = i_{D\ p} = 0.683$ A_p,
$v_G = 10$ V_p, $P_{load} = 12.2$ W, $P_Q = 0.05$ W,
no heat sink

6-9   omitted

6-11 $v_{each\ Q} = 19$ mV_p, $P_{each\ Q} = 2.4$ mW,
no sink

6-13 $R_{load} = 150$ Ω, $R_B < 300$ Ω, $t_{rise} = 277$ ns

6-15 R1 > 1.4 $k\Omega$,   R2 = 330 Ω,   R3 = 10 Ω
$t_{rise} = 2.2$ μs, $R_{Q\ on} = 0.24$ Ω,
$P_{load} = 70.9$ W, $P_Q = 1.1$ W,
$\Theta_{SA} < 76$ °C/W

6-17 Polarity is reversed, $v_L = 400$ V_p

6-19 R1 = 470 Ω,   R2 = 10 Ω, $t_{rise} = 693$ ns,
$R_{on\ @\ 140°C} = 0.23$ Ω, $P_{load} = 3.90$ W,
$P_Q = 0.22$ W,   no heat sink needed

6-21 Use Figure 6-22, replace R2 with a short,
R1 = 150 Ω,   R3 = 10 Ω, t_{rise} = 303 ns,
$R_{on \, @ \, 140°C}$ = 0.55 Ω, P_{load} = 4.36 W,
P_Q = 0.59 W,   no heat sink is needed

6-23 Use Figure 6-25, but omit the 7406,
R1 < 9.3 kΩ,   pick R2 ≈ 1 kΩ,
P_{load} = 196 mW, P_Q = 4 mW

6-25 omitted

6-27 f < 17.4 kHz

6-29 omitted

Chapter 7

7-1 D = 55.6%, t_{on} = 3.7 μs,
$\Delta I_{pp \, L}$ = 148 mA$_{pp}$, $i_{p \, L}$ = 874 mA$_p$,
$\Delta v_{pp \, load}$ = 14.8 mV$_{pp}$, $I_{rms \, C}$ = 42.8 mV$_{pp}$,
P_Q = 690 mW, P_{diode} = 187 mW,
$I_{in \, dc}$ = 494 mA$_{dc}$

7-3 L > 74 μH,   C > 400 μF, ESR_C < 0.5 Ω,
$I_{in \, dc}$ = 494 mA$_{dc}$, P_{in} = 4.44 W

7-5 $V_{load \, dc}$ = 4.99 V$_{dc}$, P_{IC} = 535 mW,
no heat sink is needed

9-29 Set R_f / R_i = 13.63

7-9 D = 44.4%, t_{on} = 4.44 μs,
$\Delta I_{pp \, L}$ = 222 mA$_{pp}$, $I_{ave \, L}$ = 1.6 A$_{dc}$,
$i_{p \, L}$ = 1.71 A$_p$, $\Delta v_{pp \, load}$ = 7.6 mV$_{pp}$,
P_Q = 125 mW, P_{diode} = 269 mW,
$I_{in \, dc}$ = 1.6 A$_{dc}$

7-11 L > 111 μH,   C > 36 μF, $I_{in \, dc}$ = 1.6 A$_{dc}$,
P_{in} = 8 W

7-13 $V_{load \, dc}$ = 9.62 V$_{dc}$, P_{LM2585} = 257 mW,
no heat sink is needed

7-15 Set R_f / R_i = 11.12

7-17 $V_{load \, dc}$ = 4.99 V$_{dc}$, D = 0.357,
$I_{dc \, pri}$ = 1.60 A$_{dc}$, $I_{dc \, sec}$ = 0.8 A$_{dc}$,
$I_{p \, sec}$ = 5.23 A$_p$, Δv_{load} = 6.07 mV$_{pp}$,
P_{IC} = 1.37 W, $I_{dc \, in}$ = 1.60 A$_{dc}$

7-19 PE-68414, N_p/N_s = 1.2, L_p = 65 μH,
Set R_f / R_i = 21.76, C_{out} > 94 μF,

$I_{in \, dc}$ = 864 mA$_{dc}$, $V_{IC \, max}$ = 41.3 V$_p$,
$I_{IC \, ave}$ = 864 mA$_{dc}$, $I_{pri \, p}$ = 2.31 A$_p$,
$V_{diode \, off}$ = 49.6 V$_p$, $I_{sec \, p} = I_{diode \, max}$ = 1.93 A$_p$,
P_{IC} = 509 mW

7-21 omitted

7-23 omitted

7-25 omitted

Chapter 8

8-1 omitted

8-3 omitted

8-5 omitted

8-7 **a.** R = 72 Ω,   **b.** I_{rms} = 1.18 A$_{rms}$,
c. V_{off} = 169.7 V$_p$ **d.** P_{load} = 100 W,
e. $V_{SCR \, on}$ = 1.6 V   **f.** P_{SCR} = 1.2 W,
g. no heat sink

9-29 omitted

8-11 omitted

8-13 omitted

8-15 omitted

8-17 **a.** L > 1.7 μH,   **b.** C > 0.02 μF,
c. R > 1.7 Ω,   **d.** I_{diode} = 1.4 mA$_p$

8-19 omitted

8-21 $t(60°)$ = 2.78 ms,   by simulation R ≈ 14 kΩ

8-23 R > 350 Ω

8-25 omitted

8-27 omitted

8-29 **a.** omitted   **b.** V_{load} = 65.7 V$_{rms}$,
c. P_{load} = 60 W,   **d.** V_{load} = 46.5 V$_{rms}$,
e. P_{load} = 30 W

8-31 R_{rate} = 25 kΩ,   use a CMOS (rail-to-rail) op
amp, U1b zero voltage change from –15 V$_{dc}$
to –5 V$_{dc}$, 10 kΩ R_i change to 5 kΩ

8-33 omitted

8-35 a. $V = 50$ V_{dc},   **b.** $V = 50.4$ V_{dc},
   c. $V = 118$ V_{rms},   **d.** $V_{rms} = 118.3$ V_{rms}

8-37 $\phi_{firing} = 40°$

Chapter 9

9-1 $D = 41.6\%$

9-3 Use Figure 9-3, $R_f / R_i = 3.1$, $L > 104$ μH,
   $C > 375$ μF, $C_f = 3.3$ nF, $D = 1N5819$

9-5 $D = 70.5\%$

9-7 Use Figure 9-6, $R_f / R_i = 8.72$,
   pick $\Delta v_{load} = 10$ mV_{pp}, $C > 47$ μF,
   pick $\Delta I_L = 188$ mA_{pp}, $L > 235$ μH

9-9 $V_{load} = 5.0$ V_{dc}, $D = 34.2\%$,
   $f_{switching} = 100$ kHz, $\Delta I_{pp\,pri} = 483$ mA_{pp}

9-11 $V_p = 15.8$ V_p, $\Delta V_{pp\,load} = 3.3$ V_{pp},
   $V_{dc\,load} = 14.2$ V_{dc}

9-13 $V_{Th} = 13.4$ V_{rms}, $\overline{Z_{Th}} = 1.65$ $\Omega + 1.49$ Ω,
   $R = 1.65$ Ω, $L = 3.96$ mH

9-15 $V_p = 14.67$ V_p, $\Delta V_{pp\,load} = 1.75$ V_{pp},
   $V_{dc\,load} = 13.3$ V_{dc}

9-17 by simulation, $P = 5.10$ W, $I_{gen} = 1.03$ A_{rms}
   $S = 8.64$ VA, $F_p = 0.59$

9-19 $E_\phi = 277$ V_{rms}, $V_{p\,load} = 677$ V_p,
   $PIV_{diode} = 677$ V_p, $V_{load\,min} = 586$ V,
   $V_{load\,dc} = 646$ V_{dc}, $I_{p\,load} = 20.5$ A_p,
   $I_{load\,dc} = 19.6$ A_{dc}, $P_{load} = 12.6$ kW,
   $S_{load} = 23.0$ kVA, $F_p = 0.548$

9-21 a. $\phi = 66°$ **b.** $V_{dc\,66°} = 265$ V_{dc} **c.** $t_{delay\,SCR1} = $
   4.45 ms, $t_{delay\,SCR2} = 10.02$ ms, $t_{delay\,SCR3} = $
   15.58 ms, $t_{delay\,SCR4} = 12.80$ ms, $t_{delay\,SCR5} = $
   18.36 ms, $t_{delay\,SCR6} = 7.29$ ms

9-23 $I_{in\,p} = 25.6$ A_p, $P_Q = 1.63$ W

9-29 $V = 205$ V_{rms}

9-27 column 1,   time in 45 μs steps to 8.333 ms,
   column 2 = 200*sin(377*column1)

9-29 $V_{harmonic\,out} = 55$ mV_p

9-31 Pick R1 = 100 kΩ,   R2 = 2.2 kΩ,
   $C > 19.5$ μF

9-33 a. no effect   **b.** $V_{p\,load} > 1000$ V_p or the
   breakdown voltage of the MOSFET
   c. no effect

9-35 omitted

9-37 $R_s = 0.25$ Ω,   pick $R_H = 1$ kΩ, $R_T = 430$ Ω

Laboratory Parts and Special Equipment

Table B-1 Laboratory Parts

	Suspended Supply Amp, Chapter 1, p59 - 63	Power Op Amp, Chapter 4, p208 - 212	Class B Amp, Chapter 5, p263 - 268	Transistor Switch, Chapter 6, p321 - 323	Buck Regulator, Chapter 7, p379 - 381	Boost Regulator, Chapter 7, p381 - 383	Triac, Chapter 8, p437 - 443
resistors							
assortment of $\frac{1}{4}$ W							
100k pot							1
capacitors 0.1	4	2	5	2		2	
0.33							1
100				1			
470					1	2	
inductor 200 μ – 400 μ					1	1	
transformer 25.2V @ 0.1A							1
diodes 8 V zener	2		2				
1N5819					1	1	1
1N400x							4
transistors 2N3055				1			
2N3904							2
2N5550*	1		1				
2N5401*	1		1				
IRF630*			1	1		1	
IRF9630*			1	1	1		
IC 7406				1	1	1	
LM741			1				2
LF411	1		1				
MOC3011							1
MOC3030							1
OPA548		1					
Heat sinks TO-3				1			
TO-220 Θ< 5C/W		1	2	1			

* indicates parts required for ± 56 V operation.

2N3904, 2N3906, IRF530, and IRF9530 may be substituted for operation from ± 28 V.

Table B-2 Special Equipment

	Suspended Supply Amp, Chapter 1, p59 - 63	Power Op Amp, Chapter 4, p208 - 212	Class B Amp, Chapter 5, p263 - 268	Transistor Switch, Chapter 6, p321 - 323	Buck Regulator, Chapter 7, p379 - 381	Boost Regulator, Chapter 7, p381 - 383	Triac, Chapter 8, p437 - 443
+/- 28 V @ 3 A supply		1		1			
+/- 56 V supply*	1		1				
8 Ω, 50 W load		1					
100 Ω, 50 W rheostat		1	1		1	1	
50 W loudspeaker		1					
OPA548 printed circuit board		1					
dBspl meter		1					
music source		1					
dc motor, 24 V @ 1 - 3 A				1			
light - safety enclosure (Figure 8-40)							1

Index